Fundamentals of Robotic Grasping and Fixturing

Second Edition

Fundamentals of Robotic Grasping and Fixturing

Second Edition

Caihua Xiong
Wenbin Chen
Han Ding
Youlun Xiong

Huazhong University of Science & Technology, China

World Scientific

NEW JERSEY · LONDON · SINGAPORE · BEIJING · SHANGHAI · TAIPEI · CHENNAI

Published by

World Scientific Publishing Co. Pte. Ltd.

5 Toh Tuck Link, Singapore 596224

USA office: 27 Warren Street, Suite 401-402, Hackensack, NJ 07601

UK office: 57 Shelton Street, Covent Garden, London WC2H 9HE

Library of Congress Cataloging-in-Publication Data
Names: Xiong, Caihua author | Chen, Wenbin author | Ding, Han, 1963–   author |
    Xiong, You-Lun author.
Title: Fundamentals of robotic grasping and fixturing / Caihua Xiong, Wenbin Chen,
    Han Ding, Youlun Xiong, Huazhong University of Science & Technology, China.
Description: Second edition. | New Jersey : World Scientific, 2025. |
    Includes bibliographical references and index.
Identifiers: LCCN 2025007203 | ISBN 9789819810277 hardcover |
    ISBN 9789819810284 ebook for institutions | ISBN 9789819810291 ebook for individuals
Subjects: LCSH: Robots--Motion--Mathematical models | Robot hands--Design and construction
Classification: LCC TJ211.4 .X56 2025 | DDC 629.8/933--dc23/eng/20250326
LC record available at https://lccn.loc.gov/2025007203

British Library Cataloguing-in-Publication Data
A catalogue record for this book is available from the British Library.

For any available supplementary material, please visit
https://www.worldscientific.com/worldscibooks/10.1142/14224#t=suppl

Desk Editors: Soundararajan Raghuraman/Steven Patt

Typeset by Stallion Press
Email: enquiries@stallionpress.com

Preface

Robotic grasping and fixturing (RGF) play a crucial role in various fields, including dexterous manipulation, manufacturing, and assembly, as they significantly impact production quality, cycle time, and cost. While the human hand serves as a highly dexterous gripper with exceptional grasping and manipulation capabilities, the goals of RGF differ in terms of forces and mechanisms. Robotic grasping focuses on closure and stability, whereas robotic fixturing emphasizes localization accuracy in addition to closure and stability. Although extensive research has been conducted on robotic grasping, fixturing, and anthropomorphic manipulation, the current state-of-the-art robotic hand still falls short of the natural human hand in terms of dexterity.

The primary objective of this book is to provide readers with a fundamental understanding of transforming experience-based design into science-based design for robotic grasping system to replicate human natural grasp and robotic fixturing systems to obtain the optimal clamping performance. Additionally, the book explores how to transfer human hand motion intelligence to robotic hands mechanically. By bridging the gap between art and science, the book delves into multifingered robot hand grasping, basic fixture design principles, evaluating and planning of robotic grasping/fixturing, mechanical replicating principle of human hand grasping, as well as the modeling and applications of RGF.

In this second edition, three new chapters have been added to cover the fundamentals of implementing human grasping function mechanically, thus keeping up with the development of dexterous robotic hands. The book serves as a reference for academic researchers, manufacturing and industrial engineers, and technicians involved in dexterous prosthesis research and development. It can also be utilized as a textbook for engineering graduate students in this field.

The authors of this book have dedicated years of work to RGF. The content primarily stems from the authors' recent research and includes contributions from other experts in the field. By offering an overall perspective and scientific foundation of RGF, the book aims to provide comprehensive information, mathematical models, and practical insights into the development and application of grippers and fixtures in the industry. It serves as a valuable reference for academic researchers interested in robotic manipulation and challenges traditional notions by introducing underactuated features in dexterous hand design.

This book is organized as follows. Chapter 1 describes the essential characteristics, similarities, and differences between both robotic grasping and fixturing and the applications of RGF. Chapter 2 provides the essential mathematical foundation for the study of robotic grasping and fixturing. Chapter 3 analyzes the grasping/fixturing closure. Chapter 4 discusses grasping stability. Chapter 5 develops a fast and efficient force planning method to obtain the desired joint torques which will allow multifingered hands to firmly grasp an object with arbitrary shape. Chapter 6 addresses the problem of grasp capability analysis of multifingered robot hands. Chapter 7 analyzes the existence of the uncontrollable grasping forces (i.e., passive contact forces) in enveloping grasp or fixturing and formulates a physical model of compliant enveloping grasp. Chapter 8 derives the kinematic equations of pure rolling contact over the surfaces of two contacting objects and develops a direct force control method based on the position control for robotic manipulation systems. Chapter 9 discusses the dynamic stability of a grasped/fixtured object and presents a quantitative measure for evaluating dynamically grasps. In Chapter 10, a mapping model between the error space of locators and the workpiece locating error space is built up for 3-D workpieces. Chapter 11 describes that deformations at contacts between the workpiece and locators/clamps resulting from large contact forces cause overall workpiece displacement and affect the localization accuracy of the workpiece. Chapter 12 describes the design and implementation of an anthropomorphic hand for replicating human grasping functions. Chapter 13 describes the mechanical implementation of kinematic synergy for continual grasping generation.

The research works in this book have been supported by the National Natural Science Foundation of China (Grant No. 52027806, 91648203, 51335004, 52075191, U1913205), the National Key R&D Program (Grant No. 2016YFE0113600, 2018YFB1307201), and the '973' National Basic Research Program of China (Grant No. 2011CB013301).

About the Authors

 **Caihua Xiong** currently is the Director of the Institute of Medical Equipment Science and Engineering at Huazhong University of Science and Technology (HUST). He is a Changjiang Scholar Distinguished Professor by the Ministry of Education. Prof. Xiong has authored over 200 high-level papers published in renowned international journals such as *National Science Review, IEEE Transactions on Robotics* (TRO), and *International Journal of Robotics Research*. Professor Xiong has been honored with the Gold Award at the Geneva International Invention Expo. He actively participates in editorial roles for several international journals, including serving as a Technical Editor for the IEEE/ASME Transactions on Mechatronics, an Associate Editor of the *International Journal of Social Robotics*, and the editorial member of the *Journal of Bionic Engineering*.

Wenbin Chen is the Huazhong Scholar Distinguished Professor at Huazhong University of Science and Technology (HUST). He specializes in exoskeletal rehabilitation robotics, humanoid robotics, and soft robotics, with a focus on the development of innovative robotic systems. He has obtained several Chinese medical device certifications for his inventions and was awarded the First Prize of the China Mechanical Industry Science and Technology Award for Technical Invention. Prof. Chen has published extensively in leading journals, contributing over 50 high-impact research papers. He has also received awards for textbook excellence, including the Ministry of Education's First Prize for Excellent Textbooks (2021).

Han Ding is an Academician of the Chinese Academy of Sciences. He is currently the Director of the Academic Committee of Huazhong University of Science and Technology (HUST), the Director of the State Key Laboratory of Intelligent Manufacturing Equipment and Technology, the Chair of the steering expert group of the National Natural Science Foundation of China's major research program "Research on the Basic Theory and Key Technology of Coexisting-Cooperative-Cognitive Robots", and the Chief Scientist of the Center for Basic Science of the National Natural Science Foundation of China on Robotized Intelligent Manufacturing. He has published four academic monographs and more than 300 papers in SCI journals. He has won the second prize in the State Natural Science Award and two times the second prize in the State Scientific and Technological Progress Award.

Youlun Xiong is an Academician of the Chinese Academy of Sciences, affiliated with the School of Mechanical Science and Engineering at Huazhong University of Science and Technology (HUST). His research expertise lies in advanced manufacturing technology, robotics, and precision measurement. He has been honored with several awards, including the National Science and Technology Progress Awards (Second and Third Class). Professor Xiong has authored influential books such as *Robotics: Modeling, Control, and Vision* (2018) and *Fundamentals of Robotics* (1995). He is currently the Director of the Academic Committee for the State Key Laboratory of Precision Manufacturing for Extreme Service Performance and serves as an editorial board member for leading journals, including *Science in China Series E: Technological Sciences* and *Science Bulletin*.

Contents

Chapter 1

Robotic Grasp and Workpiece-Fixture Systems

1.1 Introduction

The human hand which has the three most important functions, to explore, to restrain objects, and to manipulate objects with arbitrary shapes (relative to the wrist and to the palm), is used in a variety of ways [1]. The first function falls within the realm of haptics, an active research area in its own merits [2]. This book will not attempt an exhaustive coverage of this area. The work in robotic grasping and fixturing has tried to understand and emulate the other two functions. The task of restraining objects sometimes is called fixturing, and the task of manipulating objects with fingers (in contrast to manipulation with the robot arm) sometimes is called dexterous manipulation.

Our fascination with creating mechanical counterparts of human hands has led us to place great expectations on robot capabilities. The earliest mechanical hands were likely prosthetic devices to replace lost limbs. However, the majority of prosthetic hands have been designed simply for gripping objects [3]. In order to investigate the mechanics and principles of object restraint and manipulation with human hands, various multifingered robot hands have been developed, such as the Stanford/JPL hand [3], the Utah/MIT hand [4], and other hands. Compared to conventional parallel jaw grippers, multifingered robot hands have three potential advantages: (1) they have higher grip stability due to multicontact points with the grasped object; (2) they can grasp objects with arbitrary shapes; (3) it is possible to impart various movements onto the grasped object. In order for these multifingered robot hands to autonomously perform grasping and fixturing tasks in industrial settings with complex environments, it is essential to study planning methods and fundamental principles of robotic grasping and fixturing, with a specific focus on anthropomorphic grasping and manipulation. Hence, the objective of this book is to develop algorithms for grasping/fixturing planning and establish the fundamentals of robotic grasping and fixturing.

1.2   Robotic Manipulation and Multifingered Robotic Hands

The vast majority of robots in operation today consist of six-jointed "arms" with simple hands or "end effectors" for grasping/fixturing objects. The applications of robotic manipulations range from pick and place operations to moving cameras and other inspection equipment and performing delicate assembly tasks. They are certainly a far cry from the wonderful fancy about the stuff of early science fiction but are useful in such diverse arenas, such as welding, painting, transportation of materials, assembly of printed circuit boards, and repair and inspection in hazardous environments [3, 5].

The hand or end effector is the bridge between the manipulator (arm) and the environment. The traditional mechanical hands are simple, out of anthropomorphic intent. They include grippers (either two- or three-jaw), pincers, tongs, as well as some compliance devices. Most of these end effectors are designed on an *ad hoc* basis to perform specific tasks with specific tools. For example, they may have suction cups for lifting glass which are not suitable for machined parts or jaws operated by compressed air for holding metallic parts but not suitable for handling fragile plastic parts. Further, a difficulty that is commonly encountered in applications of robotic manipulations is the clumsiness of a robot equipped only with these simple hands, which is embodied in lacking of dexterity because simple grippers enable the robot to hold parts securely, but they cannot manipulate the grasped object, a limited number of possible grasps resulting in the need to change end effectors frequently for different tasks, and lacking of fine force control which limits assembly tasks to the most rudimentary ones [5].

Experience with manipulators has pointed to a need for hands that can adapt to a variety of grasps and augment the arm's manipulative capacity with fine position and force control. Multifingered or articulated hands with two or more powered joints appear to offer some solutions to the problem of endowing a robot with dexterity and versatility. The ability of a multifingered hand to reconfigure itself for performing various grasps for objects with arbitrary shapes reduces the need for changing specialized grippers. The large number of lightweight actuators associated with the degrees of freedom of the hand allows for fast, precise, and energy-efficient motions of the object held in the hand. Fine motion/force control at a high bandwidth is also facilitated for similar reasons. Indeed, multifingered hands are truly anthropomorphic analogs of human hands for grasping/fixturing objects with arbitrary shapes and implementing dexterous manipulation tasks.

There have been many attempts to devise multifingered hands for research use and extend our understanding of how articulated hands may be used to securely grasp objects and apply arbitrary forces and small motions to these objects. The Stanford/JPL hand (also known as the Salisbury Hand) is such a multifingered robot hand, as shown in Fig. 1.1. It is a three-fingered hand, each finger has three degrees of freedom and the joints are all cable-driven. The placement of the fingers consists of one finger (thumb) opposing the other two. The Utah/MIT hand is another multifingered robot hand, as shown in Fig. 1.2. It has four fingers (three fingers and a thumb) in a very anthropomorphic configuration, each finger has four degrees of freedom and the hand is cable-driven. The difference in actuation between the Salisbury Hand and the Utah/MIT hand is in how the cables (tendons) are driven: The first uses electric motors and the second pneumatic pistons.

Fig. 1.1. The Stanford/JPL hand.

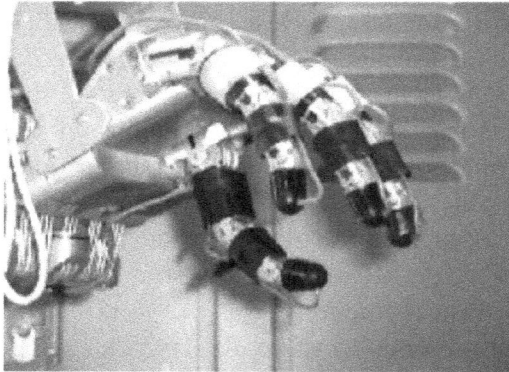

Fig. 1.2. The Utah/MIT hand.

The multifingered grasping/fixturing can be classified into two types: fingertip grasp and enveloping grasp. For the fingertip grasp, we expect the manipulation of an object to be dexterous since the active fingertip can exert an arbitrary contact force onto the object. Generally, all contact forces can be controlled actively in fingertip grasps. Figure 1.3 shows the fingertip grasps of a high-speed multi-fingered robotic hand [6]. The hand has three fingers, and the index finger has 2 degrees of freedom (DOF), the left thumb and right

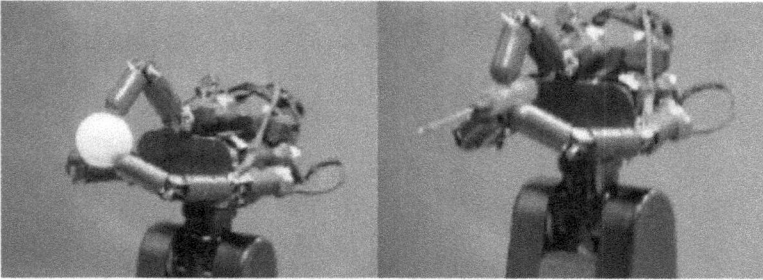

Fig. 1.3. The fingertip grasp of the high-speed multifingered robotic hand.

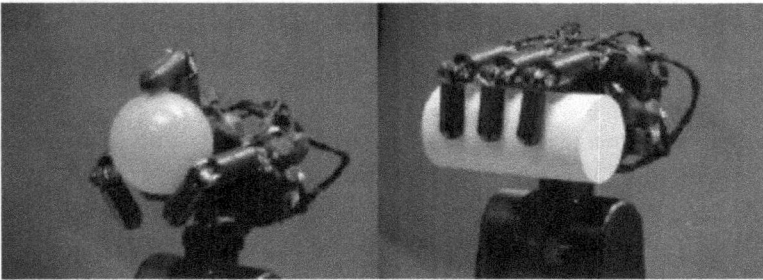

Fig. 1.4. The enveloping grasp of the high-speed multifingered robotic hand.

thumb have 3 DOF, so that the hand has 8 DOF total. In contrast to fingertip grasps, enveloping grasps are formed by wrapping the fingers (and the palm) around the object to be grasped. They are, similar to fixtures, almost exclusively used for restraint and for fixturing, and not for dexterous manipulation. We expect the grasp to be robust against an external disturbance. In the enveloping grasps, the number of actuators is commonly much less than the relative freedom of motion allowed by contacts between the object and links of fingers, thus, from a viewpoint of controllability, not all the contact forces are controllable actively, which is the main issue of the grasp force analyses in enveloping grasping and fixturing. Figure 1.4 shows the enveloping grasps of the high-speed multifingered robotic hand [6]. In fact, this is easily seen in human grasping where fingertips and distal phalanges are used in fingertip grasps for fine manipulation, while the inner parts of the hand (palm and proximal

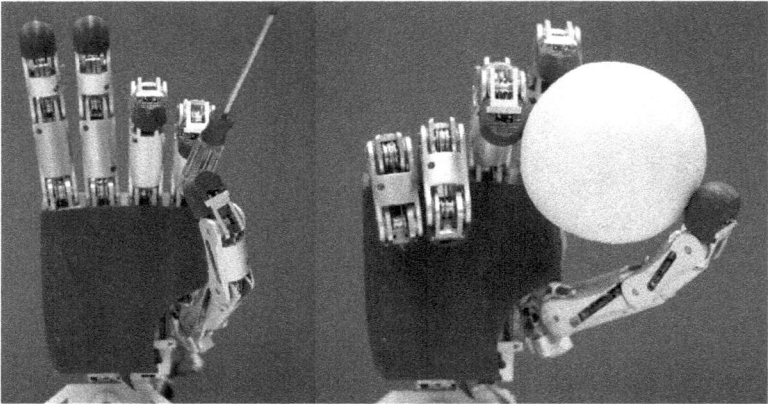

Fig. 1.5.   The precision grasp of the underactuated anthropomorphic hand.

phalanges) are used in enveloping grasps for restraint. The so-called whole arm grasps [7] and power grasps [8] belong to the enveloping grasps. Figure 1.5 shows the fingertip grasps of the underactuated anthropomorphic hand [18]. Unlike the fingertip grasp shown in Fig. 1.3, the underactuated robotic hand's fingertip grasping posture is very similar to the human hand. Its finger joints are not independently controlled but coupled together in a certain proportional relationship. Due to its human-like grasping performance and the advantages of fewer motors, such a robotic hand is widely used in prosthetic hand design.

1.3   AMT and Fixtures

Advanced Manufacturing Technology (AMT) is a key enabler to help manufacturers meet the productivity, quality, and cost reduction demands of competitive global markets [10]. It involves new manufacturing techniques and machines combined with information technology, microelectronics, and new organizational practices in the manufacturing process. AMT is viewed as providing the basis that enables firms to exploit competitive advantages fostered by the technology. The prime motivation for installing AMT is to increase the competitiveness of the firm [11].

In view of the current trend toward advanced manufacturing techniques, such as flexible manufacturing systems (FMS) and group technology (GT), the requirement for an efficient fixture design system is becoming increasingly very important. With the aid of such a system, the process of fixture design can be automated and integrated with other manufacturing modules, which will lead to higher productivity and shorter manufacturing lead times.

The fixture is a kind of gripper used to locate and hold the workpiece with locators and clamps respectively during the machining, assembling, and inspection processes. Fixtures can be classified into two types: dedicated and reconfigurable [8]. Dedicated fixtures generally imply that they have been designed for specific workpiece geometry. These types of fixtures are most suitable for mass production environments, where they can be discarded at the end of the production life and their costs can be absorbed by a large number of products. Reconfigurable fixtures, on the other hand, are designed for a family of workpiece geometries [9]. Pressures on the manufacturing industry during the 1980s have led to the development of many new techniques which come under the general description AMT. Automatically reconfigurable fixtures play a crucial role in these new technologies and they have been the subject of intensive research.

Although there are multifarious fixtures in the industry, especially in FMS, the functions of such fixtures are equivalent to a set of contact point constraints on workpieces. For example, the locating with three plane datums is equivalent to six contact point constraints on the workpiece, as shown in Fig. 1.6, where the planes A, B, and C are the so-called primary, secondary, and tertiary locating datums in the 3–2–1 locating principle [12,13], respectively. In fact, it is not easy, and not necessary to use three planes to locate a workpiece during machining, assembling, and inspecting.

The short diamond pins, short cylindrical pins, short V-blocks, and long V-blocks are usually used to locate the workpiece in fixtures. Their functions of locating may be similarly equivalent to a set of contact point constraints on workpieces. The common locating types usually used in machining and their equivalents are shown in Table 1.1.

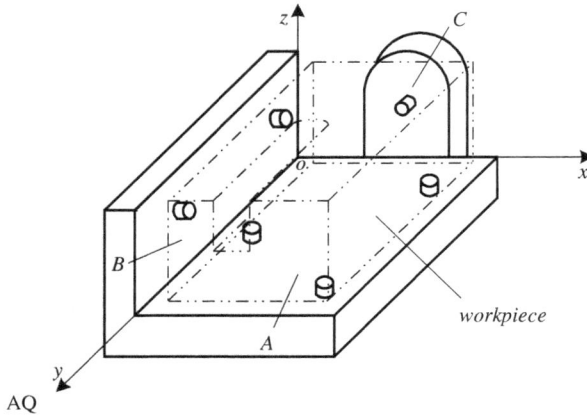

Fig. 1.6.   Example of plane locators.

The fixture design process can be divided into the following [12, 14]:

- *Fixture setup planning*: This aims to determine the number of setups, the orientation of the workpiece in each setup, and the machining surface in each setup.
- *Fixture configuration planning*: This aims to determine a set of locating and clamping points on workpiece surfaces such that the workpiece is completely restrained.
- *Fixture construction*: This aims to select fixture elements and place them into a final configuration to locate and clamp the workpiece.
- *Fixture assembly*: This aims to assemble the fixture components in strict accordance with the previous stage. Some progress has been made toward using robots for automating the assembly of the fixture.

1.4   Comparison between Grasping and Fixturing

The kinematics of workpiece fixturing is similar to that of object grasping. The analysis of motion and force constraints for multifingered grasps can be extended to fixtures. The goal of both fixturing and grasping is to immobilize an object kinematically by means of a

Table 1.1. Equivalent locating.

Locating type	Equivalent locating
 Short Diamond Pin	 1 Locator Pin
 Short Cylindrical Pin	 2 Locator Pins
 Short V-Block	 2 Locator Pins
 Long V-Block	 4 Locator Pins

suitable set of contacts. Thus, the quasi-static stability and dynamic stability are important indices to evaluate both robotic grasping and fixturing.

Contacts can be equally treated in robotic grasping because all fingers are active. However, locators and clamps cannot be equally treated in fixtures due to the passiveness of locators and the active-ness of clamps, which means that passive contact forces exist in fixtures [15]. Thus, the clamping force planning in fixtures is dif-ferent from the grasping force planning in robotic grasps. It should be noted that the passive contact forces exist in enveloping grasps as well [16].

In fixtures, the configuration of locators depends on the machining requirements, such as under-deterministic, fully deterministic, and over-deterministic locating. More importantly, the configuration of locators affects the machining quality, accessibility, and detachability (namely, loading and unloading capability), and the configuration of clamps affects the fixturing closure. Locating error is related to the configuration of locators and the errors of locators and independent of the configuration of clamps and the errors of clamps [17].

Thus, robotic grasping mainly concerns holding feasibility, compliance, and stability. In contrast, because the position and orien-tation precision of the workpiece to be fixtured depends on the pas-sive locators' tolerances and configuration, fixturing for machining emphasizes on the accurate localization of the workpiece besides the fixturing closure.

The main part of this book will focus on the mathematical model-ing of robotic grasping and fixturing to develop methodologies for the automated design of grasping and fixturing systems. What's more, the multivariate statistics methodology is also introduced to explore the dexterous grasping mechanism of the human hand. As the appli-cation of the designing methodologies, two anthropomorphic hands are provided in the final two chapters.

References

[1] Bicchi A. and Kumar V. Robotic grasping and contact: A review. *Proceedings of IEEE International Conference on Robotics and Automation*, 1, pp. 348–353, 2000.
[2] Klatzky R. and Lederman S. *Intelligent Exploration by the Human Hand.* Chapter 4, Dextrous robot manipulation, S. T. Venkataraman and T. Iberall, eds., Springer-Verlag, 1990.

[3]   Mason M. T. and Salisbury J. K. *Robot Hands and the Mechanics of Manipulation*. MIT Press, Cambridge, MA, USA, 1985.

[4]   Jacobsen S., Wood J., Bigger K., and Iverson E. The Utah/MIT hand: Work in progress. *International Journal of Robotics Research*, 4(3), pp. 21–50, 1984.

[5]   Murray R. M., Li Z. X., and Sastry S. S. *A Mathematical Introduction to Robotic Manipulation*. CRC Press, Boca Ration, Florida, USA, 1994.

[6]   Namiki A., Imai Y., Ishikawa M., and Kaneko M. Development of a high-speed multifingered hand system and its application to catching. *Proceedings of the 2003 IEEE/RSJ International Conference on Intelligent Robots and Systems*, Las Vegas, Nevada, USA, pp. 2666–2671, 2003.

[7]   Salisbury J. K. Whole-arm manipulation. *Proceedings of International Symposium on Robotics Research*, 1987.

[8]   Mirza K. and Orin D. E. Force distribution for power grasp in the digits system. *CISM-IFToMM Symposium on Theory and Practice of Robots and Manipulators*, 1990.

[9]   Benhabib B., Chan K. C., and Dai M. Q. A modular programmable fixturing system. *Transactions of ASME Journal of Engineering for Industry*, 113, pp. 93–100, 1991.

[10] Industry Canada. (2007, July 21). What is AMT? from https://www.hitechcanada.ca/amt/.

[11] DeRuntz B. D. and Turner R. M. Organizational considerations for advanced manufacturing technology. *The Journal of Technology Studies*, 29(1), pp. 1–7, 2003.

[12] Rong Y., Huang S. H., and Hou Z. *Advanced Computer-Aided Fixture Design*. Elsevier Academic Press, Burlington, MA 01803, USA, 2005.

[13] Nee A. Y. C., Whybrew K., and Senthil Kumar A. *Advanced Fixture Design for FMS*. London: Springer-Verlag, 1995.

[14] Segal L., Romanescu C., and Gojinetchi N. Methodologies for automated design of modular fixtures. *Proceedings of International Conference on Manufacturing Systems*, Iasi, Romania. pp. 151–156, 2001.

[15] Xiong C. H., Wang M. Y., Tang Y., and Xiong Y. L. On prediction of passive contact forces of workpiece-fixture systems. *Proceedings of the Institution of Mechanical Engineers Part B-Journal of Engineering Manufacture*, 219(3), pp. 309–324, 2005.

[16] Xiong C. H., Wang M. Y., Tang Y., and Xiong Y. L. Compliant grasping with passive forces. *Journal of Robotic Systems*, 22(5), pp. 271–285, 2005.

[17]   Xiong C. H., Rong Y., Tang Y., and Xiong Y. L. Fixturing model and analysis. *International Journal of Computer Applications in Technology*, 28(1), pp. 34–45, 2007.

[18]   Xiong C. H., Chen W. R., Sun B. Y., Liu M. J., Yue S. G., and Chen W. B. Design and implementation of an anthropomorphic hand for replicating human grasping functions, *IEEE Transactions on Robotics*, 32(3), pp. 652–671, 2016.

Chapter 2

Singular-Value Decomposition and Its Application

2.1 Introduction

Human hand has a number of elements related to grasping movement, for example the muscles, bones, and joints. Adjusting so many elements is a complex task that the Central Nervous System (CNS) must deal with. Recent advances in cognitive science have shown that coordination of the human hand during grasping is dominated by movement in a configuration space of highly reduced dimensionality. To better understand the grasping of human hand and scientifically design the robotic grasping and fixturing system, it is necessary to explore the constitution of the low dimension space and how the complex hand movement is constructed by the low dimensional movement, or called eigengrasp. The mathematical fundamentals of analyzing the complicated coordination of finger joints are multivariate statistical analysis, especially the singular-value decomposition and its deduction.

2.2 The Organization of Data

Without loss of generality, the joint angle of the hand or fixture can be viewed as a state variable of the system. The characteristics of several variables can be analyzed by measurements which is commonly called data. These data must frequently be arranged and displayed in various ways. Multivariate data arise whenever an investigator, seeking to understand a physical phenomenon, selects a number $p \geq 1$ of *variables* to record. The values of these variables are all recorded for each distinct experimental unit.

We will use the notation y_{jk} to indicate the particular value of the kth variable that is observed on the jth item or trial. That is,

$$y_{jk} = \text{measurement of the } k\text{th variable on the } j\text{th item} \quad (2.1)$$

Consequently, n measurements on p variables can be displayed as follows:

	Variable 1	Variable 2	$\cdots$	Variable k	$\cdots$	Variable p
Item 1:	y_{11}	y_{12}	$\cdots$	y_{1k}	$\cdots$	y_{1p}
Item 2:	y_{21}	y_{22}	$\cdots$	y_{2k}	$\cdots$	y_{2p}
$\vdots$	$\vdots$	$\vdots$	$\cdots$	$\vdots$	$\cdots$	$\vdots$
Item j:	y_{j1}	y_{j2}	$\cdots$	y_{jk}	$\cdots$	y_{jp}
$\vdots$	$\vdots$	$\vdots$	$\cdots$	$\vdots$	$\cdots$	$\vdots$
Item n:	y_{n1}	y_{n2}	$\cdots$	y_{nk}	$\cdots$	y_{np}

Or we can display these data as a rectangular array, called $\mathbf{Y}$, of n rows and p columns:

$$\mathbf{Y} = \begin{bmatrix} y_{11} & y_{12} & \cdots & y_{1k} & \cdots & y_{1p} \\ y_{21} & y_{22} & \cdots & y_{2k} & \cdots & y_{2p} \\ \vdots & \vdots & \cdots & \vdots & \cdots & \vdots \\ y_{j1} & y_{j2} & \cdots & y_{jk} & \cdots & y_{jp} \\ \vdots & \vdots & \cdots & \vdots & \cdots & \vdots \\ y_{n1} & y_{n2} & \cdots & y_{nk} & \cdots & y_{np} \end{bmatrix}$$

The array $\mathbf{Y}$, then, contains the data consisting of all of the observations on all of the variables.

Let $y_{11}, y_{21}, \ldots, y_{n1}$ be n measurements on the first variable. Then, the arithmetic average of these measurements is

$$\bar{y}_1 = \frac{1}{n} \sum_{j=1}^{n} y_{j1}$$

If the n measurements represent a subset of the full set of measurements that might have been observed, then y_1 is also called the sample *mean* for the first variable. The sample mean can be computed from the n measurements on each of the p variables so that, in general, there will be p sample means:

$$\bar{y}_k = \frac{1}{n} \sum_{j=1}^{n} y_{jk} \quad k = 1, 2, \ldots, p$$

A measure of spread is provided by the sample *variance*, defined for n measurements on the first variable as

$$s_1^2 = \frac{1}{n} \sum_{j=1}^{n} (y_{j1} - \bar{y}_1)^2$$

where $\bar{y}_1$ is the sample mean of the $\bar{y}_{j1}$. In general, for p variables, we have

$$s_k^2 = \frac{1}{n} \sum_{j=1}^{n} (y_{jk} - \bar{y}_k)^2$$

It is convenient to use double subscripts on the variances in order to indicate their positions in the array. Therefore, we introduce the notation s_{kk} to denote the same variance computed from measurements on the kth variable, and we have the notational identities $s_k^2 = s_{kk}$. The square root of the sample variance, $\sqrt{s_{kk}}$, is known as the sample *standard deviation*. This measure of variation uses the same units as the observations.

A measure of linear association between the measurements of variables i and k is provided by the sample covariance

$$s_{ik} = \frac{1}{n} \sum_{j=1}^{n} (y_{ji} - \bar{y}_i)(y_{jk} - \bar{y}_k)$$

or the average product of the deviations from their respective means. If large values for one variable are observed in conjunction with large values for the other variable, and the small values also occur together, s_{ik} will be positive. If large values from one variable occur with small values for the other variable, s_{ik} will be negative. If there is no particular association between the values for the two variables,

s_{ik} will be approximately zero. We note that the covariance reduces to the sample variance when $i = k$. Moreover, $s_{ik} = s_{ki}$ for all i and k.

The final descriptive statistic is the sample *correlation coefficient* (or *Pearson's product-moment correlation coefficient*). This measure of the linear association between two variables does not depend on the units of measurement. The sample correlation coefficient for the ith and kth variables is defined as

$$r_{ik} = \frac{s_{ik}}{\sqrt{s_{ii}}\sqrt{s_{kk}}} = \frac{\sum_{j=1}^{n}(y_{ji} - \bar{y}_i)(y_{jk} - \bar{y}_k)}{\sqrt{\sum_{j=1}^{n}(y_{ji} - \bar{y}_i)^2}\sqrt{\sum_{j=1}^{n}(y_{jk} - \bar{y}_k)^2}}$$

for $i = 1, 2, \ldots, p$ and $k = 1, 2, \ldots, p$. Note that $r_{ik} = r_{ki}$ for all i and k.

Although the signs of the sample correlation and the sample covariance are the same, the correlation is ordinarily easier to interpret because its magnitude is bounded. To summarize, the sample correlation r has the following properties:

(1) The value of r must be between -1 and $+1$ inclusive.
(2) Here, r measures the strength of the linear association. If $r = 0$, this implies a lack of linear association between the components. Otherwise, the sign of r indicates the direction of the association: $r < 0$ implies a tendency for one value in the pair to be larger than its average when the other is smaller than its average and $r > 0$ implies a tendency for one value of the pair to be large when the other value is large and also for both values to be small together.
(3) The value of r_{ik} remains unchanged if the measurements of the ith variable are changed to $\tilde{y}_{ji} = ay_{ji} + b$, $j = 1, 2, \ldots, n$ and the values of the kth variable are changed to $\tilde{y}_{jk} = cy_{jk} + d$, $j = 1, 2, \ldots, n$, provided that the constants a and c have the same sign.

Thus, the descriptive statistics computed from n measurements on p variables can also be organized into arrays:

$$\text{Sample means } \overline{\mathbf{y}} = \begin{bmatrix} \overline{y}_1 & \overline{y}_2 & \cdots & \overline{y}_p \end{bmatrix}$$

$$\text{Sample variances and covariances } \mathbf{S}_n = \begin{bmatrix} s_{11} & s_{12} & \cdots & s_{1p} \\ s_{21} & s_{22} & \cdots & s_{2p} \\ \vdots & \vdots & \cdots & \vdots \\ s_{p1} & s_{p2} & \cdots & s_{pp} \end{bmatrix}$$

$$\text{Sample correlations } \mathbf{R} = \begin{bmatrix} 1 & r_{12} & \cdots & r_{1p} \\ r_{21} & 1 & \cdots & r_{2p} \\ \vdots & \vdots & \cdots & \vdots \\ r_{p1} & r_{p2} & \cdots & 1 \end{bmatrix}$$

The sample mean array is denoted by $\overline{\mathbf{y}}$, the sample variance and covariance array by the capital letter $\mathbf{S}_n$, and the sample correlation array by $\mathbf{R}$. The subscript n on the array $\mathbf{S}_n$ is a mnemonic device used to remind you that n is employed as a divisor for the elements s_{ik}. The size of all of the arrays is determined by the number of variables, p.

2.3 Matrix Decomposition

Multivariate data can be conveniently displayed as an array of numbers. In general, a rectangular array of numbers with, for instance, n rows and p columns is called a matrix of dimension $n \times p$. The study of multivariate methods is greatly facilitated by the use of matrix algebra. The study of multivariate methods is greatly facilitated by the use of matrix algebra. We begin by introducing some very basic concepts that are essential to both geometrical interpretations and algebraic explanations of subsequent statistical techniques.

2.3.1   Singular Value Decomposition

Let $\mathbf{A}$ be an $m \times k$ matrix of real numbers. Then, there exist an $m \times m$ orthogonal matrix $\mathbf{U}$ and a $k \times k$ orthogonal matrix V such that

$$\mathbf{A} = \mathbf{U}\mathbf{\Lambda}\mathbf{V}^T \tag{2.2}$$

where the $m \times k$ matrix $\mathbf{A}$ has (i, i) entry $\lambda_i \geq 0$ for $i = 1, 2, \ldots, \min(m, k)$ and the other entries are zero. The positive constants λ_i are called the singular values of $\mathbf{A}$.

The singular value decomposition can also be expressed as a matrix expansion that depends on the rank r of $\mathbf{A}$. Specifically, there exist r positive constants $\lambda_1, \lambda_2, \ldots \lambda_r$, r orthogonal $m \times 1$ unit vectors $\mathbf{u}_1, \mathbf{u}_2, \ldots \mathbf{u}_r$ and r orthogonal $k \times 1$ unit vectors $\mathbf{v}_1, \mathbf{v}_2, \ldots \mathbf{v}_r$, such that

$$\mathbf{A} = \sum_{i=1}^{r} \lambda_i \mathbf{u}_i \mathbf{v}_i^T = \mathbf{U}_r \mathbf{\Lambda}_r \mathbf{V}_r^T$$

where $\mathbf{U}_r = \lceil \mathbf{u}_1, \mathbf{u}_2, \ldots, \mathbf{u}_r \rceil$, $\mathbf{V}_r = \lceil \mathbf{v}_1, \mathbf{v}_2, \ldots, \mathbf{v}_r \rceil$, and $\mathbf{\Lambda}_r$ is an $r \times r$ diagonal matrix with diagonal entries λ_i.

Here, $\mathbf{A}\mathbf{A}^T$ has eigenvalue-eigenvector pairs $(\lambda_i^2, \mathbf{u}_i)$, so

$$\mathbf{A}\mathbf{A}^T \mathbf{u}_i = \lambda_i^2 \mathbf{u}_i$$

with $\lambda_1^2, \lambda_2^2, \ldots, \lambda_r^2 > 0 = \lambda_{r+1}^2, \lambda_{r+2}^2, \ldots, \lambda_m^2$ (for $m > k$). Then, $\mathbf{v}_i = \lambda_i^{-1} \mathbf{A}^T \mathbf{u}_i$. Alternatively, the $\mathbf{v}_i$ are the eigenvectors of $\mathbf{A}\mathbf{A}^T$ with the same non-zero eigenvalues λ_i^2.

Proof.   According to Eq. (2.2), we have

$$\mathbf{A}\mathbf{V} = \mathbf{U}\mathbf{\Lambda} \Rightarrow \mathbf{A}^T \mathbf{A}\mathbf{V} = \mathbf{A}^T \mathbf{U}\mathbf{\Lambda} \tag{2.3}$$

and

$$\mathbf{A}\mathbf{V} = \mathbf{U}\mathbf{\Lambda} \tag{2.4}$$

Combining Eqs. (2.3) and (2.4), we have

$$\mathbf{A}^T \mathbf{A}\mathbf{V} = \mathbf{V}\mathbf{\Lambda}^T \mathbf{\Lambda} \tag{2.5}$$

that is,

$$\mathbf{A}^T \mathbf{A} \mathbf{v}_i = \mathbf{\Lambda}_i^2 \mathbf{v}_i \tag{2.6}$$

with $\mathbf{\Lambda}_1^2, \mathbf{\Lambda}_2^2, \ldots, \mathbf{\Lambda}_r^2 > 0 = \mathbf{\Lambda}_{r+1}^2, \mathbf{\Lambda}_{r+2}^2, \ldots, \mathbf{\Lambda}_m^2$ (for $m > k$). In other words, $\mathbf{A}^T \mathbf{A}$ has eigenvalue-eigenvector pairs $(\mathbf{\Lambda}_i^2, \mathbf{v}_i)$, where $i = 1, 2, \ldots, m$. According to Eq. (2.4), we have

$$\mathbf{AV} = \mathbf{U\Lambda} \Rightarrow \mathbf{A} [\mathbf{v}_1, \mathbf{v}_2, \ldots, \mathbf{v}_k]$$

$$= \begin{cases} [\mathbf{u}_1, \mathbf{u}_2, \ldots, \mathbf{u}_m] \left[\text{diag}\,\{\mathbf{\Lambda}_1, \mathbf{\Lambda}_2, \ldots, \mathbf{\Lambda}_m\}\, \mathbf{0}_{m \times (k-m)} \right], & k \geq m \\ [\mathbf{u}_1, \mathbf{u}_2, \ldots, \mathbf{u}_m] \begin{bmatrix} \text{diag}\,\{\mathbf{\Lambda}_1, \mathbf{\Lambda}_2, \ldots, \mathbf{\Lambda}_k\} \\ \mathbf{0}_{(m-k) \times k} \end{bmatrix}, & k < m \end{cases}$$

$$\Rightarrow \mathbf{u}_i = \mathbf{\Lambda}_i^{-1} \mathbf{A} \mathbf{v}_i \tag{2.7}$$

According to Eq. (2.2), we also have

$$\mathbf{A}^T = \mathbf{V\Lambda}^T \mathbf{U}^T \Rightarrow \mathbf{A}^T \mathbf{U} = \mathbf{V\Lambda}^T \Rightarrow \mathbf{A A}^T \mathbf{U} = \mathbf{A V \Lambda}^T \tag{2.8}$$

Combining Eqs. (2.4) and (2.8), we have

$$\mathbf{A A}^T \mathbf{U} = \mathbf{U \Lambda \Lambda}^T \tag{2.9}$$

that is,

$$\mathbf{A A}^T \mathbf{u}_i = \mathbf{\Lambda}_i^2 \mathbf{u}_i \tag{2.10}$$

with $\mathbf{\Lambda}_1^2, \mathbf{\Lambda}_2^2, \ldots, \mathbf{\Lambda}_r^2 > 0 = \mathbf{\Lambda}_{r+1}^2, \mathbf{\Lambda}_{r+2}^2, \ldots, \mathbf{\Lambda}_m^2$ (for $m > k$). In other words, $\mathbf{A A}^T$ has eigenvalue-eigenvector pairs $(\mathbf{\Lambda}_i^2, \mathbf{u}_i)$ where $i = 1, 2, \ldots, m$. According to Eq. (2.8), we have

$$\mathbf{A}^T \mathbf{U} = \mathbf{V\Lambda}^T \Rightarrow \mathbf{A}^T [\mathbf{u}_1, \mathbf{u}_2, \ldots, \mathbf{u}_m]$$

$$= \begin{cases} [\mathbf{v}_1, \mathbf{v}_2, \ldots, \mathbf{v}_k] \left[\text{diag}\{\mathbf{\Lambda}_1, \mathbf{\Lambda}_2, \ldots, \mathbf{\Lambda}_m\}\, \mathbf{0}_{m \times (k-m)} \right]^T, & k \geq m \\ [\mathbf{v}_1, \mathbf{v}_2, \ldots, \mathbf{v}_k] \begin{bmatrix} \text{diag}\,\{\mathbf{\Lambda}_1, \mathbf{\Lambda}_2, \ldots, \mathbf{\Lambda}_k\} \\ \mathbf{0}_{(m-k) \times k} \end{bmatrix}^T, & k < m \end{cases}$$

$$\Rightarrow \mathbf{v}_i = \mathbf{\Lambda}_i^{-1} \mathbf{A}^T \mathbf{u}_i \tag{2.11}$$

Obviously, matrices $\mathbf{A}^T \mathbf{A}$ and $\mathbf{A A}^T$ has the same non-zero eigenvalues $\mathbf{\Lambda}_i^2$. $\qquad \square$

The matrix expansion for the singular-value decomposition written in terms of the full-dimensional matrices $\mathbf{U}, \mathbf{V}, \mathbf{\Lambda}$ is

$$\underset{(m \times k)}{\mathbf{A}} = \underset{(m \times m)}{\mathbf{U}} \ \underset{(m \times k)}{\mathbf{\Lambda}} \ \underset{(k \times k)}{\mathbf{V}}^T$$

where $\mathbf{U}$ has m orthogonal eigenvectors of $\mathbf{A}\mathbf{A}^T$ as its columns and $\mathbf{V}$ has k orthogonal eigenvectors of $\mathbf{A}^T\mathbf{A}$ as its columns.

2.3.2   Low-rank Matrix Approximation

The singular value decomposition is closely connected to a result concerning the approximation of a rectangular matrix by a lower-rank or lower-dimensional matrix. If an $m \times k$ matrix $\mathbf{A}$ is approximated by $\mathbf{B}$, having the same dimension but a lower rank, the sum of squared differences

$$\sum_{i=1}^{m} \sum_{j=1}^{k} (a_{ij} - b_{ij})^2 = \mathrm{tr}\left[(\mathbf{A} - \mathbf{B})(\mathbf{A} - \mathbf{B})^T\right]$$

Let $s < k = rank(\mathbf{A})$, applying singular value decomposition to the matrix $\mathbf{B}$, then

$$\mathbf{B} = \sum_{i=1}^{s} \mathbf{\Lambda}_i \mathbf{u}_i \mathbf{v}_i^T$$

is the rank-s least squares approximation to $\mathbf{A}$, and

$$\min(\mathrm{tr}[(\mathbf{A} - \mathbf{B})(\mathbf{A} - \mathbf{B})^T]) = \sum_{i=s+1}^{k} \mathbf{\Lambda}_i^2$$

over all $m \times k$ matrices $\mathbf{B}$ having rank no greater than s.

Proof.   To establish this result, we use $\mathbf{U}\mathbf{U}^T = \mathbf{I}_m$ and $\mathbf{V}\mathbf{V}^T = \mathbf{I}_k$ to write the sum of squares as

$$\mathrm{tr}\left[(\mathbf{A} - \mathbf{B})(\mathbf{A} - \mathbf{B})^T\right]$$

$$= \mathrm{tr}\left[\mathbf{U}\mathbf{U}^T(\mathbf{A} - \mathbf{B})\mathbf{V}\mathbf{V}^T(\mathbf{A} - \mathbf{B})^T\right]$$

$$= \mathrm{tr}\left[\mathbf{U}^T(\mathbf{A} - \mathbf{B})\mathbf{V}\mathbf{V}^T(\mathbf{A} - \mathbf{B})^T\mathbf{u}\right]$$

$$= \operatorname{tr}\left[(\boldsymbol{\Lambda} - \mathbf{C})(\boldsymbol{\Lambda} - \mathbf{C})^T\right]$$

$$= \sum_{i=1}^{m}\sum_{j=1}^{k}(\boldsymbol{\Lambda}_{ij} - c_{ij})^2 = \sum_{i=1}^{m}(\boldsymbol{\Lambda}_i - c_{ii})^2 + \sum\sum_{i\neq j}c_{ij}^2$$

where $\mathbf{C} = \mathbf{U}^T\mathbf{B}\mathbf{V}$. Clearly, the minimum occurs when $c_{ij} = 0$ for $i \neq j$ and $c_{ii} = \boldsymbol{\Lambda}_i$ for the s largest singular values. The other $c_{ij} = 0$. That is,

$$\sum_{i=1}^{m}(\boldsymbol{\Lambda}_i - c_{ii})^2 + \sum\sum_{i\neq j}c_{ij}^2 = \sum_{i=s+1}^{k}\boldsymbol{\Lambda}_i^2$$

and

$$\mathbf{C} = \left[\begin{bmatrix}\operatorname{diag}\{\boldsymbol{\Lambda}_1, \boldsymbol{\Lambda}_2, \ldots, \boldsymbol{\Lambda}_s\} & \\ & \mathbf{0}_{(m-s)\times(m-s)}\end{bmatrix}\right],$$

$$\mathbf{0}_{(m-k)\times k}$$

so that $\mathbf{B} = \mathbf{U}\mathbf{C}\mathbf{V}^T = \sum_{i=1}^{s}\boldsymbol{\Lambda}_i\mathbf{u}_i\mathbf{v}_i^T$.

2.3.3 Principal Component Analysis

Algebraically, principal components are particular linear combinations of the p random variables $y_1, y_2, \ldots, y_p$. Geometrically, these linear combinations represent the selection of a new coordinate system obtained by rotating the original system, with $y_1, y_2, \ldots, y_p$ as the coordinate axes. The new axes represent the directions with maximum variability and provide a simpler and more parsimonious description of the covariance structure.

As we shall see, principal components depend solely on the covariance matrix $\boldsymbol{\Sigma}$ (or the correlation matrix $\boldsymbol{\rho}$) of $y_1, y_2, \ldots, y_p$. Their development does not require a multivariate normal assumption. On the other hand, principal components derived for multivariate normal populations have useful interpretations in terms of the constant density ellipsoids. Further, inferences can be made from the sample components when the population is multivariate normal.

Let the random vector $\mathbf{y}^T = [y_1, y_2, \ldots, y_p]$ have the covariance matrix $\boldsymbol{\Sigma}$ with eigenvalues $\boldsymbol{\Lambda}_1 \geq \boldsymbol{\Lambda}_2 \geq \cdots \geq \boldsymbol{\Lambda}_p \geq 0$. Consider the linear combinations

$$
\begin{aligned}
z_1 &= \mathbf{a}_1^T \mathbf{y} = a_{11} y_1 + a_{12} y_2 + \cdots + a_{1p} y_p \\
z_2 &= \mathbf{a}_2^T \mathbf{y} = a_{21} y_1 + a_{22} y_2 + \cdots + a_{2p} y_p \\
&\vdots \\
z_p &= \mathbf{a}_P^T \mathbf{y} = a_{p1} y_1 + a_{p2} y_2 + \cdots + a_{pp} y_p
\end{aligned}
\tag{2.12}
$$

Then, the variance and covariance of z_i are

$$
\mathrm{Var}(z_i) = \mathbf{a}_i^T \boldsymbol{\Sigma} \mathbf{a}_i \quad i = 1, 2, \ldots, p \tag{2.13}
$$

$$
\mathrm{Cov}(z_i, z_k) = \mathbf{a}_i^T \boldsymbol{\Sigma} \mathbf{a}_k \quad i, k = 1, 2, \ldots, p \tag{2.14}
$$

The principal components are those uncorrelated linear combinations $z_1, z_2, \ldots, z_p$ whose variances in (2.13) are as large as possible. The first principal component is the linear combination with maximum variance. That is, it maximizes $\mathrm{Var}(z_1) = \mathbf{a}_1^T \boldsymbol{\Sigma} \mathbf{a}_1$. It is clear that $\mathrm{Var}(z_2)$ can be increased by multiplying any $\mathbf{a}_1$ by some constant. To eliminate this indeterminacy, it is convenient to restrict attention to coefficient vectors of unit length. We therefore define the following:

First principal component $=$ linear combination $\mathbf{a}_1^T \mathbf{y}$ that maximizes $\mathrm{Var}(\mathbf{a}_1^T \mathbf{y})$ subject to $\mathbf{a}_1^T \mathbf{a}_1 = 1$

Second principal component $=$ linear combination $\mathbf{a}_2^T \mathbf{y}$ that maximizes $\mathrm{Var}(\mathbf{a}_2^T \mathbf{y})$ subject to $\mathbf{a}_2^T \mathbf{a}_2 = 1$ and $\mathrm{Cov}(\mathbf{a}_1^T \mathbf{y}, \mathbf{a}_2^T \mathbf{y}) = 0$

At the ith step, the ith principal component $=$ linear combination $\mathbf{a}_i^T \mathbf{x}$ that maximizes $\mathrm{Var}(\mathbf{a}_i^T \mathbf{y})$ subject to $\mathbf{a}_1^T \mathbf{a}_1 = 1$ and $\mathrm{Cov}(\mathbf{a}_i^T \mathbf{y}, \mathbf{a}_k^T \mathbf{y}) = 0$ for $k < i$

Let $\boldsymbol{\Sigma}$ be the covariance matrix associated with the random vector $\mathbf{y}^T = [y_1, y_2, \ldots, y_p]$. Let $\boldsymbol{\Sigma}$ have the eigenvalue–eigenvector pairs $(\boldsymbol{\Lambda}_1, \mathbf{e}_1), (\boldsymbol{\Lambda}_2, \mathbf{e}_2), \ldots, (\boldsymbol{\Lambda}_p, \mathbf{e}_p)$, where $\boldsymbol{\Lambda}_1 \geq \boldsymbol{\Lambda}_2 \geq \cdots \geq \boldsymbol{\Lambda}_p \geq 0$. Then, the ith principal component is given by

$$
z_i = \mathbf{e}_1^T \mathbf{y} = e_{i1} y_1 + e_{i2} y_2 + \cdots + e_{ip} y_p, \quad i = 1, 2, \ldots, p \tag{2.15}
$$

With these choices,

$$
\begin{aligned}
\mathrm{Var}(z_i) &= \mathbf{e}_i^T \boldsymbol{\Sigma} \mathbf{e}_i = \boldsymbol{\Lambda}_i \quad i = 1, 2, \ldots, p \\
\mathrm{Cov}(z_i, z_k) &= \mathbf{e}_i^T \boldsymbol{\Sigma} \mathbf{e}_k = 0 \quad i \neq k
\end{aligned}
\tag{2.16}
$$

If some Λ_i are equal, the choices of the corresponding coefficient vectors e_i are not unique. The principal components are uncorrelated and have variances equal to the eigenvalues of Σ.

Let vector $Y^T = [y_1, y_2, \ldots, y_p]$ have covariance matrix Σ, with eigenvalue–eigenvector pairs (Λ_1, e_1), (Λ_2, e_2), $\ldots$, (Λ_p, e_p), where $\Lambda_1 \geq \Lambda_2 \geq \cdots \geq \Lambda_p \geq 0$. Let $z_1 = e_1^T y$, $z_2 = e_2^T y, \ldots, z_p = e_p^T y$ be the principal components. Then,

$$\sigma_{11} + \sigma_{22} + \cdots + \sigma_{pp} = \sum_{i=1}^{p} \text{Var}(y_i)$$

$$= \Lambda_1 + \Lambda_2 + \cdots + \Lambda_p = \sum_{i=1}^{p} \text{Var}(z_i) \tag{2.17}$$

Thus, the total population variance equals $\Lambda_1 + \Lambda_2 + \cdots + \Lambda_p$ and consequently, the proportion of total variance due to (explained by) the kth principal component is

$$\begin{pmatrix} \text{Proportion of total} \\ \text{population variance} \\ \text{due to } k\text{th principal} \\ \text{component} \end{pmatrix} = \frac{\Lambda_k}{\Lambda_1 + \Lambda_2 + \cdots + \Lambda_p} \quad k = 1, 2, \ldots, p$$

$$\tag{2.18}$$

If most (for instance, 80–90%) of the total population, variance, for large p, can be attributed to the first one, two, or three components, then these components can "replace" the original p variables without much loss of information.

Each component of the coefficient vector $e_i^T = [e_{i1}, \ldots, e_{ik}, \ldots e_{ip}]$ also merits inspection. The magnitude of e_{ik} measures the importance of the kth variable to the ith principal component, irrespective of the other variables. In particular, e_{ik} is proportional to the correlation coefficient between z_i and y_k.

Thus, we can see that if $z_1 = e_1^T y$, $z_2 = e_2^T y, \ldots, z_p = e_p^T y$ are the principal components obtained from the covariance matrix Σ, then

$$\rho_{y_i, x_k} = \frac{e_{ik} \sqrt{\Lambda_i}}{\sqrt{\sigma_{kk}}} \quad i, k = 1, \ldots, p \tag{2.19}$$

are the correlation coefficients between the components z_i and the variables y_k. Here, $(\Lambda_1, \mathbf{e}_1), (\Lambda_2, \mathbf{e}_2), \ldots, (\Lambda_p, \mathbf{e}_p)$ are the eigenvalue–eigenvector pairs for $\boldsymbol{\Sigma}$.

Although the correlations of the variables with the principal components often help interpret the components, they measure only the univariate contribution of an individual y to a component z. That is, they do not indicate the importance of an y to a component z in the presence of the other y's. For this reason, some statisticians recommend that only the coefficients e_{ik} and not the correlations be used to interpret the components. Although the coefficients and the correlations can lead to different rankings as measures of the importance of the variables to a given component, it is our experience that these rankings are often not appreciably different. In practice, variables with relatively large coefficients (in absolute value) tend to have relatively large correlations, so the two measures of importance, the first multivariate and the second univariate, frequently give similar results. We recommend that both the coefficients and the correlations be examined to help interpret the principal components.

2.4   Summary

This chapter gives the basic multivariate statistical method for analyzing the characteristics of state variables from the data matrix. The basis vectors, which can be called by eigengrasps, especially for grasping data, can be found through dimensionality reduction, i.e., matrix decomposition. The basis vector is helpful to reveal the hand posture regulation mechanism on the shape of the hand. Further, the control mechanism of the contact force that is used to grasp an object can be suggested by the postures. The dimensionality of reduced space provides an alternative strategy to perform online dexterous grasp planning both for robots needing to find a correct grasp for an object and allowing the planner to work in real time to find a stable grasp.

References

[1]   Anderson, T. W. *An Introduction to Multivariate Statistical Analysis* (3rd ed.). New York: John Wiley, 2003.

[2]   Richard A., and Dean W. *Applied Multivariate Statistical Analysis* (6th ed). Pearson Prentice Hall, 2007.

[3]   Bellman R. *Introduction to Matrix Analysis* (2nd ed.) Philadelphia: Soc for Industrial & Applied Math (SIAM), 1997.

[4]   Eckart C. and G. Young. The approximation of one matrix by another of lower rank. *Psychometrika*, 1, pp. 211–218, 1936.

[5]   Graybill F. A. *Introduction to Matrices with Applications in Statistics*. Belmont, CA: Wadsworth, 1969.

[6]   Halmos P.R. *Finite-Dimensional Vector Spaces*. New York: Springer-Verlag, 1993.

[7]   Johnson R. A. and G. K. *Bhattacharyya. Statistics: Principles and Methods* (5th ed.) New York: John Wiley, 2005.

[8]   Noble B. and J. W. Daniel. *Applied Linear Algebra* (3rd ed.). Englewood Cliffs, NJ: Prentice Hall, 1988.

[9]   Anderson, T. W. Asymptotic theory for principal components analysis. *Annals of Mathematical Statistics*, 34(1), pp. 122–148, 1963.

[10] Dawkins, B. Multivariate analysis of national track records. *The American Statistician*, 43(2), pp. 110–115, 1989.

[11] Girschick, M. A. On the sampling theory of roots of determinantal equations. *Annals of Mathematical Statistics*, 10(3), pp. 203–224, 1939.

[12] Hotelling, H. Simplified calculation of principal components. *Psychometrika*, 1, pp. 27–35, 1936.

[13] Lawley, D. N. On Testing a set of correlation coefficients for equality. *Annals of Mathematical Statistics*, 34(1), pp. 149–151, 1963.

[14] Rencher, A. C. Interpretation of canonical discriminant functions, canonical variates and principal components. *The American Statistician*, 46(3), pp. 217–225, 1992.

Chapter 3

Qualitative Analysis and Quantitative Evaluation of Form-Closure Grasping/Fixturing

Form closure is considered as a purely geometric property of a set of unilateral contact constraints such as those applied on a workpiece by a mechanical fixture. This chapter provides a qualitative analysis of form-closure grasping/fixturing. The necessary and sufficient condition for form-closure grasping/fixturing is derived. Some fundamental problems related to form closure are solved, such as minimum number of frictionless contact points and the way to arrange them to achieve form closure. On the basis of qualitative analysis, the quantitative evaluation of form closure is investigated. To assess quantitatively the form-closure grasping/fixturing, two quantitative indices, one to minimize the sum of all normal contact forces and the other to minimize the maximum normal contact force, are presented. Finally, the given example verifies the analytical method and evaluating indices.

3.1 Introduction

Modern production systems must provide flexibility and rapid response to market demands. When manufacturing products, workpiece locating and clamping are important for safe and accurate machining. With the current advances in tool control techniques, the most time-consuming and labor-extensive part in a machining process is usually the process of fixturing. Flexible fixturing is one of the key technologies in the integration of such systems. Therefore, there have been tremendous efforts on the automation of fixture design [1–11].

Fixtures can be classified into two types: dedicated and reconfigurable [2]. Dedicated fixtures generally imply that they have been designed for specific workpiece geometry. These types of fixtures are most suitable for mass production environments, where they can be discarded at the end of the production life and their costs can be absorbed by a large number of products. Reconfigurable fixtures, on the other hand, are designed for a family of workpiece geometries. The automatically reconfigurable fixture [1,2] was specifically designed for flexible assembly. In previous research, the conditions for a workpiece to be accessible [1,9] to the fixture as well as detachable [1] from the fixture were derived. Yu and Goldberg [11] geometrically formalized robotic fixture loading as a sensor-based compliant assembly problem and gave a planning algorithm for generating fixture-loading plans. Cai *et al.* [5] developed a variational method to conduct robust fixture design to minimize the workpiece positional errors due to workpiece surface and fixture set-up errors. Hockenberger and De Meter [8] presented a model for the quasi-static analysis of workpiece displacement during machining. Lin *et al.* [12] proposed an approach to plan minimum deflection grasps and fixtures. The approach was based on a quality measure that characterizes the grasped/fixtured object's worst-case deflection due to disturbing wrenches lying in the unit wrench ball.

The kinematics of workpiece fixturing are similar to object grasping. The analysis [13, 14] of the motion and force constraints during grasping and manipulation of rigid bodies can be extended to fixtures. Research in grasp/fixture inevitably involves closure analysis, which can be dated back to 1885 when Reuleaux studied the form-closure mechanism for 2-D and 3-D objects. Form closure is a set of mechanical constraints that are placed around a rigid body so that the rigid body motion is not allowed in any direction.

Lakshminarayana [15] provided an algebraic proof that a minimum of seven points of contact is needed to form-close an object. Meyer [16] proved that a robot hand whose fingers make frictionless contact with a convex polyhedral object will be able to find a grasp, where the hand can exert any desired force torque on the object provided the hand has seven fingers. Xiong *et al.* [17, 34] also obtained similar results using geometric approach. Markenscoff *et al.* [18] proved that at least four frictional point contacts are needed to force close an object. Bicchi [19] clarified the differences between form closure and force closure. The form-closure grasp is a desired grasp in that the object is constrained independent of frictional properties. The object is surrounded by fingertips located at appropriate positions for constraining the object's motion in all directions. The object is constrained by purely geometrical means. Frictional constraints are not necessary to hold the object. Therefore, the form-closure grasp is an assured way of constraining objects, and is especially effective for fragile, slippery objects. However, it seems that all of the work gave the necessary, but not the sufficient conditions for form closure.

The linear programming techniques to examine the form-closure grasp conditions were studied [7, 20–22]. Liu [23] proposed an algorithm to compute grasp positions of n fingers on polygonal objects. However, the algorithm is only applicable to the planar form-closure grasps. These approaches did not attempt to seek an optimal solution. Markenscoff and Papadimitriou [24] proposed a criterion for determining the optimal grip points on 2-D polygons, disregarding friction at the contact points. The quantitative tests for force-closure grasps have been proposed by Hershkovitz *et al.* [25, 26], Salunkhe *et al.* [27], Xiong [28], Mantriota [29], and Xiong *et al.* [30, 34]. Hershkovitz *et al.* [25, 26] proposed three grasping quality measures. However, in Hershkovitz's grasping quality model, the finger contact locations are fixed and predetermined. To investigate the effects of these parameters and to obtain optimal finger contact locations, Salunkhe *et al.* [27] extended Hershkovitz's grasping quality model to configuration planning of planar objects. Up to now, the quantitative evaluation for form closure, which is important for grasp/fixture design, is hardly investigated.

The emerging computer-integrated manufacturing system requires automated grasp/fixture design. Traditionally, the design of grasping/fixturing system has been regarded as a manual process relying on human skills and experiences. In order for flexible grasping/fixturing to become a science rather than an art, this chapter describes a systematic approach to the kinematic and static analysis of manipulative tasks performed through mechanical contacts.

3.2   Qualitative Analysis

3.2.1   Kinematic Characteristics of Grasping/Fixturing

Consider an object B, as shown in Fig. 3.1. Choose reference frame $\{\mathbf{O}\}$ fixed relative to the object. Let the frames $\{\mathbf{P}\}$ and $\{\mathbf{C}_i\}$ be the inertial base frame and contact frame at the ith point of contact, respectively. The instantaneous configuration of the object can be described by the orientation and the position of the object frame $\{\mathbf{O}\}$ in terms of the inertial frame $\{\mathbf{P}\}$. We define the configuration manifold [31] $\mathcal{M}$ of the object to be the configuration space of the object. Since three parameters are needed to specify an orientation and three parameters for a position, the configuration manifold $\mathcal{M}$ is 6-D. The nominal configuration of the object and the tangent space to $\mathcal{M}$ at the nominal configuration are represented by $\xi \in \mathcal{M}$ and $\mathbf{T}_\xi \mathcal{M}$, respectively. A generalized velocity (twist) $\mathbf{V} \in \mathbf{T}_\xi \mathcal{M}$ can be

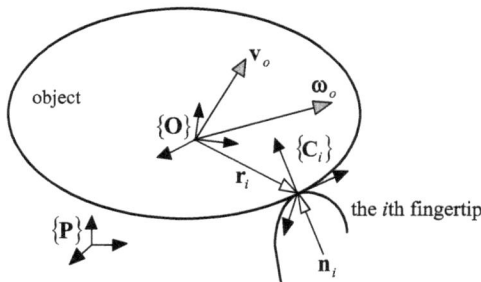

Fig. 3.1.   Grasping/Fixturing system.

written as $\mathbf{V} = (\mathbf{v}_o^T\ \omega_o^T)^T \in \Re^{6\times 1}$, where $\mathbf{v}_o \in \Re^{3\times 1}$ and $\omega_o \in \Re^{3\times 1}$ are called the linear and the angular velocities of the object, respectively. Similarly, we denote the set of generalized forces (wrench) that can be exerted on the object at configuration ξ by $\mathbf{T}_\xi^*\mathcal{M}$. $\mathbf{T}_\xi^*\mathcal{M}$ is the cotangent space to $\mathcal{M}$ at ξ and is the space of all linear functionals of $\mathbf{T}_\xi\mathcal{M}$. An element $\mathbf{F}_e \in \mathbf{T}_\xi^*\mathcal{M}$ is a combination of a force $\mathbf{F} \in \Re^{3\times 1}$ and a moment $\mathbf{M} \in \Re^{3\times 1}$ about the origin O of the inertial frame and can be written as $\mathbf{F}_e = (\mathbf{F}^T\ \mathbf{M}^T)^T$. Assuming that m fingertips grasp the object, there exists only one point of contact between each fingertip and the object to be grasped which is frictionless. Let $\mathbf{n}_i$ and $\mathbf{r}_i = (x_i\ y_i\ z_i)^T$ be the unit inner normal vector of the object and the contact position at the ith point of contact, then the motion constraint of the object at the ith point of contact can be represented as

$$\mathbf{n}_i^T\mathbf{v}_{ci} \geq 0, \quad i = 1,\ldots,m \tag{3.1}$$

where $\mathbf{v}_{ci} \in \Re^{3\times 1}$ is the velocity of the object at the ith point of contact.

Equation (3.1) implies that the object can only move in the closed half-space

$$[\mathcal{H}_i^+] = \{\mathbf{v}_{ci}|\mathbf{n}_i^T\mathbf{v}_{ci} \geq 0,\ \mathbf{v}_{ci} \in \Re^{3\times 1}\}$$

which is divided by the plane

$$\mathcal{H}_i = \{\mathbf{v}_{ci}|\ \mathbf{n}_i^T\mathbf{v}_{ci} = 0,\ \mathbf{v}_{ci} \in \Re^{3\times 1}\}$$

from the entire space $\Re^3$, as shown in Fig. 3.2.

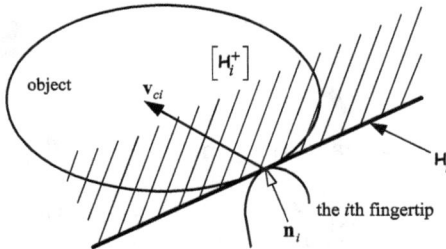

Fig. 3.2.   Closed-half space and hyperplane.

Since

$$\mathbf{v}_{ci} = \mathbf{v}_o + \omega_o \times \mathbf{r}_i \tag{3.2}$$

Eq. (3.1) can be rewritten as

$$\mathbf{n}_i^T \begin{pmatrix} \mathbf{I} \\ \mathbf{R}_i \end{pmatrix}^T \mathbf{V} \geq 0 \tag{3.3}$$

where $\mathbf{I} \in \Re^{3 \times 3}$ is an identity matrix and

$$\mathbf{R}_i = \mathbf{r}_i \times = \begin{pmatrix} 0 & -z_i & y_i \\ z_i & 0 & -x_i \\ -y_i & x_i & 0 \end{pmatrix} \in so(3)$$

$so(3)$ is the Lie algebra of the special orthogonal group $\mathbf{SO}(3)$. Equation (3.3) implies that the object can only move in the closed half-space

$$[\mathcal{H}_i^+] = \left\{ \mathbf{V} \,\middle|\, \mathbf{n}_i^T \begin{pmatrix} \mathbf{I} \\ \mathbf{R}_i \end{pmatrix}^T \mathbf{V} \geq 0, \ \mathbf{V} \in \mathbf{T}_\xi \mathcal{M} \right\}$$

which is divided by the hyperplane

$$\mathcal{H}_i = \left\{ \mathbf{V} \,\middle|\, \mathbf{n}_i^T \begin{pmatrix} \mathbf{I} \\ \mathbf{R}_i \end{pmatrix}^T \mathbf{V} = 0, \ \mathbf{V} \in \mathbf{T}_\xi \mathcal{M} \right\}$$

from the entire space $\Re^6$.

Then the motion constraint of the object with m points of contact can be represented as

$$\mathbf{N}^T \mathbf{G}^T \mathbf{V} \geq \mathbf{0} \tag{3.4}$$

where

$$\mathbf{N} = \mathrm{diag}(\,\mathbf{n}_1 \cdots \mathbf{n}_m\,) \in \Re^{3m \times m}$$

$$\mathbf{G} = \begin{pmatrix} \mathbf{I} & \cdots & \mathbf{I} \\ \mathbf{R}_1 & \cdots & \mathbf{R}_m \end{pmatrix} \in \Re^{6 \times 3m}$$

and the matrix $\mathbf{G}$ is referred to as grasp matrix.

Equation (3.4) means that the object can only move in a convex polyhedron in the space $\Re^6$ as follows:

$$\mathcal{K} = \left\{ \mathbf{V} \mid \mathbf{N}^T \mathbf{G}^T \mathbf{V} \geq 0, \ \mathbf{V} \in \mathbf{T}_\xi \mathcal{M} \right\} \tag{3.5}$$

The convex polyhedron $\mathcal{K}$ is generated by intersecting m closed half-spaces

$$[\mathbf{H}_i^+] = \left\{ \mathbf{V} \mid \mathbf{n}_i^T \begin{pmatrix} \mathbf{I} \\ \mathbf{R}_i \end{pmatrix}^T \mathbf{V} \geq 0, \ \mathbf{V} \in \mathbf{T}_\xi \mathcal{M} \right\} (i = 1, \ldots, m)$$

3.2.2 Discriminances of Form-Closure Grasping/Fixturing

If there does not exist any non-zero feasible motion direction in the entire convex polyhedron $\mathcal{K}$, that is,

$$\mathcal{K} = \{\mathbf{0}\} \tag{3.6}$$

then the corresponding grasping/fixturing is referred to as form closure.

Equation (3.6) shows that the convex polyhedron $\mathcal{K}$ does not contain any elements other than $\mathbf{0}$ for a form-closure grasp, that is, form-closure grasping/fixturing completely immobilizes an object. In contrast, if the convex polyhedron $\mathcal{K}$ contains non-zero element, then the object will be able to move in one or some related directions, which means the grasping/fixturing is not form-closure. Thus, Eq. (3.6) can be used as a qualitative measure for judging form-closure grasping/fixturing.

The problem of judging form closure can be changed to the following linear programming problem. If the virtual displacement generated by exerting external wrench $\mathbf{F}_e \in \mathbf{T}_\xi^* \mathcal{M}$ on the object is represented by $\mathbf{v} = \left(\mathbf{v}_o^T \ \boldsymbol{\omega}_o^T \right)^T \in \mathbf{T}_\xi \mathcal{M}$, then the virtual work of the grasping/fixturing can be described as $\mathbf{F}_e^T \mathbf{V}$. Thus, the problem of judging form closure can be changed into the problem of verifying whether the following linear programming has solutions or not:

$$\begin{aligned} & \text{maximize } \mathbf{F}_e^T \mathbf{V} \\ & \text{subject to } \mathbf{N}^T \mathbf{G}^T \mathbf{V} \geq 0 \end{aligned} \tag{3.7}$$

When the linear programming problem (3.7) has non-zero solutions, which means that the virtual work is not zero, the object can move. Thus, the grasping/fixturing is not form closure. On the contrary, when the linear programming problem (3.7) has zero solutions only, the virtual work is zero. Then, the object is in its equilibrium state, with the corresponding grasping/fixturing being form closure.

According to the duality theory of linear programming, we can obtain the necessary and sufficient conditions for form closure as follows:

If and only if the constraint matrix $\mathbf{GN}$ is full column rank, there exists a vector $\mathbf{0} \leq \mathbf{y} \in \Re^{m \times 1}$ such that $\mathbf{GNy} = \mathbf{0}$, then the grasping/fixturing with m frictionless contact points is form closure.

3.2.3   Minimum Number of Contacts with Frictionless for Form-Closure Grasping/Fixturing

Let f_{ciz} be the normal force of the ith fingertip/fixel exerted on the object, and the wrench generated by the force f_{ciz} can be represented as

$$\begin{pmatrix} \mathbf{f}_i \\ \mathbf{m}_i \end{pmatrix} = \begin{pmatrix} \mathbf{n}_i \\ \mathbf{r}_i \times \mathbf{n}_i \end{pmatrix} f_{ciz} \tag{3.8}$$

Thus, the wrench $\mathbf{F}_e = (\mathbf{F}^T \ \mathbf{M}^T) \in \mathbf{T}_\xi^* \mathcal{M}$ generated by m contact forces can be described as

$$\mathbf{F}_e = \begin{pmatrix} \mathbf{F} \\ \mathbf{M} \end{pmatrix} = \mathbf{G} \cdot \mathbf{N} \cdot^c \mathbf{f}_c \tag{3.9}$$

where $^c\mathbf{f}_c = \begin{pmatrix} f_{c1z} & \cdots & f_{cmz} \end{pmatrix}^T \in \Re^{m \times 1}$ is referred to as grasping/fixturing force, $\mathbf{F} = \sum_{i=1}^m \mathbf{f}_i$, and $\mathbf{M} = \sum_{i=1}^m \mathbf{m}_i$.

If the force $\mathbf{F}_e$ represents the external wrench exerted on the object, then Eq. (3.9) describes the equilibrium constraint of the grasping/fixturing forces. In addition, since the fingertip/fixel can only push, not pull the object, the unilateral constraint of the contact force f_{ciz} must be satisfied which can be described as

$$f_{ciz} \geq 0, \quad i = 1, \ldots, m \tag{3.10}$$

Now, we rewrite Eq. (3.9) as

$$\mathbf{G}^{\#} \cdot {}^{c}\mathbf{f}_c = \mathbf{F}_e \qquad (3.11)$$

where $\mathbf{G}^{\#} = \mathbf{GN} = \begin{bmatrix} \mathbf{g}_1, \ldots, \mathbf{g}_m \end{bmatrix}$, $\mathbf{g}_i \in \varphi \subset \Re^6$ $(i = 1, \ldots, m)$ is the ith column of the matrix $\mathbf{G}^{\#}$, that is, the contact wrench exerted on the object at the ith contact.

Let

$$\lambda_i = \frac{f_{ciz}}{f_{c1z} + \cdots + f_{cmz}}$$

$$\widetilde{\mathbf{F}}_e = \frac{1}{f_{c1z} + \cdots + f_{cmz}} \mathbf{F}_e$$

Then, Eq. (3.11) can be rewritten as

$$\widetilde{\mathbf{F}}_e = \sum_{i=1}^{m} \lambda_i \mathbf{g}_i, \ \sum_{i=1}^{m} \lambda_i = 1 \quad \text{and} \quad 0 \leq \lambda_i \in \Re \qquad (3.12)$$

Equation (3.12) means that the external wrench $\widetilde{\mathbf{F}}_e$ is the convex combination [32] of a finite number of contact wrenches $\mathbf{g}_i \in \varphi \subset \Re^6$ $(i = 1, \ldots, m)$. The convex combination is in fact a point in the external wrench space $\mathbf{T}_{\xi}^{*}\mathcal{M}$.

The collection of all convex combinations of contact wrenches from φ is the convex hull $co(\varphi)$ in $\Re^6$, that is,

$$co(\varphi) = co\left\{ \mathbf{g}_1, \ldots, \mathbf{g}_m \right\} = \left\{ \widetilde{\mathbf{F}}_e \ \middle| \ \begin{array}{l} \widetilde{\mathbf{F}}_e = \sum_{i=1}^{m} \lambda_i \mathbf{g}_i, \ \mathbf{g}_i \in \varphi, \\ \sum_{i=1}^{m} \lambda_i = 1, \ 0 \leq \lambda_i \in \Re \\ \text{for all } i, \text{ and} \\ m \text{ is an positive integer} \end{array} \right\}$$

$$(3.13)$$

From the viewpoint of force, we have the following definition.

Definition 1. If any external wrench applied at the grasped/fixtured object can be balanced by grasping/fixturing forces, then such grasping/fixturing is referred to as form-closure grasping/fixturing.

Definition 1 means that a necessary and sufficient condition for form closure is that the convex hull $co(\varphi)$ spans the entire wrench space $\Re^b$ ($b = 3$ for plane grasping/fixturing, $b = 6$ for space grasping/fixturing). Thus, we can derive the following theorem. Before stating Theorem 1, we give an additional definition.

Definition 2. If $\forall \mathbf{x} \in \Re^n$, $\exists \alpha > 0$ (α is a constant) such that $\mathbf{x} \in \beta \mathcal{M}$ for $\forall \beta \geq \alpha$, then the set $\mathcal{M}$ is referred to as an attractive set.

Theorem 1. *If a grasping/fixturing is said to be form closure, then*

(1) *the convex hull of the contact wrenches $co(\varphi)$ is an attractive set;*
(2) *the origin of the wrench space lies strictly in the interior of the convex hull of the contact wrenches, that is $\mathbf{0} \in int(co(\varphi))$;*
(3) *at least seven points of contact which are frictionless are needed for 3-D grasping/fixturing and 4 points for 2-D grasping/fixturing.*

Proof.   We first prove Property (1). From the definition of form closure, given any external wrench $\widetilde{\mathbf{F}}_e \in \Re^b$, there exists the positive normal contact force $\lambda_i > 0$ ($i = 1, \ldots, m$) such that $\widetilde{\mathbf{F}}_e = \sum_{i=1}^{m} \lambda_i \mathbf{g}_i \in co(\varphi)$. It is clear that there exists $\alpha > 0$ such that $\widetilde{\mathbf{F}}_e \in \beta(co(\varphi))$ for any $\beta \geq \alpha$ (α is a constant).

Then, we prove Property (2). When $\mathbf{0} \in int(co(\varphi))$, it is clear that the convex hull $co(\varphi)$ is attractive. Now, we assume that $\mathbf{0} \notin int(co(\varphi))$. It is known that the set $co(\varphi)$ is a convex set. Then, we can draw a hyperplane through the origin $\mathbf{0}$ such that the convex hull $co(\varphi)$ is located on one side of the hyperplane, that is, there exists a vector $\mathbf{z} \in \Re^b$ such that

$$\langle \mathbf{x}, \mathbf{z} \rangle \leq 0, \quad \forall \mathbf{x} \in co(\varphi) \tag{3.14}$$

When a vector $\mathbf{y} \in \Re^b$ satisfies $\langle \mathbf{y}, \mathbf{z} \rangle > 0$, we have $\langle \alpha \mathbf{y}, \mathbf{z} \rangle > 0$ for any $\alpha > 0$, as shown in Fig. 3.3. Thus, $\mathbf{y} \notin \beta(co(\varphi))$ for any $\beta > 0$, which shows that the set $co(\varphi)$ is not attractive. This result is contradictory to the fact that the set $co(\varphi)$ is attractive. Therefore, a necessary and sufficient condition for form closure is that $\mathbf{0} \in int(co(\varphi))$.

The proof of Property (3) is straightforward. From Definition 1, a necessary and sufficient condition for form closure is that the convex

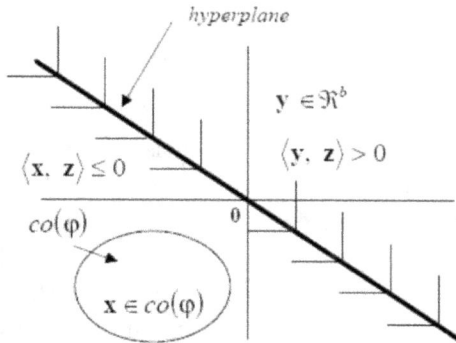

Fig. 3.3. Non-attractive set.

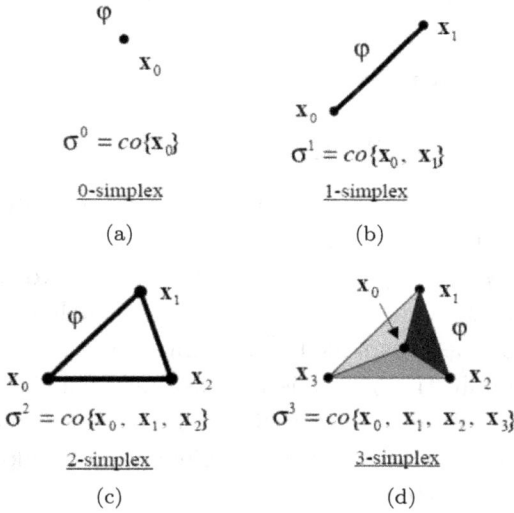

$$\sigma^0 = co\{x_0\}$$

0-simplex

(a)

$$\sigma^1 = co\{x_0, x_1\}$$

1-simplex

(b)

$$\sigma^2 = co\{x_0, x_1, x_2\}$$

2-simplex

(c)

$$\sigma^3 = co\{x_0, x_1, x_2, x_3\}$$

3-simplex

(d)

Fig. 3.4. Some simplexes.

hull $co(\varphi)$ spans the entire wrench space $\Re^b$ which means that the convex hull $co(\varphi)$ must be a b-dimensional simplex σ^b, that is, a convex hull of a set of $b + 1$ affinely independent vectors (see Fig. 3.4). Examples of k-dimensional simplexes for $k = 0, 1, 2, 3$ are illustrated in Fig. 3.4(a–d). Thus, to satisfy form closure, at least $b + 1$ points of frictionless contact are needed, that is, at least 7 and 4 points of frictionless contact are needed for 3-D and 2-D grasping/fixturing, respectively. Moreover, at least $b+1$ contact wrenches in a finite number of contact wrenches $\mathbf{g}_i \in \varphi \subset \Re^b$ $(i = 1, \ldots, m)$ are affinely independent. □

In Theorem 1, Property (3) is consistent with the most fundamental principle in the fixture design [33]. We know that any rigid workpiece has 6 degrees of freedom and 12 directions of motion in the space $\Re^6$. Locators can prevent motions in one direction only; therefore, for complete location, exactly 6 locators are required. Using a clamping device restricts the remaining motion directions. That is the principle: 6 locators and 1 clamp in the fixture design. However, any rigid workpiece has only 3 degrees of freedom in 2-D space $\Re^3$, thus 4 contact points are enough for constraining the workpiece.

3.3   Quantitative Evaluation

3.3.1   Evaluation Criteria

In Section 3.2, we analyzed qualitatively the form closure of grasping/fixturing. However, just judging by form closure is insufficient for grasping/fixture planning. Here, we answer such a question of how to construct a form closure such that the contact configuration of the grasp/fixture becomes more reasonable.

From the qualitative analysis of grasping/fixturing, we find that there are many contact configurations that satisfy the necessary and sufficient condition of form closure. Generally speaking, given the external wrench exerted on the object, the necessary unisense contact forces that balance the external wrench vary with contact configurations. The grasping/fixturing effect on the object varies with different contact forces as well. Hard squeezing may deform or damage the object. On the contrary, the reasonable smaller contact forces can guarantee stable grasping/fixturing without damaging the object. To save energy, and not to damage the object, one always expects that the best grasp/fixture should be the one that has small contact forces. Thus, the objective of planning grasp/fixture is to seek the contact configuration corresponding to the smaller contact forces.

Here, we define an objective function as follows:

$$\text{minimize} \sum_{i=1}^{m} f_{ciz} \qquad (3.15)$$

The procedure of seeking the optimal contact configuration is changed to the following nonlinear programming problem which minimizes the sum of all normal contact force under the constraints of Eqs. (3.9) and (3.10):

$$
\begin{aligned}
&\text{minimize} \ \left(\Phi_1(\eta) = \sum_{i=1}^{m} f_{ciz} \right) \\
&\text{subject to } \mathbf{G} \cdot \mathbf{N} \cdot {}^c\mathbf{f}_c = \mathbf{F}_e, \ f_{ciz} \geq f_{i\_limit} > 0, \\
&\mathbf{r}_i = \{ (x_i \ \ y_i \ \ z_i) \in (x \ \ y \ \ z) | \ S(x, \, y, \, z) = 0 \}, \\
&(x_i \ \ y_i \ \ z_i) \in g(x \ \ y \ \ z), \\
&i = 1, \ldots, m,
\end{aligned}
\qquad (3.16)
$$

where $\eta = (f_{c1z}, \ldots, f_{cmz}, x_1, y_1, z_1, \ldots, x_m, y_m, z_m)^T \in \Re^{4m \times 1}$ is the design variable, f_{i_limit} is the lower limit of the ith normal contact force that ensures the object to be grasped/fixtured stably, $S(x, y, z) = 0$ represents the profile equation of the object surface, and $g(x \ \ y \ \ z)$ denotes the feasible grasp/fixture domain.

However, in planning the contact configuration, we must consider that the contact force of each fingertip/fixture is generally not the same. If the capability of withstanding the normal contact force for the grasped/fixtured object is the same at each contact point, then when the squeezing damage occurs, it will occur at the contact point where the normal contact force exerted on the object is the largest. Based on the fact mentioned above, we give a modified nonlinear programming method as follows:

$$
\begin{aligned}
&\text{minimize} \ (\Phi_2(\eta) = \text{maximize} \ \{ f_{c1z}, \ldots, f_{cmz} \}) \\
&\text{subject to } \mathbf{G} \cdot \mathbf{N} \cdot {}^c\mathbf{f}_c = \mathbf{F}_e, f_{ciz} \geq f_{i\_limit} > 0, \\
&\mathbf{r}_i = \{ (x_i \ \ y_i \ \ z_i) \in (x \ \ y \ \ z) | S(x, y, z) = 0 \}, \\
&(x_i \ \ y_i \ \ z_i) \in g(x \ \ y \ \ z), \\
&i = 1, \ldots, m,
\end{aligned}
\qquad (3.17)
$$

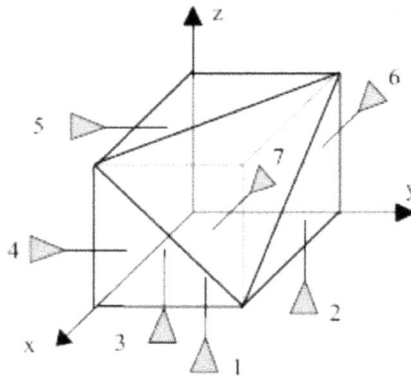

Fig. 3.5. Seven contact points constrain an object (only one contact point on the incline plane).

In the nonlinear programming method (3.17), the objective of planning the contact configuration is to minimize the maximum normal contact force under the constraints of Eqs. (3.9) and (3.10) so that the possibility of squeezing damage can be minimized.

3.3.2   Numerical Example

Consider an example where an object is grasped with seven frictionless contact points (see Fig. 3.5); the object is in fact a cubic rigid body with one corner cut out.

Assume that the side length of the cubic object is 100 unit and the weight of the object (after cutting a corner) is 10 unit. The coordinates of the weight center can be calculated which are (125/3, 125/3, 125/3). At the same time, assume that there are three contact points on the bottom face of the object, two contact points on the left side face, and one contact point on the back face and the incline, respectively. In order to withstand the external disturbing wrench, the lower limits of all normal contact forces are set as 5 unit, i.e., $f_{i\_limit} = 5$, $i = 1, \ldots, 7$. Using the nonlinear programming method described in (3.16), we obtain the optimal contact configuration, the corresponding coordinates, and the normal contact forces of the seven contact points, as shown in Table 3.1.

Table 3.1. Optimal contact configuration corresponding to Fig. 3.3 (obtained using nonlinear programming (3.16)).

| Contact point | Coordinates of contact points | | | Contact forces |
	x	y	z	
1	56.114	22.831	0	8.474
2	48.002	58.513	0	5.000
3	86.389	63.384	0	6.526
4	66.407	0	63.856	5.000
5	85.042	0	50.395	5.000
6	0	31.676	50.039	10.000
7	65.646	63.265	71.090	17.321

Table 3.2. Optimal contact configuration corresponding to Fig. 3.3 (obtained using nonlinear programming (3.17)).

| Contact point | Coordinates of contact points | | | Contact forces |
	x	y	z	
1	57.753	23.226	0	8.690
2	44.186	66.568	0	5.017
3	85.712	63.528	0	6.293
4	60.159	0	56.586	5.000
5	80.897	0	48.234	5.000
6	0	40.378	53.998	10.000
7	59.149	70.666	70.185	17.321

Now, we plan the contact configuration corresponding to Fig. 3.5 using the nonlinear programming (3.17) again. The results are shown in Table 3.2.

Comparing Tables 3.1 and 3.2, we can find that when there are 3 contact points on the bottom face of the object, 2 contact points on the left side face, and 1 contact point on the back face and the incline respectively, the optimal contact configurations are almost similar, no matter whether we use the nonlinear programming (3.16) or (3.17). Moreover, in both cases, the difference between the obtained maximum and minimum contact forces is significant, which indicates that we must adjust the contact configuration further so that the contact forces can be distributed more evenly.

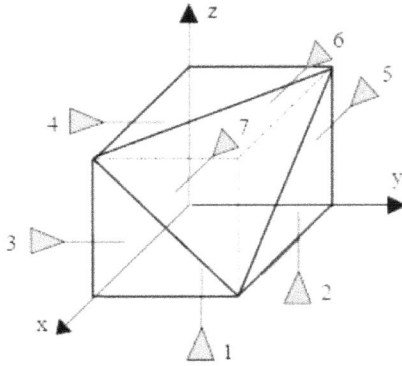

Fig. 3.6.   Seven contact points constrain an object (two contact points on the incline plane).

Table 3.3.   Optimal contact configuration corresponding to Fig. 3.6 (obtained using nonlinear programming (3.16)).

| Contact point | Coordinates of contact points | | | Contact forces |
	x	y	z	
1	51.888	48.924	0	13.427
2	10.830	11.601	0	6.573
3	15.696	0	18.379	5.000
4	57.662	0	58.098	5.000
5	0	0.000	4.031	10.000
6	83.672	73.031	43.297	5.000
7	52.472	63.802	83.726	12.320

Now we assume that there are only 2 contact points on the bottom face of the object, 2 contact points on the incline, and the number of contact points on the other faces does not change, as shown in Fig. 3.6. Meanwhile, assume that the weight of the object and constrained conditions are similar to those in Fig. 3.3.

Using the nonlinear programming method described in Eq. (3.16), we obtain the optimal contact configuration, and the normal contact forces of the 7 contact points corresponding to Fig. 3.6 are shown in Table 3.3.

Table 3.4. Optimal contact configuration corresponding to Fig. 3.6 (obtained using nonlinear programming (3.17)).

Contact point	Coordinates of contact points			Contact forces
	x	y	z	
1	56.558	54.073	0	10.000
2	13.694	13.688	0	10.000
3	15.794	0	16.990	5.000
4	56.408	0	55.065	5.000
5	0	0.002	1.420	10.000
6	78.347	70.685	50.968	8.543
7	42.097	60.542	97.361	8.778

Then we plan the contact configuration corresponding to Fig. 3.6 using the nonlinear programming (3.17) again. The related results are shown in Table 3.4.

From Tables 3.3 and 3.4, we note that when the contact points are arranged as in Fig. 3.6, the difference between the maximum and minimum contact forces corresponding to the optimal contact configuration is much smaller than that in Fig. 3.3, which shows that the contact configuration in Fig. 3.6 is more reasonable than that in Fig. 3.3.

Furthermore, comparing Tables 3.3 and 3.4, we can find that the difference between the maximum and minimum contact forces in Table 3.4 is much smaller than that in Table 3.3. That is, the contact configuration in Table 3.4 is more reasonable than that in Table 3.3, which implies that the nonlinear programming method described in Eq. (3.17),which is used to determine the optimal contact configuration for form-closure grasping, is a more reasonable method.

3.4 Summary

In this chapter, the motion and the force constraints of grasped/fixtured objects are analyzed qualitatively from the viewpoint of manifold. A qualitative measure for judging form closure is presented. The necessary and sufficient conditions for formclosure are given. Some properties related to formclosure are proved. Using the duality theory of linear programming, an efficient method to judge formclosure is developed.

On the basis of qualitative analysis, the evaluation of grasping/fixturing is investigated. Two quantitative indices for evaluating formclosure are presented: one to minimize the sum of all normal contact forces and the other to minimize the maximum normal contact force. Using the two quantitative indices, the optimal contact configuration can be determined in the feasible grasping/fixturing domain. The presented example verified the proposed two quantitative indices. The obtained results showed that the second index, which minimizes the maximum normal contact force, is more reasonable.

In short, this chapter first analyzed qualitatively the form-closure grasping/fixturing. It derived the minimum number of frictionless contact points needed and the way to arrange them to achieve form-closure. To evaluate quantitatively form-closure grasping/fixturing, two quantitative indices are presented. The given example verified the analysis method and evaluating indices, which can be applied to diverse areas, such as reconfigurable fixtures, multifingered robotic grasp, robotic assembly, and manufacturing automation.

References

[1] Asada H., By A. B. Kinematic analysis of workpart fixturing for flexible assembly with automatically reconfigurable fixtures. *IEEE Transactions on Robotics and Automation*, 1(2), pp. 86–94, 1985.

[2] Benhabib B., Chan K. C., and Dai M. Q. A Modular programmable fixturing system. *Transactions of ASME Journal of Engineering for Industry*, 113, pp. 93–100, 1991.

[3] Brost R. C. and Goldberg K. Y. A. Complete algorithm for designing planar fixtures using modular components. *IEEE Transactions on Robotics and Automation*, 12(1), pp. 31–46, 1996.

[4] Brost R. C. and Peters R. R. Automatic design of 3-D fixtures and assembly pallets. *International Journal of Robotics Research*, 17(12), pp. 1243–1281, 1998.

[5] Cai W., Hu S. J., and Yuan J. X. A variational method of robust fixture configuration design for 3-D workpieces. *Transactions of ASME*

Journal of Manufacturing Science and Engineering, 119, pp. 593–602, 1997.

[6] Chen Y. C. On the fixturing of non-prismatic workpieces under frictionless contact models. *Proceedings of the 1995 IEEE International Conference on Robotics and Automation*, pp. 121–126, 1995.

[7] DeMeter E. C. Restraint analysis of fixtures which rely on surface contact. *Transactions of ASME Journal of Engineering for Industry*, 116, pp. 207–215, 1994.

[8] Hockenberger M. J. and DeMeter E. C. The application of meta functions to the quasi-static analysis of workpiece displacement within a machining fixture. *Transactions of ASME Journal of Manufacturing Science and Engineering*, 118, pp. 325–331, 1996.

[9] Li J., Ma W., and Rong Y. Fixturing surface accessibility analysis for automated fixture design. *International Journal of Production Research*, 37(13), pp. 2997–3016, 1999.

[10] Wallack A. S. and Canny J. F. Planning for modular and hybrid fixtures. *Algorithmica*, 19, pp. 40–60, 1997.

[11] Yu K. and Goldberg K. Y. A complete algorithm for fixture loading. *International Journal of Robotics Research*, 17(11), pp. 1214–1224, 1998.

[12] Lin Q., Burdick J., and Rimon E. Minimum-deflection grasps and fixtures. *Proceedings of the 1998 IEEE International Conference on Robotics and Automation*, pp. 3322–3328, 1998.

[13] Hirai S. and Asada H. Kinematics and statics of manipulation using the theory of polyhedral convex cones. *International Journal of Robotics Research*, 12(5), pp. 434–447, 1993.

[14] Salisbury J. K. and Craig J. J. Articulated hands: Force control and kinematic issues. *International Journal of Robotics Research*, 1(1), pp. 4–17, 1982.

[15] Lakshminarayana K. Mechanics of form closure. Paper No. 78-DET-32, pp. 2–8, *New York: American Society of Mechanical Engineering*, 1978.

[16] Meyer W. Seven fingers allow force–torque closure grasps on any convex polyhedron. *Algorithmica*, 15, pp. 278–292, 1993.

[17] Xiong Y. L., Sanger D. J., and Kerr D. R. Geometric modelling of boundless grasps. *Robotica*, 11, pp. 19–26. 1993.

[18] Markenscoff X. and Ni L. Papadimitriou C. H. The geometry of grasping. *International Journal of Robotics Research*, 9(1), pp. 61–72, 1990.

[19] Bicchi A. On the closure properties of robotics grasping. *International Journal of Robotics Research*, 14(4), pp. 319–334, 1995.

[20]   Asada H. and Kitagawa M. Kinematic analysis and planning for form closure grasps by robotic hands. *Robotics and Computer Integrated Manufacturing*, 5(4), pp. 293–299, 1989.

[21]   Liu Y. H. Qualitative test and force optimization of 3D frictional form-closure grasps using linear programming. *IEEE Transactions on Robotics and Automation*, 15(1), pp. 1630–1673, 1999.

[22]   Trinkle J. C. On the stability and instantaneous velocity of grasped frictionless objects. *IEEE Transactions on Robotics and Automation*, 8(5), pp. 560–572, 1992.

[23]   Liu Y. H. Computing n-finger form-closure grasps on polygonal objects. *International Journal of Robotics Research*, 19(2), pp. 149–158, 2000.

[24]   Markenscoff X. and Papadimitriou C. H. Optimal grip of a polygon. *International Journal of Robotics Research*, 8(2), pp. 17–29, 1989.

[25]   Hershkovitz M., Tasch U. and Teboulle M. Toward a Formulation of the human grasping quality sense. *Journal of Robotic Systems*, 12(4), pp. 249–256, 1995.

[26]   Hershkovitz M., Tasch U., Teboulle M., and Tzelgov J. Experimental validation of an optimization formulation of human grasping quality sense. *Journal of Robotic Systems*, 14(11), pp. 753–766, 1997.

[27]   Salunkhe B., Mao W. X., and Tasch U. Optimal grasping formulations that result in high quality and robust configurations. *Journal of Robotic Systems*, 15(12), pp. 713–729, 1998.

[28]   Xiong C. H. and Xiong Y. L. Stability index and contact configuration planning for multifingered grasp. *Journal of Robotic Systems*, 15(4), pp. 183–190, 1998.

[29]   Mantriota G. Communication on optimal grip points for contact stability. *International Journal of Robotics Research*, 18(5), pp. 502–513, 1999.

[30]   Xiong C. H., Li Y. F., Ding H., and Xiong Y. L. On the dynamic stability of grasping. *International Journal of Robotics Research*, 18(9), pp. 951–958, 1999.

[31]   Lang S. *Differential and Riemannian Manifolds*. Berlin: Springer, 1996.

[32]   Rockafellar R. T. *Convex Analysis*. Princeton: Princeton University Press, 1970.

[33]   Nee A. Y. C., Whybrew K., Senthil A. *Advanced Fixture Design for FMS*. London: Springer, 1995.

[34]   Xiong C. H., Li Y. F., Rong Y. K., and Xiong Y. L. Qualitative analysis and quantitative evaluation of fixturing. *Robotics and Computer Integrated Manufacturing*, 18(5-6), pp. 335–342, 2002.

Chapter 4

Stability Index and Contact Configuration Planning of Force-Closure Grasping/Fixturing

It is necessary to plan the contact configuration to guarantee a stable grasp. This chapter discusses the grasping stability of multifingered robot hands. The fingers are assumed to be point contacts with friction. A stability index for evaluating a grasp, which is proportional to the ellipsoidal volume in the grasping task space, is proposed. The invariance of the index is proved under an object linear coordinate transformation and under a change of the torque origin. The similar invariance of the index is also proved under a change of the dimensional unit. The optimal grasping of an object by a multifingered robot hand can be obtained using the stability index to plan the grasp configurations. The index is applicable to plan adaptable fixtures as well. A nonlinear programming method to plan configurations is addressed. Several examples are given using the index to evaluate a grasp, in which the obtained optimal grasping is consistent with what human beings expect. The sensibility of optimal grasping is analyzed in these examples.

4.1   Introduction

A number of articles deal with grasping stability [1–7]. A grasp on an object is in force closure (which is referred to as stable grasp) if and only if we can exert, through the set of contacts, arbitrary force, and moment on this object. Equivalently, any motion of the object is resisted by a contact force [2]. In the contact configuration space, there are many configurations which satisfy the force closure conditions. How to put the fingertips such that the grasp is stable largely depends on the experience. It is, in general, a complex decision process, requiring tradeoffs between many grasp evaluation measures.

Li and Sastry defined a grasp map and gave three task-oriented quality measures for grasping [1]. Nguyen defined the grasp stability on polygonal and polyhedral objects as the minimum distance from any point of contact to the edge of its independent regions [2]. Montana formulated a model of the dynamics of two-fingered grasps, where the state of a grasp is the position of the contact points. On the basis of this model, Montana derived a quantitative measure of contact grasp stability [3]. Xiong *et al.* proposed quantitative tests for force-closure grasps [10]. Nakamura *et al.* developed a method of evaluating contact stability [4]. The concept of contact stability was defined as the ability to keep contact with an object without slipping for a class of unexpected disturbing forces [4]. Xiong *et al.* described the geometric approach to the analysis and synthesis of the force closure grasp with frictional contact on the basis of the concepts of constraint cone and freedom cone [5]. Bicchi [6] investigated the closure properties of robotic grasping. As described in Chapter 3, form closure was considered as a purely geometric property of a set of unilateral constraints, while force closure was related to the capability of the particular robotic device being considered to apply forces through contacts [6]. Varma described a graphical representation of the quality of a grasp [7]. Xiong *et al.* [11] investigated the dynamic stability of grasping/fixturing which is discussed in Chapter 9.

Grasp stability is one of the main problems discussed in most articles mentioned above. However, the frictional force constraints and unisense constraints of the normal contact force are considered in the articles [1–7], selecting the positions of contact points is poor. Randomly chosen contact points may render situations where the frictional constraints between the fingertips and the grasped object can never be satisfied, which results in failure in achieving a stable grasp. Moreover, we note that grasp stability varies for different contact configurations. These facts indicate the need for studies on the quantitative index of grasp stability related to contact configurations.

It will be beneficial for grasp planning to have a quantitative index of grasp stability of evaluating a grasp. In this chapter, we define a grasp map and propose such a quantitative index. Some properties of the index are discussed. A nonlinear programming method for determining the optimal grasping is presented.

4.2 Description of Contacts with Friction

First, we consider the hard finger contact with friction. In this situation, the fingertip contact force is represented in the contact frame $\{\mathbf{C}_i\}$ as follows:

$$^c\mathbf{f}_{ci} = (f_{cix}\, f_{ciy}\, f_{ciz})^T \in \Re^{3\times 1}, \quad i = 1, \ldots, m \quad (4.1)$$

where f_{cix}, f_{ciy}, and f_{ciz} are the orthogonal tangent elements, normal element of the ith fingertip contact force at the contact.

Then, we consider the soft finger contact with friction. In this situation, the fingertip can exert a torque m_{ciz} on the object around the normal direction at the ith contact besides the force f_{cix}, f_{ciy}, and f_{ciz}. The corresponding fingertip contact force is represented as

$$^c\mathbf{f}_{ci} = (f_{cix}\, f_{ciy}\, f_{ciz}\, 0\, 0 m_{ciz})^T \in \Re^{6\times 1}, \quad i = 1, \ldots, m \quad (4.2)$$

The relationship between the external wrench $\mathbf{F}_e \in \Re^{6\times 1}$ (described in the object frame $\{\mathbf{O}\}$) and the contact force $^c\mathbf{f}_c$ is represented as

$$\mathbf{G}_c^o\mathbf{R}^c\mathbf{f}_c = \mathbf{F}_e \tag{4.3}$$

where

$$^c\mathbf{f}_c = \begin{pmatrix} ^c\mathbf{f}_{c1}^T & ^c\mathbf{f}_{c2}^T & \cdots & ^c\mathbf{f}_{cm}^T \end{pmatrix}^T \in \Re^{3m \times 1}$$

$$\mathbf{G} = \begin{pmatrix} \mathbf{I} & \cdots & \mathbf{I} \\ \mathbf{R}_1 & \cdots & \mathbf{R}_m \end{pmatrix} \in \Re^{6 \times 3m}$$

$$_c^o\mathbf{R} = \text{block diag}\begin{pmatrix} _{c1}^o\mathbf{R} & _{c2}^o\mathbf{R} & \cdots & _{cm}^o\mathbf{R} \end{pmatrix} \in \Re^{3m \times 3m}$$

for the hard finger contact with friction model;

$$^c\mathbf{f}_c = \begin{pmatrix} ^c\mathbf{f}_{c1}^T & ^c\mathbf{f}_{c2}^T & \cdots & ^c\mathbf{f}_{cm}^T \end{pmatrix}^T \in \Re^{6m \times 1}$$

$$\mathbf{G} = \begin{pmatrix} \mathbf{I} & \mathbf{0} & \cdots & \mathbf{I} & \mathbf{0} \\ \mathbf{R}_1 & \mathbf{I} & \cdots & \mathbf{R}_m & \mathbf{I} \end{pmatrix} \in \Re^{6 \times 6m}$$

$$_c^o\mathbf{R} = \text{block diag}\begin{pmatrix} _{c1}^o\mathbf{R} & _{c1}^o\mathbf{R} & _{c2}^o\mathbf{R} & _{c2}^o\mathbf{R} & \cdots & _{cm}^o\mathbf{R} & _{cm}^o\mathbf{R} \end{pmatrix} \in \Re^{6m \times 6m}$$

for the soft finger contact with friction model. $_{ci}^o\mathbf{R} \in \Re^{3 \times 3}$ is the orientation matrix of the ith contact frame $\{\mathbf{C}_i\}$ with respect to the object frame $\{\mathbf{O}\}$.

If the number of columns of the grasp matrix $\mathbf{G}$ is more than 6, then the corresponding grasp is called the over-constrained grasp, that is, given the external wrench $\mathbf{F}_e$, we cannot uniquely determine the contact force $^c\mathbf{f}_c$ by using Eq. (4.3). However, the contact force $^c\mathbf{f}_c$ can be represented as

$$^c\mathbf{f}_c = (\mathbf{G}_c^o\mathbf{R})^+ \mathbf{F}_e + \left[\mathbf{I}_L - (\mathbf{G}_c^o\mathbf{R})^+ (\mathbf{G}_c^o\mathbf{R})\right]\boldsymbol{\lambda} \tag{4.4}$$

where $\mathbf{I}_L$ is a $3m \times 3m$ identity matrix, $\boldsymbol{\lambda} \in \Re^{3m \times 1}$ is an arbitrary vector for the hard finger contact with friction model, $\mathbf{I}_L$ is a $6m \times 6m$ identity matrix, and $\boldsymbol{\lambda} \in \Re^{6m \times 1}$ is an arbitrary vector for the soft finger contact with friction model.

Since $_{ci}^o\mathbf{R}$ is the orientation matrix, the Moore–Penrose generalized inverse matrix of the matrix $\mathbf{G}$ can be represented as

$$(\mathbf{G}_c^o\mathbf{R})^+ = {}_c^o\mathbf{R}^T\mathbf{G}^T \left(\mathbf{G}\mathbf{G}^T\right)^{-1} \tag{4.5}$$

In Eq. (4.4), $\mathbf{f}_e = (\mathbf{G}_c^o\mathbf{R})^+ \mathbf{F}_e$ denotes the set of finger forces that can resist the external wrench, while $\mathbf{f}_N = \left[\mathbf{I}_L - (\mathbf{G}_c^o\mathbf{R})^+ (\mathbf{G}_c^o\mathbf{R})\right]\boldsymbol{\lambda}$

denotes the set of internal forces in the null space of the matrix $\mathbf{G}_c^o \mathbf{R}$. The internal forces $\mathbf{f}_N$ can be exerted at the contact points without causing a net wrench on the object and can be used to modify the finger force $\mathbf{f}_c$ to avoid sliding at the contacts (this implies we can plan the internal forces to ensure that the contact forces lie within the friction cone at each contact point). The grasp force planning is discussed in Chapter 5.

4.3 Conditions of Force Closure Grasp

When a multifingered robotic hand grasps an object, the normal elements of the contact force must be positive so that the object can be grasped stably whether the contact is hard contact with friction or soft contact with friction, that is, the constraint (3.10) described in Chapter 3 must be satisfied.

Moreover, the finger contact force must be within the friction cone at each contact for the hard finger contact with friction, that is, the tangential force f_{ci}^t of the finger contact force must satisfy the following constraint:

$$f_{ci}^t = \sqrt{f_{cix}^2 + f_{ciy}^2} \le \mu f_{ciz}, \quad i = 1, \ldots, m \tag{4.6}$$

where μ is the static friction coefficient.

However, the constraints (3.10) and (4.6) are not enough for the soft finger contact with friction. Since there exists interrelationship between the contact force and torque for the soft contact with friction, the friction limit surface [12] constraint must be satisfied so that the sliding and rotation at contacts between the object and fingertips can be avoided, which is represented as follows:

$$\left(\frac{f_{ci}^t}{\mu f_{ciz}} \right)^2 + \left(\frac{m_{ciz}}{m_{\max}} \right)^2 \le 1, \quad i = 1, \ldots, m \tag{4.7}$$

where $m_{\max}$ is the allowable maximum torque without rotation at contacts between the object and fingertips.

It should be noted that the constraint (4.7) is transformed into the constraint (4.6) when $m_{ciz} = 0$ (that is the fingertip cannot exert torque on the object at the ith contact, i.e., the fingertip is hard contact, not soft contact with the object.), which means that the constraint (4.7) is the general one for the grasp with friction.

Particularly, if a grasp satisfies the constraints (3.10) and (4.7), and the rank of the grasp matrix is 6, that is,

$$R\left(\mathbf{G}\right) = 6 \tag{4.8}$$

then we can choose appropriately the grasp force within the friction cone or friction limit surface to withstand the external wrench exerted on the object. Such a grasp is called stable grasp [1] or force closure grasp [13].

4.4   Grasp Stability Index

4.4.1   Definition of the Grasp Stability Index

Without loss generality, assume that the contacts between the fingertips and the object are modeled as the point contacts with friction. From Eq. (4.4), we can make use of the equation $\|\mathbf{f}_e\|^2 = 1$ to define a supersphere in the finger force space. The supersphere will be mapped into a superellipsoid in the workspace. The superellipsoid is referred to as force superellipsoid which is similar to the manipulability ellipsoid defined by Yoshikawa [8]. The force superellipsoid equation is defined by

$$\mathbf{F}_e^T(\mathbf{G}\mathbf{G}^T)^{-1}\mathbf{F}_e = 1 \tag{4.9}$$

The principal axes of the superellipsoid are $\mathbf{e}_1\sigma_1, \ldots, \mathbf{e}_6\sigma_6$, where $\mathbf{e}_i \in \Re^6$ is the ith eigenvector of the matrix $\mathbf{G}\mathbf{G}^T$ and σ_i is the ith singular value of the matrix $\mathbf{G}$ which corresponds to the length of the ith principal axis of the superellipsoid ($i = 1, \ldots, m$). Here we define the product of the singular values as the grasp stability index W, that is,

$$W = \sigma_1\sigma_2\cdots\sigma_6 \tag{4.10}$$

Obviously, the index is related to the profile of the grasped object and the contact configurations (i.e., the grasp matrix $\mathbf{G}$). Since

$det(\mathbf{G}\mathbf{G}^T) = (\sigma_1\sigma_2\cdots\sigma_6)^2$, the index also can be expressed as

$$W = \sqrt{det(\mathbf{G}\mathbf{G}^T)} \qquad (4.11)$$

Particularly, Eq. (4.11) can be simplified further for the trifingered grasp [9, 10]. When three fingers grasp an object, we obtain

$$\mathbf{G}\mathbf{G}^T = \begin{bmatrix} 3\mathbf{I} & \mathbf{R}_1^T + \mathbf{R}_2^T + \mathbf{R}_3^T \\ \mathbf{R}_1 + \mathbf{R}_2 + \mathbf{R}_3 & \mathbf{R}_1\mathbf{R}_1^T + \mathbf{R}_2\mathbf{R}_2^T + \mathbf{R}_3\mathbf{R}_3^T \end{bmatrix}$$

$$= 3 \begin{bmatrix} \mathbf{I} & \frac{\mathbf{R}_1^T + \mathbf{R}_2^T + \mathbf{R}_3^T}{3} \\ \frac{\mathbf{R}_1 + \mathbf{R}_2 + \mathbf{R}_3}{3} & \frac{\mathbf{R}_1\mathbf{R}_1^T + \mathbf{R}_2\mathbf{R}_2^T + \mathbf{R}_3\mathbf{R}_3^T}{3} \end{bmatrix}$$

Let

$$\mathbf{B} = \frac{1}{3}\left(\mathbf{R}_1^T + \mathbf{R}_2^T + \mathbf{R}_3^T\right)$$

$$\mathbf{H} = \frac{1}{3}\left(\mathbf{R}_1\mathbf{R}_1^T + \mathbf{R}_2\mathbf{R}_2^T + \mathbf{R}_3\mathbf{R}_3^T\right)$$

$$\mathbf{E} = \frac{1}{3}\left(\mathbf{R}_1 + \mathbf{R}_2 + \mathbf{R}_3\right)$$

then

$$\mathbf{G}\mathbf{G}^T = 3 \begin{bmatrix} \mathbf{I} & \mathbf{B} \\ \mathbf{E} & \mathbf{H} \end{bmatrix} \qquad (4.12)$$

Let

$$\mathbf{D} = \begin{bmatrix} \mathbf{I} & -\mathbf{B} \\ \mathbf{0} & \mathbf{I} \end{bmatrix} \in \Re^{6\times6}$$

Right-multiplying both sides of Eq. (4.12) with the matrix $\mathbf{D}$ yields

$$(\mathbf{G}\mathbf{G}^T)\mathbf{D} = 3 \begin{bmatrix} \mathbf{I} & \mathbf{B} \\ \mathbf{E} & \mathbf{H} \end{bmatrix} \begin{bmatrix} \mathbf{I} & -\mathbf{B} \\ \mathbf{0} & \mathbf{I} \end{bmatrix} = 3 \begin{bmatrix} \mathbf{I} & \mathbf{0} \\ \mathbf{E} & \mathbf{H} - \mathbf{E}\mathbf{B} \end{bmatrix}$$

Consequently,

$$det(\mathbf{G}\mathbf{G}^T) = 3det(\mathbf{H} - \mathbf{E}\mathbf{B})$$

i.e.,

$$W = \sqrt{3 det(\mathbf{H} - \mathbf{EB})} \qquad (4.13)$$

The volume of the superellipsoid is given by

$$V = K(\sigma_1 \sigma_2 \cdots \sigma_6) = KW \qquad (4.14)$$

where K is a constant; for plane grasp, $K = 4\pi/3$; for spatial grasp, $K = \pi^3/6$.

Therefore, the physical meaning of the index is very clear. The index is proportional to the volume of the force superellipsoid. The larger the volume of the force superellipsoid, the greater ability the grasping has to withstand any disturbance wrench on the grasped object. The volume is 0 when W equals 0 which indicates that the contact configuration becomes singularity; at this time, the grasp will lose the capability of withstanding the external wrench in a direction or several directions. Thus, the grasp is not stable. On the contrary, the volume is the biggest when the index reaches the biggest value which implies the grasp has the best ability to withstand the external wrench in all directions. The superellipsoid becomes a super-sphere when the smallest singular value equals the biggest one (i.e., the condition number of the matrix $\mathbf{G}$ is 1), the contact configuration has been termed isotropic. Hence, we can judge the contact configuration by the index.

4.4.2   Some Properties of the Index

We can see from the representation of the grasp matrix $\mathbf{G}$ that changes the body coordinate, the torque origin, and the dimensional unit results in a different grasp matrix. Can these changes influence the stability index? To answer the problem, we give a theorem as follows:

Theorem. *Stability index is invariant under a body linear coordinate transformation and a change of the torque origin. The index is similar invariant under a change of the dimensional unit.*

Proof. At first, prove the invariance of the index under the linear coordinate transformation. Let the change of the body coordinate from $oxyz$ to $o'x'y'z'$ be denoted by $\mathbf{T}$ as follows:

$$\mathbf{T} = \begin{bmatrix} \mathbf{C} & \mathbf{P} \\ \mathbf{0} & 1 \end{bmatrix}$$

A point $\mathbf{r} = \begin{pmatrix} x & y & z \end{pmatrix}^T$ in the old coordinate is transformed to a point $\tilde{\mathbf{r}} = \mathbf{Cr} + \mathbf{P}$ in the new coordinate. The representation of the matrix $\mathbf{G}$ is transformed under $\mathbf{T}$ (assuming the torque origin coincides with the origin of each body coordinate) to

$$\tilde{\mathbf{G}} = \begin{bmatrix} \mathbf{I} & \cdots & \mathbf{I} \\ (\tilde{\mathbf{r}}_1 \times) & \cdots & (\tilde{\mathbf{r}}_m \times) \end{bmatrix}$$

$$= \begin{bmatrix} \mathbf{I} & \cdots & \mathbf{I} \\ ((\mathbf{Cr}_1 + \mathbf{P}) \times) & \cdots & ((\mathbf{Cr}_m + \mathbf{P}) \times) \end{bmatrix}$$

$$= \begin{bmatrix} \mathbf{I} & \mathbf{0} \\ (\mathbf{P} \times) & \mathbf{C} \end{bmatrix} \begin{bmatrix} \mathbf{I} & \cdots & \mathbf{I} \\ (\mathbf{r}_1 \times) & \cdots & (\mathbf{r}_m \times) \end{bmatrix}$$

$$= \mathbf{T}_1 \mathbf{G}$$

where $\mathbf{C}$ is a 3×3 orthogonal orientation matrix with determinant 1, $\mathbf{P} \in \Re^{3 \times 1}$ is the position vector,

$$\mathbf{T}_1 = \begin{bmatrix} \mathbf{I} & \mathbf{0} \\ (\mathbf{P} \times) & \mathbf{C} \end{bmatrix}$$

and $det\,(\mathbf{T}_1) = 1$. Thus,

$$\tilde{W} = \sqrt{det\,(\tilde{\mathbf{G}}\tilde{\mathbf{G}}^T)} = \sqrt{det\,(\mathbf{T}_1 \mathbf{G} \mathbf{G}^{\mathbf{T}} \mathbf{T}_1^T)} = \sqrt{det\,(\mathbf{G} \mathbf{G}^T)} = W \tag{4.15}$$

Then, prove the invariance of the index W under the change of the torque origin. Assume the torque origin changes from o to s, and $p_1, p_2, \ldots, p_m$ are contact points. The representation of the matrix $\mathbf{G}$

is transformed to

$$\tilde{\mathbf{G}} = \begin{bmatrix} \mathbf{I} & \cdots & \mathbf{I} \\ (\overline{sp_1}\times) & \cdots & (\overline{sp_m}\times) \end{bmatrix}$$

$$= \begin{bmatrix} \mathbf{I} & \cdots & \mathbf{I} \\ ((\overline{so}+\mathbf{r}_1)\times) & \cdots & ((\overline{so}+\mathbf{r}_m)\times) \end{bmatrix}$$

$$= \begin{bmatrix} \mathbf{I} & 0 \\ (\overline{so}\times) & \mathbf{I} \end{bmatrix} \begin{bmatrix} \mathbf{I} & \cdots & \mathbf{I} \\ (\mathbf{r}_1\times) & \cdots & (\mathbf{r}_m\times) \end{bmatrix}$$

$$= \mathbf{T}_2\mathbf{G}$$

where

$$\mathbf{T}_2 = \begin{bmatrix} \mathbf{I} & 0 \\ (\overline{so}\times) & \mathbf{I} \end{bmatrix}$$

and $det\,(T_2) = 1$. Thus,

$$\tilde{W} = \sqrt{det\,(\tilde{\mathbf{G}}\tilde{\mathbf{G}}^T)} = \sqrt{det\,(\mathbf{T}_2\mathbf{G}\mathbf{G}^T\mathbf{T}_2^T)} = \sqrt{det\,(\mathbf{G}\mathbf{G}^T)} = W \tag{4.16}$$

Finally, prove the similar invariance of the index $\mathbf{W}$ under the change of the dimensional unit. Assuming the position vector is changed from $\mathbf{r}_i$ to $\tilde{\mathbf{r}}_i = k\mathbf{r}_i$, where k is a positive constant factor. The new representation of the matrix $\mathbf{G}$ is transformed to

$$\tilde{\mathbf{G}} = \begin{bmatrix} \mathbf{I} & \mathbf{I} & \cdots & \mathbf{I} \\ (\tilde{\mathbf{r}}_1\times) & (\tilde{\mathbf{r}}_2\times) & \cdots & (\tilde{\mathbf{r}}_m\times) \end{bmatrix}$$

$$= \begin{bmatrix} \mathbf{I} & \mathbf{I} & \cdots & \mathbf{I} \\ k(\mathbf{r}_1\times) & k\,(\mathbf{r}_2\times) & \cdots & k(\mathbf{r}_m\times) \end{bmatrix}$$

$$= \begin{bmatrix} \mathbf{I} & 0 \\ 0 & k\mathbf{I} \end{bmatrix} \begin{bmatrix} \mathbf{I} & \mathbf{I} & \cdots & \mathbf{I} \\ (\mathbf{r}_1\times) & (\mathbf{r}_2\times) & \cdots & (\mathbf{r}_m\times) \end{bmatrix}$$

$$= \mathbf{T}_3\mathbf{G}$$

where

$$det\,(T_3) = det \begin{bmatrix} \mathbf{I} & 0 \\ 0 & k\mathbf{I} \end{bmatrix} = k^3$$

Thus,

$$\tilde{W} = \sqrt{det\left(\tilde{\mathbf{G}}\tilde{\mathbf{G}}^{\mathbf{T}}\right)} = \sqrt{det\left(\mathbf{T}_3\mathbf{GG}^T\mathbf{T}_3^T\right)} = k^3\sqrt{det\left(\mathbf{GG}^T\right)} = k^3W \tag{4.17}$$

4.4.3 Contact Configuration Planning

Generally speaking, we cannot locate the contact points of fingertips on any place of the surface of the object to be grasped, such as a hole on the surface of the object, thus, the objective of the contact configuration planning is to select a better configuration in the feasible grasp space so that the grasp has a better stability. Finding optimal contact configuration can be expressed as a nonlinear programming problem as follows:

$$\text{minimize} \quad -W = -\sqrt{det\left(\mathbf{GG}^T\right)}$$
$$\text{subject to}$$
$$\mathbf{r}_i = \{(x_i\ y_i\ z_i)^T \in (x\ y\ z)^T|S(x,\ y,\ z) = 0\} \tag{4.18}$$
$$(x_i\ y_i\ z_i) \in g(x,\ y,\ z)$$
$$i = 1,\ 2,\ \ldots,\ m$$

4.5 Examples

In this section, we calculate the stability index for various grasps and determine the optimal contact configuration in the feasible grasp space from the viewpoint of grasp stability.

Example 1. Consider a trifingered hand grasping a triquetrous object whose cross section is an equilateral triangle, as shown in Fig. 4.1. Assume that the first contact point is fixed at the middle center of a side of the triquetrous object, while the second and third contact points moves on the other two sides in a parallel direction with the first side. The value of W is obtained in Fig. 4.2. The maximum W is obtained for $y = 6$ which shows that the grasp is very stable. A hexagon is a special case of the triquetrous object. It is interesting to note that the optimal contact configuration of the

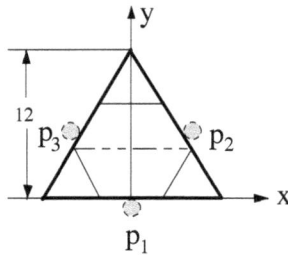

Fig. 4.1.   A triquetrous object grasped by a trifingered hand.

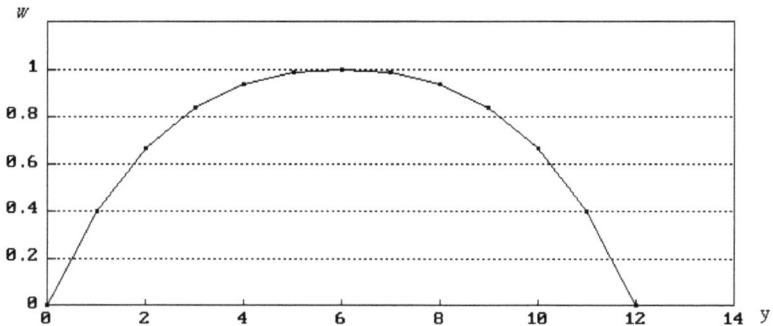

Fig. 4.2.   Stability index W for triquetrous object.

hexagon is same as that of the triquetrous object. In Fig. 4.2, we can see that the value of W varies slowly near the optimal configuration which means that we have a high degree of flexibility in the configuration planning.

Example 2. Consider a trifingered hand grasping a ball, as shown in Fig. 4.3. Without loss generality, assume that the first contact point is located at the North Pole and the second contact point can be moved on the latitude of the ball, while the third contact point can be moved on the surface of the ball. The value of W is obtained for different contact configurations and the results are given in Fig. 4.4. The maximum $W(=1)$ is obtained for $\varphi = 120°$, $\theta = 180°$, and $\gamma = 120°$. The $W = 1$ shows that the contact configuration is optimal. Therefore, the grasp stability is also best. The result is almost coincidence with the grasping way of a human being. In Fig. 4.4, it

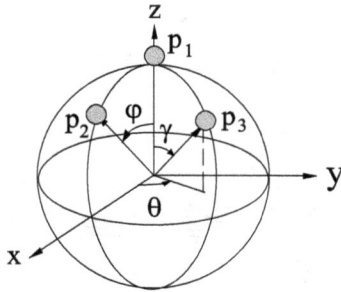

Fig. 4.3. A ball grasped by a trifingered hand.

is easy to see that the value of W varies slowly near the optimal configuration ($W = 1$), which means that we have a high degree of flexibility in determining the contact positions.

Example 3. Let us consider a trifingered hand grasping an object, as shown in Fig. 4.5. The object can be regarded as an ellipsoid. We assume that the parametric equation of the ellipsoid surface is expressed as

$$
\begin{aligned}
x &= 40 \cos \alpha \cos \beta \\
y &= 50 \cos \alpha \sin \beta \\
z &= 30 \sin \alpha
\end{aligned}
\tag{4.19}
$$

where α and β are the angular coordinates on the ellipsoid.

However, as shown in Fig. 4.5, the feasible grasp space is not the whole of the surface of the ellipsoid because there are two holes at the top and bottom of the ellipsoid, respectively. The parametric equation of the elliptical hole at the top is given by

$$
\begin{aligned}
x &= 40 \cos(\pi/3) \cos \beta \\
y &= 50 \cos(\pi/3) \sin \beta \\
z &= 30 \sin(\pi/3)
\end{aligned}
\tag{4.20}
$$

Similarly, the parametric equation of the hole at the bottom is given by

$$
\begin{aligned}
x &= 40 \cos(-2\pi/9) \cos \beta \\
y &= 50 \cos(-2\pi/9) \sin \beta \\
z &= 30 \sin(-2\pi/9)
\end{aligned}
\tag{4.21}
$$

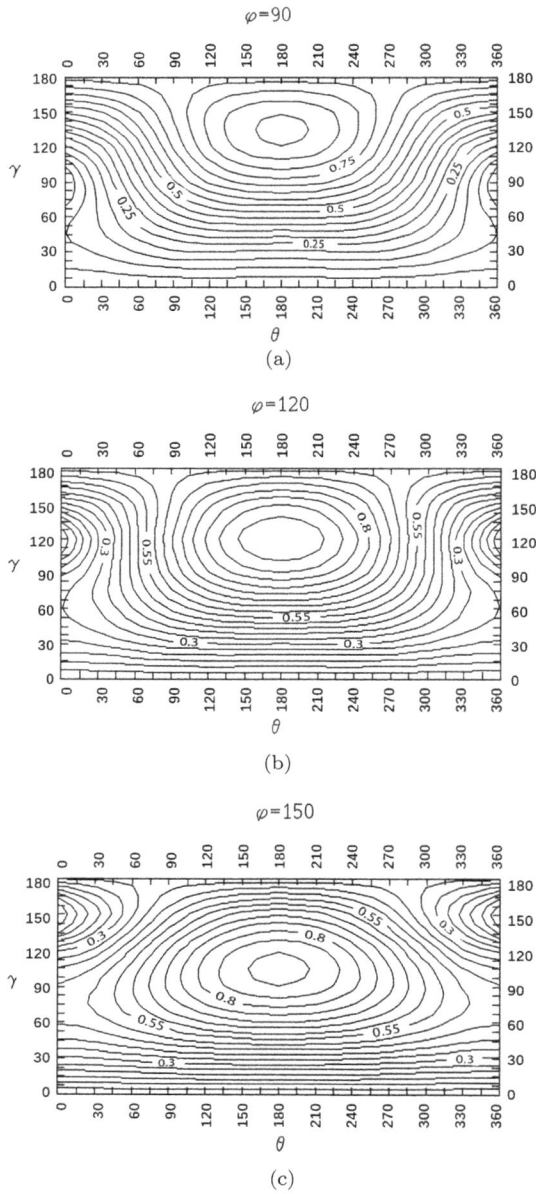

Fig. 4.4.   The equiscalar curve of stability index for the ball.

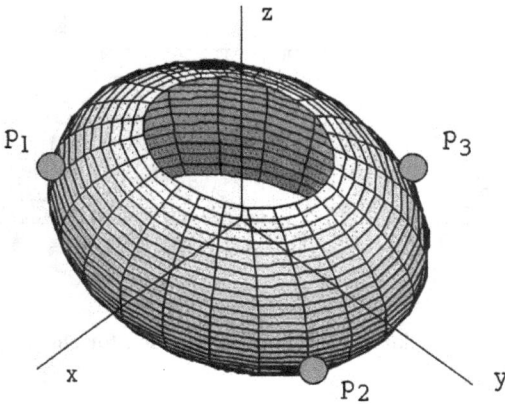

Fig. 4.5. An ellipsoidal object grasped by a trifingered hand.

Since we leave the internal grasp out of consideration, that is, the fingertip contacts cannot be fixed in the interior of the object, the feasible grasp domain can be represented as

$$-2\pi/9 \leq \alpha \leq \pi/3$$
$$0 \leq \beta \leq 2\pi \tag{4.22}$$

Using the nonlinear programming method as mentioned in Section 4.4.3, we obtain the optimal contact configuration, namely $\alpha_1 = 0°$, $\beta_1 = 0°$ for the first contact, $\alpha_2 = 0°$, $\beta_2 = 117°$ for the second contact, and $\alpha_3 = 0°$, $\beta_3 = 243°$ for the third contact. Thus, at this time, the grasp stability is also the best. The result is almost consistent with our expectations.

Given the first and the second contact locations, namely $\alpha_1 = 0°$, $\beta_1 = 0°$, $\alpha_2 = 0°$ and $\beta_2 = 117°$, we can obtain the value of W for different α_3 and β_3, and the results are given in Fig. 4.6. As can be seen, when $\alpha_3 = 0°$, $\beta_3 = 0°$; $\alpha_3 = 0°$, $\beta_3 = 360°$; and $\alpha_3 = 0°$, $\beta_3 = 117°$, that is, only two fingers grasping the ellipsoid, W equals 0, which indicates that the contact configuration becomes singularity; to be more exact, the grasp will lose the capability of withstanding the external moment about the connecting line of the two contact points. The result is clear because we have assumed that the contacts between the fingertips and the object are hard contacts with friction (this means that each finger can only transmit any force to the object

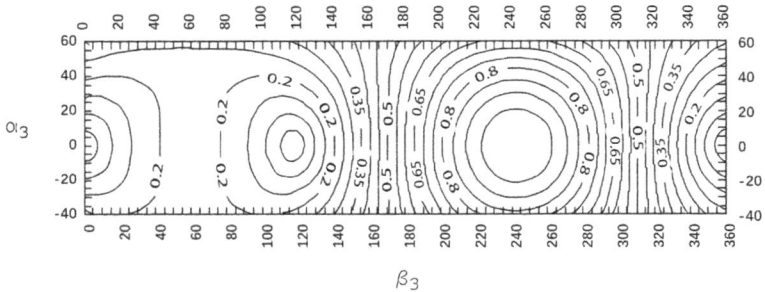

Fig. 4.6.   The equiscalar curve of stability index for the ellipsoidal object.

through the contact, but it cannot transmit any torque). Thus, at this time, the grasp is not stable.

Similar to the two examples above, Fig. 4.6 also shows that the value of W varies slowly near the optimal contact configuration ($W = 1$), which implies that we have a high degree of flexibility in selecting contact locations.

4.6   Summary

In this chapter, a quantitative index for evaluating a grasp is derived. Some properties of the index are discussed. A nonlinear programming method for finding optimal contact configuration is presented. The test of several examples illustrates the effectiveness of the index. The index is applicable not only to evaluate the configuration of the multifingered robot hand grasp but also to plan adaptable fixtures.

References

[1]   Li Z. X. and Sastry S. S. Task-oriented optimal grasping by multifingered robot hands. *IEEE Journal of Robotics and Automation*, 4, pp. 32–43, 1988.

[2]   Nguyen V. Constructing force-closure grasps. *International Journal of Robotics Research*, 7, pp. 3–16, 1988.

[3]   Montana D. Contact stability for two-fingered grasps. *IEEE Transactions on Robotics and Automation*, 8, pp. 421–430, 1992.

[4]   Nakamura Y., Nagai K., and Yoshikawa T. Dynamics and stability in coordination of multiple robotic mechanisms. *International Journal of Robotics Research*, 8, pp. 44–61, 1989.

[5]   Xiong Y. L., Sanger D. J., and Kerr D. R. Geometric modelling of bounded and frictional grasps. *Robotica*, 11, pp. 185–192, 1993.

[6]   Bicchi A. On the closure properties of robotic grasping. *International Journal of Robotics Research*, 14, pp. 319–334, 1995.

[7]   Varma V. K. and Tasch U. A new representation for a robot grasping quality measure. *Robotica*, 13, pp. 287–295, 1995.

[8]   Yoshikawa T. Manipulability of robotic mechanism. *International Journal Robotics Research*, 4, pp. 3–9, 1985.

[9]   Xiong C. H. and Xiong Y. L. The determination of fingertip contact positions of a multifingered robot hand for 3-D object. *Proc. 2nd Asian Conf. on Robotics and Its Application*, Oct., Beijing, pp. 342–347, 1994.

[10] Xiong C. H. and Xiong Y. L. Stability index and contact configuration planning for multifingered grasp. *Journal of Robotic Systems*, 15(4), pp. 183–190, 1998.

[11] Xiong C. H., Li Y. F., Ding H., and Xiong Y. L. On the dynamic stability of grasping. *International Journal of Robotics Research*, 18(9), pp. 951–958, 1999.

[12] Goyal S. and Ruina A. planar sliding with dry friction-part 1. limit surface and moment function. *Wear*, 143, pp. 307–330, 1991.

[13] Cole A., Hauser J., and Sastry S. Kinematics and control of a multifingered robot hand with rolling contact. *IEEE Transactions on Automation and Control*, 34(4), pp. 398–403, 1989.

Chapter 5

Active Grasp Force Planning

The real-time control of multifingered grasp involves the problem of the force distribution which is usually underdetermined. It is known that the results of the force distribution are used to provide force or torque set-points to the actuators, so they must be obtained in real time. The objective of this chapter is to develop a fast and efficient force planning method to obtain the desired joint torques which will allow multifingered hands to firmly grasp an object with arbitrary shape. In this chapter, the force distribution problem in a multifingered hand is treated as a nonlinear mapping from the object size to joint torques. We represent the nonlinear mapping using artificial neural networks (ANNs), which allow us to deal with the complicated force planning strategy. A nonlinear programming method, which optimizes the contact forces of fingertips under the friction constraints, unisense force constraints and joint torque constraints, is presented to train the ANNs. The ANNs used for this research are based on the functional link (FL) network and the popular backpropagation (BP) network. It is found that the FL network converges more quickly to the smaller error by comparing the training process of the two networks. The results obtained by simulation show that the FL network is able to learn the complex nonlinear mapping to an acceptable level of accuracy and can be used as a real-time grasp planner.

5.1   Introduction

Since multifingered robot hands provide more flexibility than simple single degree-of-freedom (DOF) grippers such as those used in industrial robots, they can perform more complex grasping and manipulating tasks. Especially when making a robot carry out dexterous works in the application area of unstructured, hazardous environments, multifingered robot hands are very useful. For example, in the application environment of nuclear waste site clean-up, the types of materials to be handled are not well known and cannot be modeled *a priori* for the robot. On the other hand, the robot does not need to handle these objects for fine manipulation. Grasping and transporting objects for sorting and packing and carrying out elementary tasks such as reorienting and separating may frequently be all that is required [16]. Here we divide the grasping process into three phases: precontact, contact, and postcontact. In the precontact phase, viewer-relative, extrinsic properties of the object such as spatial location and orientation are used to guide hand/wrist orientation and a ballistic reach toward the object. At the same time, object-relative, intrinsic properties such as shape and size are used to preshape the hand in anticipation of the grasp.

In highly unstructured environments, uncertainties and noises can degrade the performance of the grasping system. Noise disturbs the signals of the sensors, thus the execution of the motion commands is inaccurate and the exact position of the grasped object is uncertain, which makes the object disaligned in the hand. In order to avoid the disalignment, it is necessary to specify adjustment parameters for a grasp. Two adjustment parameters are used by Stansfield [16]. Adjustment parameters are used once a preshape has been invoked to adjust the grasp to the specific object. In the contact phase, the shape and size of an object to be grasped are used to plan joint torques so that they generate the required external object force, including object weight, to avoid sliding or breaking contact. When multifingered hands are used for fine manipulating objects, like the peg-in-hole insertion, torques applying to the motors have to compensate for the gravity of the peg and the reaction forces by grasp-force-adaptation while inserting the peg into the hole. An adaptive control strategy is needed which handles the uncertain information about the state of a peg [6, 18]. In the final or postcontact phase, which occurs after the fingers are in contact with the object, the major objective may be to control a hand-arm system to follow a given trajectory. The orientation of a hand/wrist, the ballistic reach toward the object, hand preshaping, and adjustment in the precontact phase are studied using computer vision and a compact set of heuristics [16, 17]. Coordinating a robot arm and multifingered hand is discussed by Roberts [14]. Our research interest focuses on the second phase, namely the grasping force distribution problem for unstructured environments.

Since multiple fingers share the object, the force distribution problem is usually underspecified so that multiple solutions exist. A solution which optimizes a certain objective function may then be obtained if a suitable optimization algorithm is applied [2,4,7,11–13]. However, searching for the optimal forces, which falls into the categories of linear and nonlinear programming, is computationally time-consuming. Hence, it is important to develop an efficient method for planning the forces of the fingers to cope with the computational hurdles in real time.

Artificial neural networks (ANNs) have the ability to perform complex nonlinear mapping between inputs and outputs. Recently, there has been a lot of interest in using neural networks to solve various problems in robotics. Asada [1] presented a method for representing and learning compliances from teaching data by using the back-propagation (BP) network [10]. ANNs have also been applied in grasping with multifingered hands. Xu *et al.* [20] have investigated the approach using the Hopfield neural network to guide the design of the three-finger gripper. Hanes *et al.* [9] used linear programming to train the BP network to control the force distribution for a model of the DIGITS System. Force control was implemented to ensure that the maximum normal force applied to the object is to provide the best possible grasp but at the same time not to exceed the desired clinch level. The work, however, focused on frictionless, symmetrical grasp and cannot be readily extended to work including soft fingertip contact or asymmetrical 3-D grasp.

In this chapter, we discuss the general force distribution problem in unstructured environments, where the contact type between the object and fingers may be a point contact with friction or a soft finger contact. The goal of our research is to develop an efficient force planning method to obtain the desired joint torques which will allow multifingered hands to firmly grasp objects with arbitrary shapes. The fundamental idea is to express the relationship between geometric object primitives and the desired joint torques to ensure that the object is grasped stably. As mentioned above, visual system is used to locate the spatial location and orientation of the object in space and to determine its size and geometric properties. Objects to be grasped are 3-D geometric forms which vary enormously in size and shape. Using a rule-based expert system to describe the relationship between each possible object and joint torques is almost impossible. In addition, in order to make real-time implementation possible, applying an adaptive ANN approach to deal with the complicated force planning strategy seems to be attractive. Two ANN models, which are the popular back-propagation (BP) network and the functional link (FL) network, are considered for solving the force distribution problem. The training and simulation of the two ANNs are accomplished on a personal microcomputer. A nonlinear programming method which is applied to train the ANNs is presented. In this method, an optimal solution may be obtained by minimizing the fingertip contact forces including forces and torques at contacts under the constraints, such as unisense force constraints, friction limit constraints, and maximum joint torque constraints. Comparison of two networks indicates that the latter converges quickly to the desired solution. The simulation results are given. In the simulation, the 9-DOF trifingered hand grasps a cylindroid object where the contacts are modeled as hard point contacts with friction. The results show that the FL network is able to learn the complex nonlinear mapping between the object size and joint torques to an acceptable level of accuracy and can be used as a real-time grasp planner.

5.2 Nonlinear Programming in Grasp

As mentioned in Section 5.1, the force distribution problem is usually under-constrained so that multiple solutions exist. An optimal solution can be obtained by a linear or nonlinear programming approach. Since the frictional constraints are nonlinear, the nonlinear programming seems better suited to the optimal force distribution problem, especially involving point contact with friction or soft finger contact.

It is quite obvious that large grasping forces and torques are not appropriate for grasping breakable objects. On the other hand, in order to increase the strength of the grasp for sturdy objects, it is advantageous to use larger grasping forces. However, Cutkosky and Wright [5] pointed out that increasing the gripping force reduced the chance of slipping but also made the grip less stable with respect to disturbances. The grasping forces depend on the objective function used in actual environments. In unstructured environments, especially when the types of materials of objects to be grasped are not well known, it is difficult to prespecify a desired normal force or clinch level as used in Ref. [9]. In order to prevent crushing of the grasped objects, we define the minimization of the norm of the finger contact forces/torques as an objective function which is similar to that used in Ref. [13]. The relatively versatile objective function is specified as follows:

$$\text{minimize } {}^c\mathbf{f}_c^T \mathbf{w}^c \mathbf{f}_c \tag{5.1}$$

where $\mathbf{w}$ is an $n \times n$ positive-definite weighing matrix (usually diagonal).

If the fingertips are modeled as springs, the sum of the square of the finger contact forces/torques indicates the potential energy stored due to the deformation of the tissues. Minimization of the norm of the finger contact forces/torques reduces the energy required to grasp an object, namely minimization of the energy supplied by the actuators to grasp an object.

In order to maintain the contact, the finger contact forces/torques must satisfy the constraints (3.10) and (4.7).

The joint torques of the ith finger are represented as

$$\tau_i = \mathbf{J}_{fi}^{T} \, {}_{ci}^{o}\mathbf{R}^c\mathbf{f}_{ci} \tag{5.2}$$

where $\mathbf{J}_{fi}$ is the Jacobian of the ith finger ($\mathbf{J}_{fi} \in \Re^{3 \times n_i}$ for the hard finger with friction contact, and $\mathbf{J}_{fi} \in \Re^{6 \times n_i}$ for the soft finger with friction contact).

The joint torque of the multifingered robotic hand with m fingers is represented as

$$\tau = \mathbf{J}_{h\,c}^{To}\mathbf{R}^c\mathbf{f}_c \tag{5.3}$$

where the Jacobian of the multifingered hand is

$$\mathbf{J}_h = \text{block diag}\left(\mathbf{J}_{f1} \; \mathbf{J}_{f2} \; \cdots \; \mathbf{J}_{fm} \right) \in \Re^{3m \times \sum_{i=1}^{m} n_i}$$

for the hard finger with friction contact and

$$\mathbf{J}_h = \text{block diag}\left(\mathbf{J}_{f1} \; \mathbf{J}_{f2} \; \cdots \; \mathbf{J}_{fm} \right) \in \Re^{6m \times \sum_{i=1}^{m} n_i}$$

for the soft finger with friction contact.

$$\tau = \left(\tau_1^T \; \tau_2^T \; \cdots \; \tau_m^T \right)^T \in \Re^{\sum_{i=1}^{m} n_i \times 1}$$

is the joint torque of the multifingered hand.

In addition, since the magnitude of each joint torque for a multi-fingered hand has its upper bound, the maximum joint torque constraints must be taken into account, which may be represented as

$$-\tau_{\max} \le \tau \le \tau_{\max} \tag{5.4}$$

Substituting Eq. (5.3) into Eq. (5.4) yields

$$-\tau_{\max} \le \mathbf{J}_{h\,c}^{To}\mathbf{R}^c\mathbf{f}_c \le \tau_{\max} \tag{5.5}$$

where $\tau_{\max}$ is the upper bound of the joint torque of the multifingered hand.

Hence, the optimal force distribution problem can be solved by a nonlinear programming method which minimizes the finger contact forces, while satisfying the unisense force constraints, friction limit surface constraints [8], joint torque constraints, etc. From Eqs. (5.1), (3.10), (4.7), and (5.5), the nonlinear programming problem becomes

$$\text{minimize } {}^c\mathbf{f}_c^T \mathbf{w}^c \mathbf{f}_c$$

s. t.

$$\mathbf{G}_c^o \mathbf{R}^c \mathbf{f}_c - \mathbf{F}_e = 0$$

$$- f_{ciz} \leq 0$$

$$\left(\frac{f_{ci}^t}{\mu f_{ciz}}\right)^2 + \left(\frac{m_{ciz}}{m_{\max}}\right)^2 - 1 \leq 0 \qquad (5.6)$$

$$\mathbf{J}_{hc}^{To} \mathbf{R}^c \mathbf{f}_c - \boldsymbol{\tau}_{\max} \leq 0$$

$$- \mathbf{J}_{hc}^{To} \mathbf{R}^c \mathbf{f}_c - \boldsymbol{\tau}_{\max} \leq 0$$

$$i = 1, 2, \ldots, m$$

5.3   Force Planning Using Neural Networks

The nonlinear programming, which is used to plan the forces in multifingered hands, generally requires a computationally expensive process. In particular, if the object to be grasped is often changed, it is necessary to plan the forces again.

ANNs have distinct advantages, such as performing complex nonlinear mapping between inputs and outputs, adaptive nature, and capability of operating in real time. It is very suitable to learn the nonlinear programming with ANNs. Thus, we take an adaptive ANN approach to plan joint torques for various robot hands and for various objects. As for different multifingered hands such as the Stanford/JPL hand, Utah/MIT hand, or Belgrade hand, we can determine the contact configurations and the grasp modes using the planning methods proposed in Chapters 4 and 5 and preshape the hand according to the intrinsic properties (namely, dimensions and shape) of an object to be grasped. For example, a cube has three dimensions (length, width, and height). Then, the optimal joint torques can be obtained by the nonlinear programming method (5.6), where the contact type may be a hard contact with friction or a soft finger contact. For each multifingered hand, we need to build different

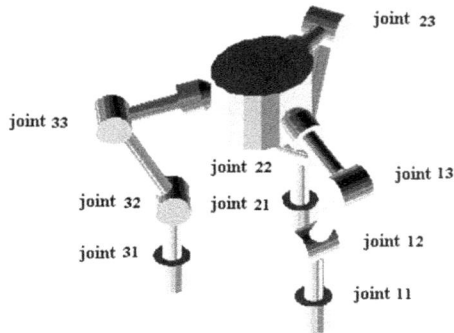

Fig. 5.1. Trifingered hand grasps the cylindroid object.

neural network structures to learn the corresponding joint torques to a given object. A complex object is considered to be the union of geometric primitives including cylinder, cube, sphere, cone, torus, etc. Here, each multifingered hand is treated as a black box. The supervised learning method is used. The networks must be taught to learn the nonlinear mapping relationship between geometric object primitives and the desired joint torques by observation of the grasp system's inputs and outputs, once the networks are trained by an example, yield a constant time solution to the complex force planning problem.

The example for training the ANNs is shown in Fig. 5.1. In the example, a trifingered hand grasps a cylindriod object. The fingers are identical in construction, each with three revolute joints. The length of each link is 40 mm, 70 mm, and 50 mm, respectively. The radius of the semisphere fingertip is 10 mm. The maximum and minimum joint-driven torques of each finger are 0.9 Nm and −0.9 Nm, respectively. To simplify the problem, we assume that the contact type between the object and the finger is a hard point contact with friction and the frictional coefficient between the fingertip and object is 0.4. This means that each of all fingertips cannot apply a moment of the object about the surface normal, i.e., $m_{ci}^n = 0$, $i = 1, 2$, and 3 in Eq. (5.6).

Assume that the weight of the object to be grasped is 1 kg. If the object is represented by two half lengths of principal axes, e.g., ra

and rb, we can construct the nonlinear mapping between the object size and the desired joint-driven torques to firmly grasp the object without slippage.

The set of training samples is collected from the nonlinear programming method. This is done by specifying different object sizes (i.e., desired input) and determining the corresponding joint torques (i.e., desired outputs) with the nonlinear programming method as stated in Section 5.2.

Now, we consider the popular BP network algorithm which allows us to plan the forces in multifingered hands via iterative learning of the network. The learning procedure of the BP network follows the delta rule [15]. Typical delta rule-based neural networks apply a gradient descent technique which attempts to minimize the mean squared error between the desired and actual network outputs by modifying connection weights based on a set of input–output pairs (known as the training set). The mean squared error (MSE) of a system is defined as

$$E = \frac{1}{2p} \sum_{u=1}^{p} \sum_{j=1}^{k} (t_{uj} - o_{uj})^2 \tag{5.7}$$

where p denotes the total number of input samples, k denotes the number of output units, o_{uj} is the output of the network (actual network output), and t_{uj} is the teacher signal (desired output). For the convenience of representation, we will omit the subscript u later on.

The networks generally consist of a layer of input units, multiple layers of hidden units, and a layer of output units. There may be as many or a few units in each layer as required.

Neurons behave as functions. Neurons transduce an unbounded input activation net_j into a bounded output signal o_j. Usually, a sigmoidal or an S-shaped curve describes the transduction. We define the sigmoidal function as

$$o_j^{(l)} = f(net_j^{(l)}) = \frac{2}{1 + e^{-net_j^{(l)}}} - 1 \tag{5.8}$$

where $o_j^{(l)}$ is the general output of the jth unit of the lth layer, $net_j^{(l)}$ is the input of jth unit of the lth layer, it has the usual form:

$$net_j^{(l)} = \sum_i w_{ji}^{(l)} o_i^{(l-1)} \qquad (5.9)$$

where $w_{ji}^{(l)}$ is the weight of connections between the ith unit on the $(l-1)$th layer and the jth unit on the lth layer. The corrective signal δ_j can be calculated as shown in the following:

$$\delta_j = \begin{cases} \frac{1}{2}(t_j - o_j)(1 + o_j)(1 - o_j) & \text{if } j \text{ is output unit,} \\ \frac{1}{2}(1 + o_j)(1 - o_j)\sum_k \delta_k w_{kj} & \text{if } j \text{ is hidden unit,} \end{cases} \qquad (5.10)$$

where t_j is the teacher signal (or desired output).

The recursive weight updating formula (learning rule) is given [15] by

$$w_{ji}(t + 1) = w_{ji}(t) + \eta \delta_j o_i + \alpha(w_{ji}(t) - w_{ji}(t - 1)), \qquad (5.11)$$

where η is the learning rate, α is the momentum coefficient.

According to Hecht-Nielsen [10], for any function F and any $\varepsilon > 0$, there exists a three-layer BP network that can approximate F to within ε MSE accuracy. Based on this result, we have constructed a three-layer BP network to plan the force in the multifingered hand.

Sizes of the input layer and output layer are decided by the problem. The input unit is 2 (corresponding to ra and rb) and output unit is 9 (corresponding to the joint-driven torques: $\tau_{11}, \tau_{12}, \tau_{13}, \ldots, \tau_{31}, \tau_{32}, \tau_{33}$) for our force planning problem. Sizes of the hidden layer are changeable.

We construct the training set by choosing 121 samples where ra and rb are uniformly distributed from 20–30 mm, respectively. We note that grasp stability varies for different shape objects. A quantitative measure of stability which is applied to determine the positions of contact on the grasped object is proposed in Chapter 3. The desired network outputs are obtained from the nonlinear programming with respect to different object sizes (i.e., ra and rb) which are shown in Fig. 5.2.

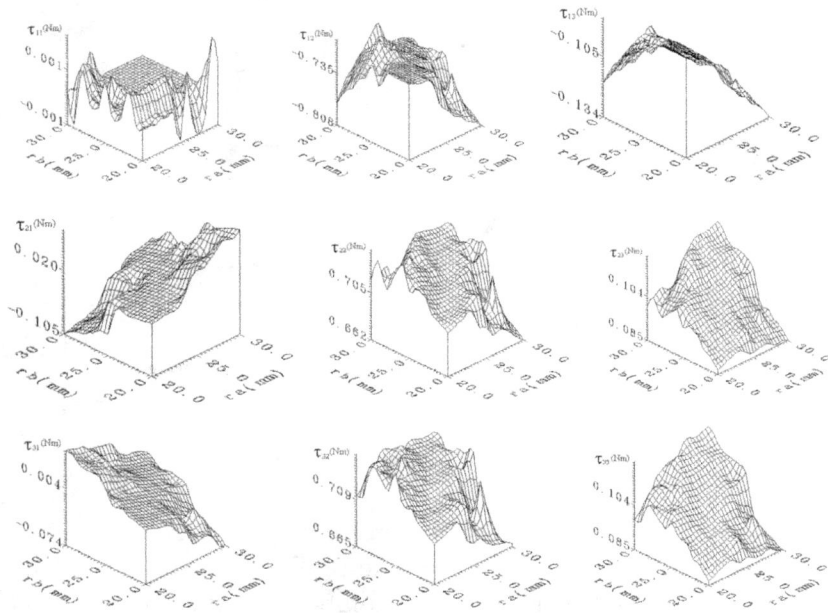

Fig. 5.2. Desired output torques.

Before the BP algorithm can be applied, one needs to specify the learning rate and momentum. Due to lack of efficient theoretical methods, the learning rate and momentum are determined empirically. The reasonable learning rate and momentum coefficient are found, both are 0.1, by testing a number of variations and selecting the ones having the best rate of convergence over the entire training set.

It is interesting to observe the effects of varying the number of hidden units on the performance of the network. Figure 5.3 displays the number of training iterations and their effect on the MSE over the 121 training samples with respect to 0 hidden unit (i.e., without hidden layer), 6 hidden units, and 10 hidden units network. The relationship between the number of training and the elapsed time with respect to different sizes of the hidden layer is shown in Figs. 5.3 and 5.4 show that the BP network with 6 hidden units converges more quickly to the smaller error. For the 6 hidden units network,

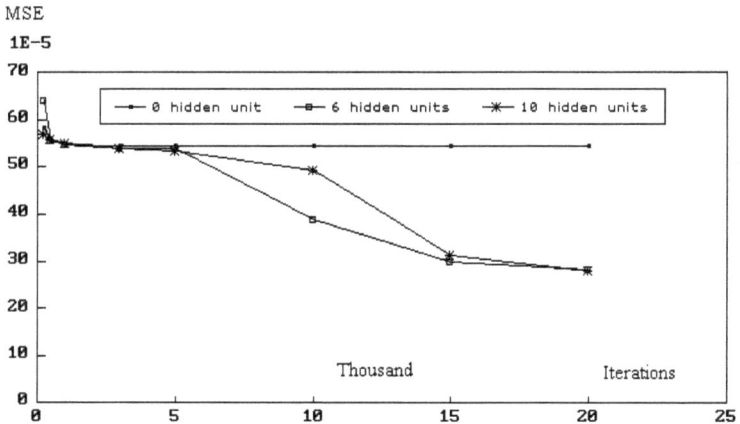

Fig. 5.3.   Training effect curve.

the MSE is 2.82×10^{-4} over the entire training set after 20 000 iterations (the elapsed time is 187 min). Figures 5.3 and 5.4 also show that continued increasing the number of hidden units cannot improve the performance of the network, in contrast, reaching the same MSE will need much more training time.

In an attempt to improve the performance of the grasp system without sacrificing accuracy, we use the composite method of outer product and functional extension to reinforce the original input patterns of BP network so that the original input patterns are mapped into much higher dimension pattern space. This means the nonlinearity of neural network is increased. Such a reinforced neural network is called FL network and has no hidden layers. The learning procedure of the FL network follows the same delta rule as the BP network. Our FL network model consists of 11 input units and 9 output units (corresponding to the joint-driven torques: $\tau_{11}, \tau_{12}, \tau_{13}, \ldots, \tau_{31}, \tau_{32}, \tau_{33}$). The components of input patterns of the FL network are described by ra, $\sin(\pi ra)$, $\cos(\pi ra)$, rb, $\sin(\pi rb)$, $\cos(\pi rb)$, $rarb$, $ra\sin(\pi rb)$, $ra\cos(\pi rb)$, $rb\sin(\pi ra)$, and $rb\cos(\pi ra)$, which are reinforced by the original input patterns (ra and rb) with the composite method of outer product and functional extension.

The learning rate and momentum coefficient are the same as those in the previous BP network. The training samples of the FL network

Time (min)

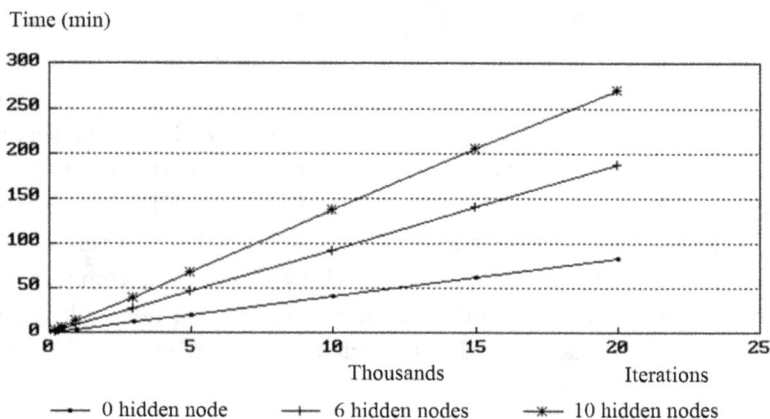

Fig. 5.4. Training time curve.

MSE

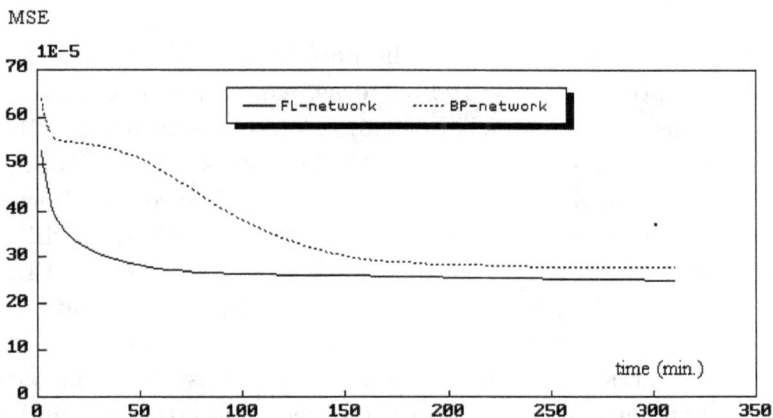

Fig. 5.5. Mean-squared error (MSE) for two networks.

are also the same as those in the BP network except for the input patterns. Figure 5.5 compares the training process of the BP network with 6 hidden units and the FL network. The graph shows the MSE of the networks with respect to different training times over the 121 training samples as iterations progress. As can be seen, the BP network appears to have reached its first stable MSE after about 10 min training. The BP network remains in the above stable MSE for a relatively short period of time. If the training continues beyond 50 min, the MSE of the BP network begins to fall rapidly until the training time reaches 187 min and then changes very little throughout the process. However, in the FL network, the MSE falls rapidly during the first 50 min training and then declines very slowly. In terms of training time, the FL network converges quickly to the smaller error. In terms of accuracy, the FL network is better. For the FL network, the MSE over the entire training set is 2.5×10^{-4} after 310 min training (i.e., 30000 iterations). But the BP network requires more training time to achieve similar accuracy as the FL network without hidden layers. Thus, the FL network is more efficient in planning forces for multifingered grasp.

5.4   Simulation

Our main objective was to test the performance of the FL network. For this purpose, we chose 100 patterns, which were not trained, for processing as the test set where ra and rb were uniformly distributed from 20.5 to 29.5 mm, respectively. All the results have been obtained by simulation on a personal microcomputer. Figure 5.6 shows the joint-driven torques of the multifingered hand required for different input patterns, which are automatically generated by the FL network. Note that the joint-driven torques vary between the admissible limits. The joint torques in Fig. 5.6 change evenly as the size of the object to be grasped varies, which means the grasp has some adaptability to the geometric error of the object. Moreover, we find that Fig. 5.6 is somewhat different from Fig. 5.2. It is important to know whether or not the network output torques which maintain grasp stability, even though the difference exists.

As mentioned in Section 5.2, in order to grasp firmly the object without slippage, the finger contact forces at each contact point must satisfy the conditions:

(1) *The normal component of the contact forces must be positive.*
(2) *The contact forces must lie within the friction cone for hard point contact with friction.*

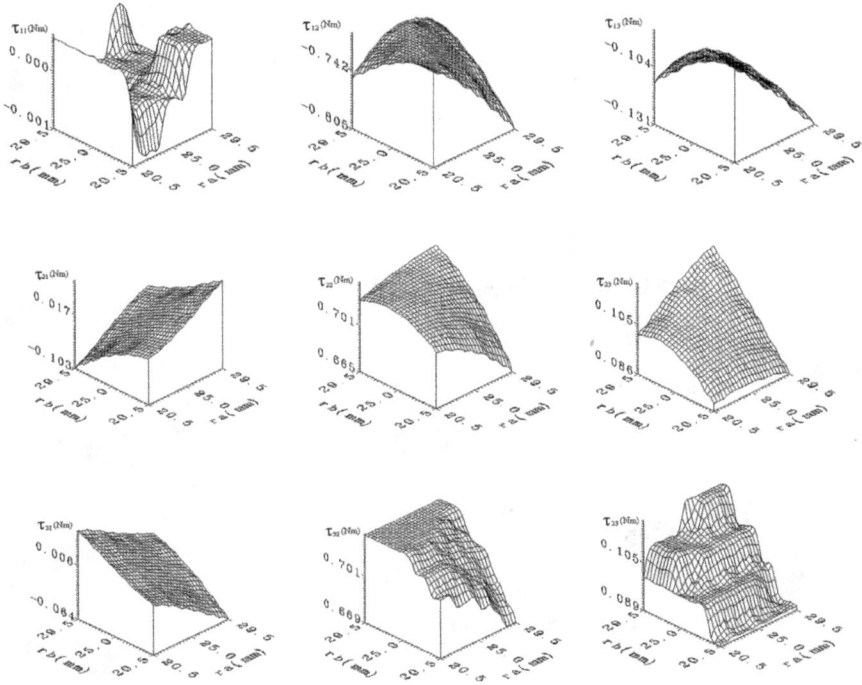

Fig. 5.6. Joint-driven torques generated by FL network.

In order to judge whether the output torques of the FL network maintain grasp stability, we define [19]

$$\Delta_i = {}^c\mathbf{n}_i^T {}^c\mathbf{f}_{ci} - \|{}^c\mathbf{f}_{ci}\| \cos\theta, \quad i = 1, 2, 3 \qquad (5.12)$$

as the performance index of sliding.

In Eq. (5.12), ${}^c\mathbf{n}_i \in \Re^{3\times 1}$ is the inner unit normal vector of the object surface at the ith contact point and θ is the frictional angle ($\cos\theta = 1/\sqrt{1 + \mu^2}$, μ is the coefficient of friction).

The meaning of Eq. (5.12) becomes apparent in Fig. 5.7, which shows the friction cone and the finger contact force on the object at the ith contact point.

It is obvious that the grasp is stable when all $\Delta_i \geq 0$ ($i = 1, 2, 3$), that is, the contact forces of each finger satisfy the above conditions (5.1) and (5.2). In contrast, the grasp is not stable when one of all Δ_i

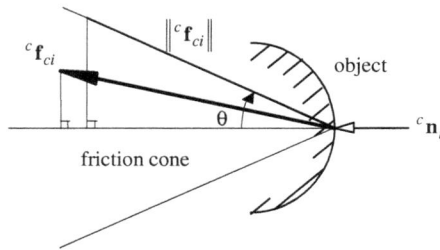

Fig. 5.7.   Finger contact force constraints.

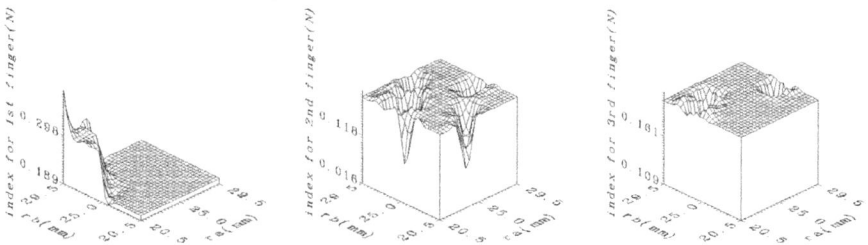

Fig. 5.8.   Stability analysis of grasp.

$(i = 1, 2, 3)$ is negative. Figure 5.8 illustrates the test results, i.e., Δ_i $(i = 1, 2, 3)$ for different ra and rb, and shows that the grasp is stable. The reason why the grasp is stable is due to redundant degrees of freedom of the hand and the adaptability of the grasp system.

5.5   Summary

In this chapter, we have developed an efficient grasp force planning method based on neural networks to be used for real-time finger force distribution in unstructured environments. Two neural network models, namely the BP network and the FL network of which input patterns are obtained by using the composite method of outer product and functional extension to reinforce the original input patterns of BP network, are considered. We first present a nonlinear programming algorithm for solving the force distribution problem in the multifingered grasp system, which is applicable to the three contact types, namely hard point contact with no friction, hard point contact with friction, and soft finger contact. Then we use the non-linear programming force solutions to train the neural networks. It is found that the FL network converges more quickly to the smaller error by comparing the training process of the two networks. The results obtained by simulation of a trifingered hand grasping a cylin-droid object show that the FL network can provide a better solu-tion to ensure that the object is grasped stably, despite slight errors existing. The reason why the grasp is stable is because of redun-dant DOF of the hand and the adaptability of the grasp system. The simulation results also indicate that the neural-network-based force planning results can be obtained fast enough to be supplied as set points for a real-time force controller. Thus, the approach can be applied in unstructured environments such as in flexible manufac-turing which requires grasping and manipulation. The approach also can be used for multiple manipulator systems and can be extended to legged walking machines.

A significant advantage of using a neural planner is that the model of the grasp system does not have to be known at the time of planner design. Unlike most other plan schemes such as linear or nonlinear programming method, here the learning is done iteratively, based only on observations of input/output relationship of grasp system. Also, a change in the physical setup of the system, such as using different multifingered hands or grasping different objects, would only involve retraining and not any major system software modifications. Since the neural planner is system-independent, the same planner could be used for different hands, and each one could be trained according to the specific objects that the multifingered hands would be required to grasp.

References

[1] Asada H. Teaching and learning of compliance using neural nets: Rep-resentation and generation of nonlinear compliance. *IEEE Interna-tional Conference on Robotics and Automation*, pp. 1237–1243, 1990.

[2] Bicchi A. On the closure properties of robotic grasping. *International Journal of Robotics Research*, 14(4), pp. 319–334, 1995.

[3]   Chevallier D. P. and Payandeh S. On computing the friction forces associated with three-fingered grasp. *International Journal of Robotics Research*, 13(2), pp. 119–126, 1994.

[4]   Choi M. H., Lee B. H., and Ko M. S. Optimal load distribution for two cooperating robots using a force ellipsoid. *Robotica*, 11, pp. 61–72, 1993.

[5]   Cutkosky M. R. and Wright P. K., Friction stability and the design of robotic fingers. *International Journal of Robotics Research*, 5(4), pp. 20–37, 1986.

[6]   Dorsam T., Fatikow S. and Streit I. Fuzzy-based grasp force-adaptation for multifingered robot hands. *Proceedings of the 3rd IEEE Conference on Fuzzy Systems (IEEE World Congress on Computational Intelligence)*, pp. 1468–1471. 1994.

[7]   Gorce P., Villard C. and Fontaine J. G. Grasping, Coordination and optimal force distribution in multifingered mechanisms. *Robotica*, 12, pp. 243–251, 1994.

[8]   Goyal S., Ruina A. and Papadopoulos J. Limit surface and moment function description of planar sliding. *IEEE International Conference on Robotics and Automation*, pp. 794–799, 1989.

[9]   Hanes M. D., Ahalt S. C., Mirza K. and Orin D. E. Neural network control of force distribution for power grasp. *IEEE International Conference on Robotics and Automation*, pp. 746–751, 1991.

[10] Hecht-Nielsen R. Theory of the back propagation network. *Proceedings of the International Joint Conference on Neural Networks*, Washington, DC, 1989.

[11] Lu W. S. and Meng Q. H. On optimal force distribution of coordinating manipulators. *International Journal of Robotics and Automation*, 7(2), pp. 70–79, 1992.

[12] Nahom M. A. and Angeles J. Real-time force optimization in parallel kinematic chains under inequality constraints. *IEEE Transactions on Robotics and Automation*, 8(4), pp. 439–450, 1992.

[13] Nakamura Y., Nagai K. and Yoshikawa T. Dynamics and stability in coordination of multiple robotic mechanisms. *International Journal of Robotics Research*, 3(2), pp. 44–61, 1989.

[14] Roberts K. S. Coordinating a robot arm and multi-finger hand using the quaternion representation. *IEEE International Conference on Robotics and Automation*, pp. 1252–1257, 1990.

[15] Rumelhart D. E., McClelland J. L., and Williams R. J. *Parallel Distributed Processing: Explorations in the Microstructure of Cognition*, Vol. 1, Chapter 8, MIT Press, Cambridge, MA, 1986.

[16] Stansfield S. A. Robotic grasping of unknown objects: A knowledge-based approach. *International Journal of Robotics Research*, 10(4), pp. 314–326, 1991.

[17]   Tomovic R., Bekey G. A., and Karplus W. J. A strategy for grasp synthesis with multifingered robot hands. *IEEE International Conference on Robotics and Automation*, pp. 83–89, 1987.

[18]   Wohlke G. A. Neuro-fuzzy-based system architecture for the intelligent control of multi-finger robot hands. *Proceedings of the 3rd IEEE International Conference on Fuzzy Systems (IEEE World Congress on Computational Intelligence)*, pp. 64–69, 1994.

[19]   Xiong C. H. and Xiong Y. L. Neural-network based force planning for multifingered grasp. *Robotics and Autonomous Systems*, 21(4), pp. 365–375, 1997.

[20]   Xu G. H., Kaspar H. and Schweitzer S. G. Application of neural networks on robot grippers. *Proceedings of International Joint Conference on Neural Networks*, 3, pp. 337–342, 1990.

Chapter 6

Grasp Capability Analysis

This chapter addresses the problem of grasp capability analysis of multifingered robot hands. The aim of the grasp capability analysis is to find the maximum external wrench that the multifingered robot hands can withstand, which is an important criterion in the evaluation of robotic systems. The study of grasp capability provides a basis for the task planning of force control of multifingered robot hands. For a given multifingered hand geometry, the grasp capability depends on the joint driving torque limits, grasp configuration, contact model, and so on. A systematic method of the grasp capability analysis, which is in fact a constrained optimization algorithm, is presented. In this optimization, the optimality criterion is the maximum external wrench, and the constraints include the equality constraints and the inequality constraints. The equality constraints are for the grasp to balance the given external wrench, and the inequality constraints are to prevent the slippage of fingertips, the overload of joint actuators, the excessive forces over the physical limits of the object, etc. The advantages of this method are the ability to accommodate diverse areas such as multiple robot arms, intelligent fixtures, and so on.

6.1   Introduction

During the past few years, the pace of research on artificial multifingered hands for robot manipulators has significantly increased [1–14]. Each finger of multifingered robot hands can be considered as an independent manipulator; thus, operation of a hand requires coordinated motion of several separate small manipulators. When holding an object with a multifingered hand, the grasp must satisfy a number of conditions, such as static equilibrium, no slippage, and the ability to resist disturbance in all directions. This means that the contact locations and the hand configuration must be carefully chosen, so as to guarantee that the grasp has the properties mentioned above. Once the contact locations and the hand configuration are chosen properly, how large external wrench can the grasp resist without violating the joint driving torque limits for multifingered hands? We define the problem of finding the maximum external wrench in any direction that the multifingered hand can withstand as the problem of grasp capability analysis. The grasp capability is an important criterion in the evaluation of robotic systems and its analysis will provide a basis for the task planning of force control of multifingered hands.

Coordinated force control of multifingered hands has been studied by many authors [3, 4, 6–11]. For example, Nakamura *et al.* [3] considered the control of force distribution in multiple robotic mechanisms, maintained the formulation of friction cones (quadratic constraints), and defined minimum norm forces (a quadratic function) as the objective function. Nonlinear programming based on a Lagrange multiplier method has been applied to solve the optimization problem. The load distribution problem in multifingered hands can be solved using linear programming. Cheng and Orin [4] proposed a compact-dual linear programming method to optimize a linear objective function for the forces exerted at contact points. Lu and Meng [6] have formulated the load-sharing tasks for multiarm coordinating robots as a least pth unconstrained minimization problem with a sufficiently large p while satisfying basic constraints on the joint forces that the robots can provide. Shin and Chung [7] presented a general force distribution method, based on weak point force minimization. Chevallier and Payandeh [8] proposed a method based on the screw geometry for computing the friction forces between the fingertips of a dexterous mechanical hand and the object as a function of the external wrench. Panagiotopoulos and Al-Fahed [9] investigated the optimal control problem related to the grasping of objects by multifingered hands. The linear complementarity problem, which governed the static unilateral and frictional contact problem, has been formulated. Sheridan *et al.* [10] presented a fuzzy controller for a robotic power grasp system which attempted to achieve three objectives: obtaining contact with all finger links and the palm, centering the object in the palm, and controlling the link and palm normal forces such that they lie within a specified clinch range. Xiong and Xiong [11] developed an artificial neural-network-based force planning method for solving the load distribution problem of multifingered grasp.

It should be noted that the problem of the grasp capability analysis differs from the load distribution problem studied in [3,4,6–11] in that we are more interested in the capability of multifingered hands to bear a load, whereas the load distribution problem concentrates more on how a load can be handled by considering how each finger and their joint actuators should behave in the handling. Li *et al.* [5] discussed the problem of finding the maximum applicable force/torque by multiple robot arms in coordinated motion and formulated the problem as an optimization problem under the constraints of the dynamic equation of the robots and their joint driving torque limits.

However, the multifingered grasp is different from that of one of multiple robot arms, in that forces can be applied in one direction only. The fingers can only push, not pull, on the object because there is no glue between the object and the fingers. In a revolute (respectively, prismatic) arm, torques (respectively, forces) can be applied in both directions at the joints of the arm. This is the reason why the grasp capability of multifingered hands is not the same as the grasp capability of multiple robot arms. At the same time, this is also the reason why the grasp capability analysis of multifingered hands is more complex than the grasp capability analysis of multiple robot arms. The purpose of this chapter is to solve the problem of the grasp capability analysis of multifingered hands.

6.2   Evaluation of Multifingered Grasp Capability

Assume that m fingers grasp an object, and the contacts between the fingertips and the object are modeled as hard-finger contacts. This means that each finger can transmit any force to the object to be grasped through the contact (as long as it is within the friction cone) but it can not transmit any torque. The external wrench exerted $\mathbf{F}_e$ on the object is represented as

$$\mathbf{F}_e = \left(\mathbf{F}^T \ \mathbf{M}^T \right)^T = \left(f_1 \ f_2 \ f_3 \ t_1 \ t_2 \ t_3 \right)^T \tag{6.1}$$

where $\mathbf{F} = \left(f_1 \ f_2 \ f_3 \right)^T$, $\mathbf{M} = \left(t_1 \ t_2 \ t_3 \right)^T$ are the forces and torques withstood jointly by the multifingered robot hand.

Assume that we are given a directional vector $\left(a_1 \ a_2 \ a_3 \ b_1 \ b_2 \ b_3 \right)^T$ in the external wrench space, along which we expect that the multifingered hand will withstand an external wrench $\mathbf{F}_e$ on the object with the largest possible magnitude, namely we expect that the multifingered hand will resist a force applied on the object with the largest possible magnitude along a directional vector $\mathbf{d}^f = \left(a_1 \ a_2 \ a_3 \right)^T$ in the object coordinate frame $\{\mathbf{O}\}$, and a torque around a directional vector $\mathbf{d}^t = \left(b_1 \ b_2 \ b_3 \right)^T$ in the object coordinate frame $\{\mathbf{O}\}$. We can obtain a unit vector $\mathbf{d}$ in the external wrench space, $\mathbf{d} = \left(a_1^f \ a_2^f \ a_3^f \ b_1^t \ b_2^t \ b_3^t \right)^T$, which is simply the normalization of the vector $\left(a_1 \ a_2 \ a_3 \ b_1 \ b_2 \ b_3 \right)^T$.

Thus, Eq. (6.1) can be rewritten as

$$\mathbf{F}_e = \|\mathbf{F}_e\| \, \mathbf{d} \tag{6.2}$$

From Eq. (4.3), the force equilibrium of grasping can be represented as

$$\mathbf{G}_c^o \mathbf{R}^c \mathbf{f}_c = \|\mathbf{F}_e\| \, \mathbf{d} \tag{6.3}$$

When the external wrench exerted on the object increases, the grasping forces required to balance the external wrench have to increase. However, once the grasp capability goes beyond the limits of the joint driving torques of the multifingered hand, if the external wrench continuously increases, then the slippage between the object and fingertip will occur, which will result in an unstable grasp.

The maximum external wrench in one direction that the grasp can withstand within the joint torque limits reflects the grasp capability in this direction. The aim of the grasp capability analysis is to find the maximum external wrench that the multifingered robotic hand can withstand. Thus, the problem of the grasp capability analysis can be transformed into the problem of finding the maximum external wrench in any direction. Here we define the magnitude of the external wrench as the objective function, that is,

$$\text{minimize} - \|\mathbf{F}_e\| \tag{6.4}$$

Hence, the problem of finding maximum external wrench can be solved by a nonlinear programming method similar to the one in Chapter 4 besides the different objective function, which maximizes the magnitude of the external wrench in any direction, while satisfying the equality constraints of the fingertip contact forces, the inequality constraints of the joint torque limits, the static friction forces, and the unisense fingertip contact forces.

From Eq. (6.3), and the nonlinear programming (5.6) in Chapter 5, the problem of the grasp capability analysis is formulated as

$$\text{minimize} - \|\mathbf{F}_e\|$$
s. t.

$$\mathbf{G}_c^o \mathbf{R}^c \mathbf{f}_c - \|\mathbf{F}_e\| \, \mathbf{d} = \mathbf{0}$$
$$\mathbf{J}_{hc}^{To} \mathbf{R}^c \mathbf{f}_c - \boldsymbol{\tau}_{\max} \le \mathbf{0}$$
$$-\mathbf{J}_{hc}^{To} \mathbf{R}^c \mathbf{f}_c - \boldsymbol{\tau}_{\max} \le \mathbf{0} \tag{6.5}$$
$$f_{ci}^t - \mu f_{ciz} \le 0$$
$$-f_{ciz} \le 0$$
$$i = 1, 2, \ldots, m$$

We note that the nonlinear programming problem described above does not consider the constraint of the upper bound of the fingertip normal contact force. The solution of the maximum external wrench, which is obtained by the nonlinear programming method mentioned above, is true for the sturdy object. However, the upper bound constraint of the fingertip normal contact force must be taken into account if the object to be grasped is breakable, such as eggs.

Assuming that the maximum normal contact force of the fingertip at the ith contact point, which guarantees that the breakable object is not damaged, is $f_{i\max}$, then we can obtain the upper bound constraint of the fingertip normal contact force as follows:

$$f_{ciz} \leq f_{i\max}, \quad i = 1, 2, \ldots, m \tag{6.6}$$

Thus, when the upper bound constraint of the fingertip normal contact force is taken into account, the nonlinear programming problem associated with the grasp capability analysis problem for multi-fingered hands is the same as the problem formulated above except for the additional constraint given by Eq. (6.6). We restate the problem here as follows:

$$
\begin{aligned}
\text{minimize} \ & -\|\mathbf{F}_e\| \\
\text{s.t.} \ \\
& \mathbf{G}_c^o \mathbf{R}^c \mathbf{f}_c - \|\mathbf{F}_e\| \, \mathbf{d} = \mathbf{0} \\
& \mathbf{J}_{hc}^{To} \mathbf{R}^c \mathbf{f}_c - \boldsymbol{\tau}_{\max} \leq \mathbf{0} \\
& -\mathbf{J}_{hc}^{To} \mathbf{R}^c \mathbf{f}_c - \boldsymbol{\tau}_{\max} \leq \mathbf{0} \\
& f_{ci}^t - \mu f_{ciz} \leq 0 \\
& -f_{ciz} \leq 0 \\
& f_{ciz} - f_{i\max} \leq 0 \\
& i = 1, 2, \ldots, m
\end{aligned}
\tag{6.7}
$$

6.3   Numerical Example

In this section, we give a 3-D grasp example to prove the effectiveness of the proposed method for the grasp capability analysis. A sphere with a radius of 41.7 mm is grasped by three fingers. The three-fingered hand has identical structures with the one in Chapter 4. Assume that the frictional coefficient between the fingertip and the object is 0.32 and the maximum joint driving torque is 0.9 Nm.

The matrices used in the nonlinear programming method are given as follows:

$$\mathbf{J}_{f1} = \begin{pmatrix} -(l_3c(\theta_{12}+\theta_{13})+l_2c\theta_{12})s\theta_{11} & -(l_3s(\theta_{12}+\theta_{13})+l_2s\theta_{12})c\theta_{11} \\ (l_3c(\theta_{12}+\theta_{13})+l_2c\theta_{12})c\theta_{11} & -(l_3s(\theta_{12}+\theta_{13})+l_2s\theta_{12})s\theta_{11} \\ 0 & l_3c(\theta_{12}+\theta_{13})+l_2c\theta_{12} \end{pmatrix}$$

$$\begin{pmatrix} -l_3c\theta_{11}s(\theta_{12}+\theta_{13}) \\ -l_3s\theta_{11}s(\theta_{12}+\theta_{13}) \\ l_3c(\theta_{12}+\theta_{13}) \end{pmatrix}$$

$$\mathbf{J}_{f2} = \begin{pmatrix} (l_3c(\theta_{22}+\theta_{23})+l_2c\theta_{22})s\theta_{21} & (l_3s(\theta_{22}+\theta_{23})+l_2s\theta_{22})c\theta_{21} \\ (l_3c(\theta_{22}+\theta_{23})+l_2c\theta_{22})c\theta_{21} & -(l_3s(\theta_{22}+\theta_{23})+l_2s\theta_{22})s\theta_{21} \\ 0 & l_3c(\theta_{22}+\theta_{23})+l_2c\theta_{22} \end{pmatrix}$$

$$\begin{pmatrix} l_3c\theta_{21}s(\theta_{22}+\theta_{23}) \\ -l_3s\theta_{21}s(\theta_{22}+\theta_{23}) \\ l_3c(\theta_{22}+\theta_{23}) \end{pmatrix}$$

$$\mathbf{J}_{f3} = \begin{pmatrix} (l_3c(\theta_{32}+\theta_{33})+l_2c\theta_{32})s\theta_{31} & (l_3s(\theta_{32}+\theta_{33})+l_2s\theta_{32})c\theta_{31} \\ -(l_3c(\theta_{32}+\theta_{33})+l_2c\theta_{32})c\theta_{31} & (l_3s(\theta_{32}+\theta_{33})+l_2s\theta_{32})s\theta_{31} \\ 0 & l_3c(\theta_{32}+\theta_{33})+l_2c\theta_{32} \end{pmatrix}$$

$$\begin{pmatrix} l_3c\theta_{31}s(\theta_{32}+\theta_{33}) \\ l_3s\theta_{31}s(\theta_{32}+\theta_{33}) \\ l_3c(\theta_{32}+\theta_{33}) \end{pmatrix}$$

$$_{c1}^{o}\mathbf{R} = \begin{pmatrix} 0 & 0 & -1 \\ 1 & 0 & 0 \\ 0 & -1 & 0 \end{pmatrix}, \quad _{c2}^{o}\mathbf{R} = \begin{pmatrix} -\frac{\sqrt{3}}{2} & 0 & \frac{1}{2} \\ -\frac{1}{2} & 0 & -\frac{\sqrt{3}}{2} \\ 0 & -1 & 0 \end{pmatrix},$$

$$_{c3}^{o}\mathbf{R} = \begin{pmatrix} \frac{\sqrt{3}}{2} & 0 & \frac{1}{2} \\ -\frac{1}{2} & 0 & \frac{\sqrt{3}}{2} \\ 0 & -1 & 0 \end{pmatrix}$$

$$\mathbf{R}_1 = \begin{pmatrix} 0 & 0 & 0 \\ 0 & 0 & -r_b \\ 0 & r_b & 0 \end{pmatrix}, \quad \mathbf{R}_2 = \begin{pmatrix} 0 & 0 & r_b\cos(\pi/6) \\ 0 & 0 & r_b\sin(\pi/6) \\ -r_b\cos(\pi/6) & -r_b\sin(\pi/6) & 0 \end{pmatrix}$$

$$\mathbf{R}_3 = \begin{pmatrix} 0 & 0 & -r_b\cos(\pi/6) \\ 0 & 0 & r_b\sin(\pi/6) \\ r_b\cos(\pi/6) & -r_b\sin(\pi/6) & 0 \end{pmatrix}$$

where $l_1 = 40\,\text{mm}$, $l_2 = 70\,\text{mm}$, and $l_3 = 50\,\text{mm}$ are the lengths of finger links. r_b is the radius of the sphere, which is 0.0417 m for this example. $sq_{ij} = \sin q_{ij}$, $cq_{ij} = \cos q_{ij}$, $s(q_{ij} + q_{kl}) = \sin(q_{ij} + q_{kl})$, $c(q_{ij} + q_{kl}) = \cos(q_{ij} + q_{kl})$, and q_{ij} is the jth joint angle of the ith finger.

The finger joint configuration of the three-fingered hand is determined by the corresponding indices and methods proposed in Chapters 2 and 3. For this example, $q_{11} = 0$, $q_{12} = p/3$, $q_{13} = p/2$, $q_{21} = p/3$, $q_{22} = p/3$, $q_{23} = p/2$, $q_{31} = p/3$, $q_{32} = p/3$, $q_{33} = p/2$.

Once the desired object position, fingertip contact locations on the object, and the finger joint configuration are given, we can analyze the grasp capability of the three-fingered hand. For simplifying the problem, we assume that the three-fingered grasp system is at a steady state, and the maximum external wrench that the three-fingered hand can withstand is the maximum force it can apply along the direction

$$\mathbf{d} = \begin{pmatrix} \cos a \cos b & \cos a \sin b & \sin a & 0 & 0 & 0 \end{pmatrix}^T$$

as shown in Fig. 6.1.

Using the nonlinear programming method, namely Eq. (6.5), we obtain the maximum external wrench in any direction that the hand can resist without the upper bound constraint of the fingertip normal contact force. the symmetrical grasp, the following figures illustrate the results with β between 0 and $2\pi/3$. Figure 6.2 shows the maximum external wrench in the whole direction that the three-fingered hand can withstand within the joint torque limits without the upper bound constraint of the fingertip normal contact force.

As can be seen, the maximum external wrench that the three-fingered hand can withstand varies along different directions. When

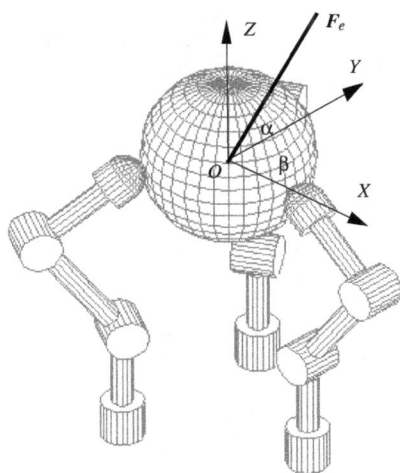

Fig. 6.1. Three-fingered hand grasps a sphere.

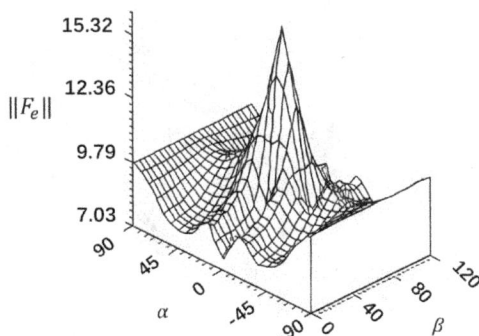

Fig. 6.2. Maximum external wrench without the constraints of the normal contact force limits.

$\alpha = 0°$ and $\beta = 60°$, the external wrench that the trifingered hand can withstand is the largest, which is 16.34 N. In other words, when the external wrench applies along the direction from the center of the sphere to the middle point of the two fingertips, the grasp has the best ability to withstand external wrench. This result can help us better select the grasp posture and plan grasp task. The corresponding normal contact forces of the three fingertips are shown in Fig. 6.3.

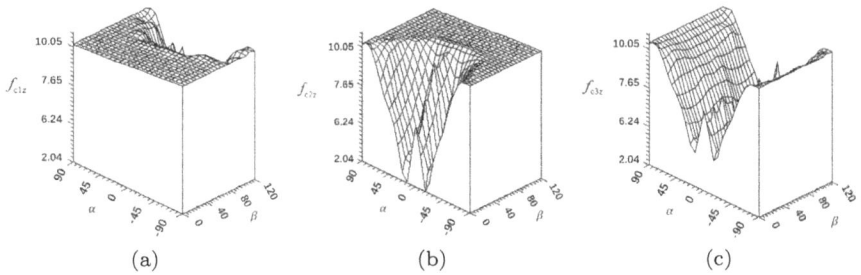

Fig. 6.3. Normal contact forces without upper bound constraints. (a) Normal contact force of the 1st fingertip. (b) Normal contact force of the 2nd fingertip. (c) Normal contact force of the 3rd fingertip.

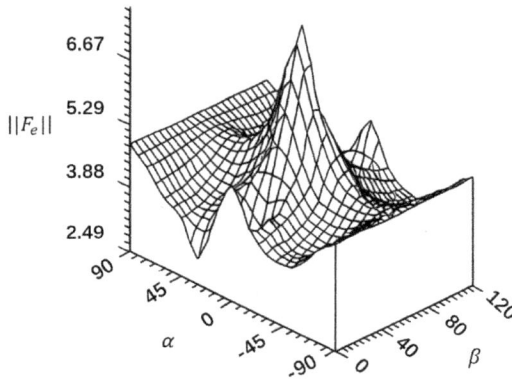

Fig. 6.4. Maximum external wrench with $f_{i\max} = 5\,\text{N}$.

It can be found that the component of the contact force normal to the object's surface at each contact is positive which implies the unisense force constraints are satisfied. However, in some directions, the component of the normal contact force is very large which means the object may be damaged when the maximum external wrench exerts along some directions.

Assuming that the maximum components of all fingertip contact forces normal to the sphere's surface at each contact, which guarantees the object not to be damaged, are $5\,\text{N}$, we analyze the grasp capability again using Eq. (6.7). Figure 6.4 shows the maximum external wrench in the whole direction that the three-fingered

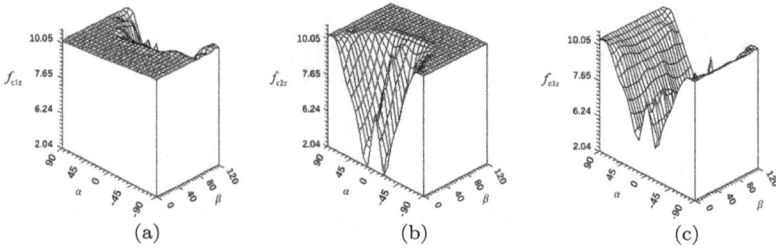

Fig. 6.5.   Normal contact forces with $f_{imax} = 5\,\text{N}$, (a) Normal contact force of the 1st fingertip. (b) Normal contact force of the 2nd fingertip. (c) Normal contact force of the 3rd fingertip.

hand can withstand with the upper bound constraint of the fingertip normal contact force. The corresponding normal contact forces of the three fingertips are shown in Fig. 6.5. Note that the normal contact forces in the whole direction vary between the admissible limits. Comparison of Figs. 6.2 and 6.4 shows that the maximum external wrench that the three-fingered hand can withstand becomes smaller, which is 7.771 N, when the upper bound constraints of the normal contact force are taken into account.

Further, assuming that the surface of the sphere to be grasped is very rough, and the frictional coefficient between the sphere and the fingertips is 0.48. The maximum components of all fingertip contact forces normal to the sphere's surface at each contact, which guarantee the object is not to be damaged, are 10 N, we analyze the grasp capability again using Eq. (6.7). Figure 6.6 shows the maximum external

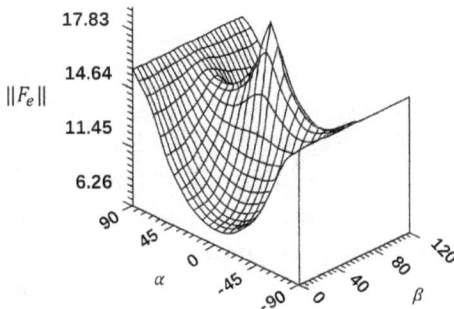

Fig. 6.6.   Maximum external wrench with $f_{imax} = 10\,\text{N}$ and $\mu = 0.48$.

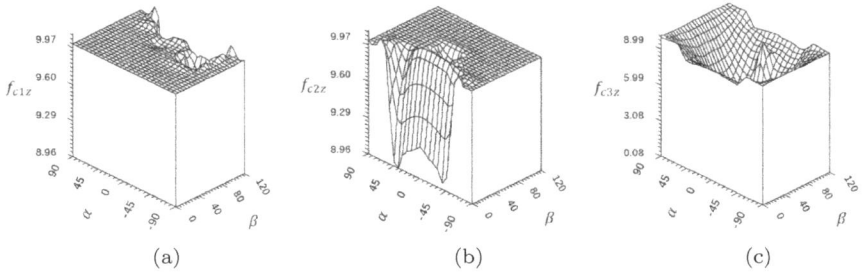

Fig. 6.7. Normal contact forces with $f_{i\max} = 10\,\mathrm{N}$ and $\mu = 0.48$. (a) Normal contact force of the 1st fingertip. (b) Normal contact force of the 2nd fingertip. (c) Normal contact force of the 3rd fingertip.

wrench in the whole direction that the three-fingered hand can with stand with $f_{i\max} = 10\mathrm{N}$. The corresponding normal contact forces of the three fingertips are shown in Fig. 6.7. Note that the normal contact forces in the whole direction vary between the admissible limits. Comparing Figs. 6.2 and 6.6 show that the maximum external wrench that the trifingered hand can withstand becomes larger when the surface of the object to be grasped is very rough. When $\alpha = 0°$ and $\beta = 60°$, the external wrench that the trifingered hand can withstand has a maximum value which is 19.01 N. In addition, as can be seen, when the external wrench applies along the direction from the center of the sphere to the North Pole or South Pole of the sphere, the grasp has also a better ability to withstand external wrench.

6.4   Summary

Grasp capability can represent an important criterion in evaluation, programming, and design of multifingered robot hands, intelligent fixtures, and similar devices. A throughout analysis of the grasp capability can not only provide additional information for the design and control of artificial multifingered hands but also help us better plan the grasp task for the existing multifingered hands.

The problem of grasp capability analysis is, in fact, the problem of finding the maximum external wrench in any direction that the multifingered hand can withstand. We formulate the problem as a nonlinear programming problem which maximized the external wrench in any direction under the constraints of the joint driving torque limits, the friction forces, and the unisense forces. We also take into account the upper bound constraints of the fingertip normal contact force in order that the object to be grasped is not to be damaged. The presented example shows that the proposed method is effective.

The method is applicable to diverse areas such as multiple robot arms and intelligent fixtures.

References

[1] Cole A., Hauser J., and Sastry S. Kinematics and control of a multifingered robot hand with rolling contact. *IEEE Transactions on Automatic Control*, 34(4), pp. 398–403, 1989.

[2] Li Z. X. and Sastry S. Task oriented optimal grasping by multifingered robot hands. *IEEE Journal of Robotics and Automation*, 14, pp. 32–43, 1988.

[3] Nakamura Y., Nagai K., and Yoshikawa T. Dynamics and stability in coordination of multiple robotic mechanisms. *International Journal of Robotics Research*, 8(2), pp. 44–61, 1989.

[4] Cheng F. T., Orin D. E. Efficient algorithm for optimal force distribution-the compact-dual LP method. *IEEE Transactions on Robotics and Automation*, 6, pp. 178–187, 1990.

[5] Li Z., Tarn T. J., Bejczy A. K. Dynamic workspace analysis of multiple cooperating robot arms. *IEEE Transactions on Robotics and Automation*, 7(5), pp. 589–596, 1991.

[6] Lu W. S. and Meng Q. H. On optimal force distribution of coordinating manipulators. *International Journal of Robotics and Automation*, 7(2), pp. 70–79, 1992.

[7] Shin Y. D. and Chung M. J. An optimal force distribution scheme for cooperating multiple robot manipulators. *Robotica*, 11, pp. 49–59, 1993.

[8] Chevallier D. P. and Payandeh S. On computing the friction forces associated with three-fingered grasp. *International Journal of Robotics Research*, 13(2), pp. 119–126, 1994.

[9] Panagiotopoulos P. D. and Al-Fahed A. M. Robot hand grasping and related problem: Optimal control and identification. *International Journal of Robotics Research*, 13(2), pp. 127–136, 1994.

[10]   Sheridan M. J., Ahalt S. C., and Orin D. E. Fuzzy control for robotic power grasp. *Advanced Robotics*, 9(5), pp. 535–546, 1995.

[11]   Xiong C. H. and Xiong Y. L. Neural-network based force planning for multifingered grasp. *Robotics and Autonomous Systems*, 21(4), pp. 365–375, 1997.

[12]   Xiong C. H., Li, Y. F., Xiong Y. L., Ding H., and Huang Q. S. Grasp capability analysis of multifingered robot hands. *Robotics and Autonomous Systems*, 27(4), pp. 211–224, 1999.

[13]   Xiong C. H. and Xiong Y. L. Stability index and contact configuration planning of multifingered grasp. *Journal of Robotic Systems*, 15(4), pp. 183–190, 1998.

[14]   Shimoga K. B. Robot grasp synthesis algorithms: A survey. *International Journal of Robotics Research*, 15(3), pp. 230–266, 1996.

Chapter 7

Compliant Grasping with Passive Forces

Since friction is central to robotic grasp, developing an accurate and tractable model of contact compliance, particularly in the tangential direction, and predicting the passive force closure are crucial to robotic grasping and contact analysis. This chapter analyzes the existence of the uncontrollable grasping forces (i.e., passive contact forces) in enveloping grasp or fixturing and formulates a physical model of compliant enveloping grasp. First, we develop a locally elastic contact model to describe the nonlinear coupling between the contact force with friction and elastic deformation at the individual contact. Further, a set of "compatibility" equations is given so that the elastic deformations among all contacts in the grasping system result in a consistent set of displacements of the object. Then, combining the force equilibrium, the locally elastic contact model, and the "compatibility" conditions, we formulate the natural compliant model of the enveloping grasp system where the passive compliance in joints of fingers is considered and investigate the stability of the compliant grasp system. The crux of judging passive force closure is to predict the passive contact forces in the grasping system, which is formulated into a nonlinear least square in this chapter. Using the globally convergent Levenberg–Marquardt method, we predict contact forces and estimate the passive force closure in the enveloping grasps. Finally, a numerical example is given to verify the proposed compliant enveloping grasp model and the prediction method of passive force closure.

7.1   Introduction

Grasping with multifingered robot hands can be divided into two types: fingertip grasps and enveloping grasps. For the fingertip grasp, we expect the manipulation of an object to be dexterous since the active fingertip can exert an arbitrary contact force onto the object. Generally, all the contact forces can be controlled actively in fingertip grasps. In contrast to fingertip grasps, enveloping grasps are formed by wrapping the fingers (and the palm) around the object to be grasped. They are, similar to fixtures, almost exclusively used for restraint and for fixturing and not for dexterous manipulation. We expect the grasp to be robust against an external disturbance. In fixtures and enveloping grasps, the number of actuators is commonly much less than the relative freedom of motion allowed by contacts between the object and links of fingers, thus, from a view point of controllability, not all the contact forces are controllable actively, which is the main issue of the grasp force analyses in enveloping grasping and fixturing. This chapter focuses on the fundamental problems of the modeling of compliant enveloping grasping or fixturing, and the prediction of the corresponding passive force closure, which are crucial to robotic grasping and contact analysis [1].

The study on grasping/fixturing with a rigid body contact model between an object and fingers has been reported [2–16]. It is not unusual in the literature that the contacts are assumed frictionless [2, 10, 12, 14, 16, 17]. In this case, the analysis of contact forces is simple. This assumption renders the contact forces to be determinate in a static equilibrium state when the number of passive contacts is 6. If the number of passive contacts is larger than 6, the grasping/fixturing is indeterminate, which means that we cannot determine the passive contact forces from the force/torque equilibrium equations of the system. However, frictional forces are important in practical cases to help prevent an object from slipping [1, 17–19].

The usual approach of frictionless assumption often yields impractical solutions. In reality, the influence of frictional forces cannot be simply neglected. When frictional forces are taken into account, one of the fundamental problems is that the equilibrium equation of the grasping system cannot determine the contact forces uniquely in general, which means that the system is indeterminate in the rigid body framework, that is, given joint torques, we cannot determine the grasp forces uniquely from the force/torque equilibrium equations of the system, therefore force closure of grasping [20] cannot be judged. When using fingertips to grasp an object, the contact forces between the object and fingertips are active. They can be actively regulated by coordinating control of the robot fingers [7, 9, 21, 22]. This is necessary for the goal of dexterous manipulation. In an enveloping grasp, a robot arm/finger may contact the object at many contacting points, in some cases even embracing it against a passive environment [3, 23] for a robust and stable grasp against an external disturbance.

As mentioned before, the number of actuators is commonly much less than the relative freedom of motion allowed by contacts between the object and the environment/links of fingers at some contacts; thus, such contact forces are passive. A new set of definitions for closure properties including active and passive closures is proposed in Ref. [24]. Harada *et al.* discussed the active force closure for the manipulation of multiple objects [4]. The passive forces result in more restrictive force closure conditions than those involving only active forces. The passive force closure in compliant-rigid grasps was studied in Ref. [25]. Xiong *et al.* formulated the problem of passive force closure in workpiece-fixture systems [26]. It is impossible to command an arbitrary set of grasp forces because the related Jacobian matrix is not full rank at such contacts although we can find bounds of the indeterminate contact forces for the enveloping grasp within the rigid body model [3, 8]. In fact, when joint torques exert on the joints, all of the contact forces including the active and passive will be "automatically" distributed among the contacts between the object and the links of fingers and the environment (palm) (see Fig. 7.1). Thus, using only rigid body models fails to reflect accurately the

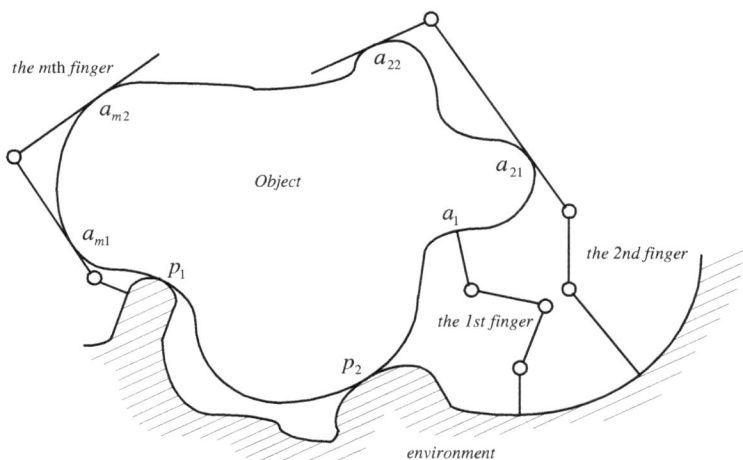

Fig. 7.1. Grasping system with passive contact constraints.

contact forces; the prediction of contact forces and force closure is a challenging and attracting research area [1, 3, 4, 23, 25–30].

If one introduces a stiffness term at each contact point, then the contact forces can be obtained uniquely for enveloping grasps [1, 25]. All force closure grasps can be made stable by constructing virtual springs at the contacts [31, 32]. On the basis of quasi-static stability analysis [15], Xiong *et al.* analyzed the dynamic stability of compliant grasps using a linear spring damper model for the fingers [33]. A stiffness quality measure for compliant grasps with frictionless contacts, which is to evaluate whether a workpiece can stay within a specified tolerance in response to machining or assembly forces, was defined in Ref. [17]. However, the contact stiffness is essentially a local property and nonlinear [18, 19]. A linear representation of a simple stiffness matrix may not be adequate or appropriate, especially in the presence of contact friction [1, 18, 19]. A comprehensive approach is to consider the grasping system as an elastic system that can be analyzed with a finite element model [28]. Such an approach often results in a large-size model and requires high computational effort. The model is also sensitive to the boundary conditions. Another approach is to use a discrete contact elasticity model to represent unidirectional contacts [29, 30]. By applying the principle of minimum total complementary energy, this model yields a constrained quadratic program for predicting the contact forces [28]. However, the discrete contact elasticity model requires prior knowledge of the contact state of each passive contact. One may have to first guess whether a particular contact is in a state of lift-off, stick, or slip. Subsequently, the general model must be assembled and solved numerically. Afterward, the inequality constraints associated with the contact states must be verified. If any of the inequality constraints is violated, a new assumption must be made and the procedure is repeated until all inequality constraints are satisfied. To simplify the model, the empirical force-deformation functions known as meta-functions were developed in Ref. [27]. Each contact is modeled with an elastic deformation region [28, 29], which increases the modeling and computational complexity considerably. Perhaps the most important is the need for a reliable estimate of contact compliance, arising with statically indeterminate grasps [1, 3, 4, 8]. Such an effective contact compliant model for enveloping grasps and fixtures with friction contacts, and the method that accurately predicts passive force closure, are currently not available [1].

In this chapter, the frictional contacts are modeled by a *locally elastic* contact model, while the links of fingers and the object to be grasped are otherwise treated as rigid bodies. Within this framework, the problem of passive and frictional contact forces in the grasping systems is resolved with the simplicity of rigid-body equilibrium and the fidelity of elastic contacts in classical mechanics.

7.2   The Model of Compliant Grasping/Fixturing

Assume that m fingers grasp an object with friction contacts, as shown in Fig. 7.1. Let $\mathbf{n}_i \in \Re^{3 \times 1}$ be the unit inner normal vector of the object at the position $\mathbf{r}_i \in \Re^{3 \times 1}$ of the ith contact. Moreover, let $\mathbf{t}_{i1} \in \Re^{3 \times 1}$ and $\mathbf{t}_{i2} \in \Re^{3 \times 1}$ be the two orthogonal unit tangential vectors of the object at the ith contact, respectively. For the ith contact, we denote by $\mathbf{f}_{ci} = (f_{in}\ f_{it1}\ f_{it2})^T \in \Re^{3 \times 1}$ the three elements of the contact force $\mathbf{f}_{ci}$ along the unit normal vector $\mathbf{n}_i$ and the unit tangential vectors $\mathbf{t}_{i1}$ and $\mathbf{t}_{i2}$, respectively. Here, $i = 1, \ldots, l$ is the number of contacts.

Thus, the force equilibrium of the grasping system is described as

$$\mathbf{G}\mathbf{f}_c - \mathbf{F}_e = 0 \tag{7.1}$$

where

$$\mathbf{G} = \begin{bmatrix} \mathbf{n}_1 & \mathbf{t}_{11} & \mathbf{t}_{12} & \cdots & \mathbf{n}_l & \mathbf{t}_{l1} & \mathbf{t}_{l2} \\ \mathbf{r}_1 \times \mathbf{n}_1 & \mathbf{r}_1 \times \mathbf{t}_{11} & \mathbf{r}_1 \times \mathbf{t}_{12} & \cdots & \mathbf{r}_l \times \mathbf{n}_l & \mathbf{r}_l \times \mathbf{t}_{l1} & \mathbf{r}_l \times \mathbf{t}_{l2} \end{bmatrix} \in \Re^{6 \times 3l}$$

is referred to as the grasping matrix,

$$\mathbf{f}_c = \left(\mathbf{f}_{c1}^T \cdots \mathbf{f}_{cl}^T \right)^T \in \Re^{3l \times 1}$$

is the contact forces of the links of the m fingers and the environment, and $\mathbf{F}_e \in \Re^{6 \times 1}$ is the external wrench exerted on the object.

Generally, given the external wrench $\mathbf{F}_e$, we cannot determine the contact forces $\mathbf{f}_c$ from Eq. (7.1) because of $3l > rank(\mathbf{G})$, which means the problem of determining contact forces from Eq. (7.1) is statically indeterminate. The degree of indeterminacy is equal to $dim(\mathrm{Ker}(\mathbf{G}))$, i.e., $3l - rank(\mathbf{G})$.

From the view point of linear space, the contact forces $\mathbf{f}_c$ in Eq. (7.1) can be represented as

$$\mathbf{f}_c = \mathbf{f}_e + \mathbf{f}_N \tag{7.2}$$

where $\mathbf{f}_e = \mathbf{G}^+ \mathbf{F}_e \in \mathrm{Im}\left(\mathbf{G}^T\right)$ denotes the set of contact forces that can balance the external wrench $\mathbf{F}_e$, while $\mathbf{f}_N = \left(\mathbf{I}_{3l} - \mathbf{G}^+ \mathbf{G}\right)\lambda \in \mathrm{Ker}(\mathbf{G})$ denotes the set of internal contact forces in the null space of the grasping matrix $\mathbf{G}$ ($dim(\mathrm{Ker}(\mathbf{G})) = 3l - 6$, and $dim\left(\mathrm{Im}\left(\mathbf{G}^T\right)\right) = rank(\mathbf{G}) = 6$), the null space $\mathrm{Ker}(\mathbf{G})$ and the row space $\mathrm{Im}\left(\mathbf{G}^T\right)$ are orthogonal complements of each other, namely $\mathrm{Im}\left(\mathbf{G}^T\right) = (\mathrm{Ker}(\mathbf{G}))^\perp$ and $\Re^{3l} = \mathrm{Im}\left(\mathbf{G}^T\right) \oplus \mathrm{Ker}(\mathbf{G})$, where $\mathbf{G}^+ = \mathbf{G}^T \left(\mathbf{G}\mathbf{G}^T\right)^{-1}$ is the Moore–Penrose generalized inverse matrix of $\mathbf{G}$, $\mathbf{I}_{3l}$ is a $3l \times 3l$ identity matrix, $\lambda \in \Re^{3l \times 1}$ is an arbitrary vector.

In the robotic fingertip grasps, grasp optimization techniques to find the optimal $\hat{\lambda}$ can be formulated by defining a cost and constraint function so as to avoid contact slippage and minimize consumption of power in the joint actuators. In non-defective systems (that is every single finger has full mobility in its task space), the optimal contact forces

$$\hat{\mathbf{f}}_c = \mathbf{G}^+ \mathbf{F}_e + \left(\mathbf{I}_{3l} - \mathbf{G}^+ \mathbf{G}\right) \hat{\lambda}$$

are applied by the fingers under some type of force control technique [23]. Thus, in this case, all the contact forces can be realized actively by controlling joint torques.

However, according to the relationship between the contact forces on the fingers and the vector $\boldsymbol{\tau}$ of joint actuator torques,

$$\boldsymbol{\tau} \in \Re^{\sum_{i=1}^{m} n_i \times 1}$$

where n_i is the joint number of the ith finger, and $\mathbf{J}^T \mathbf{f}_c = \boldsymbol{\tau}$ when the grasping system is defective. The $\mathbf{J}$ is the Jacobian of the grasping system which satisfies

$$\mathbf{J} \in \Re^{3l \times \sum_{i=1}^{m} n_i}$$

where

$$\sum_{i=1}^{m} n_i < 3l$$

is the number of actuators which is less than the relative freedom of motion allowed by contacts between the object and the fingers (the relative freedom of motion is 3 for every single contact with friction), there is no guarantee that the optimal contact forces can actually be realized by the joint actuators, which means some of the contact forces contain passive elements. In other words, the contact forces generated by joint actuators cannot span the entire 3-Dl contact space at every contact in those cases. For example, in Fig. 7.1, the number of joint actuators is 2 for the contacts a_{21} (in the 2nd finger) and a_{m2} (in the mth finger), therefore, there exists one passive element in the contact forces $\mathbf{f}_{c21}$ and $\mathbf{f}_{cm2}$, respectively. The number of actuators is 1 for the contact a_{m1} (in the mth finger), so there are 2 passive elements in the contact force $\mathbf{f}_{cm1}$. Especially, no actuator exists for the contacts p_1 and p_2 in Fig. 7.1, thus, the corresponding contact forces $\mathbf{f}_{cp1}$ and $\mathbf{f}_{cp2}$ are passive, which depend on the other active contact forces, the external wrench exerted on the object, and the material properties of the object and the grasping system.

Once the joint torques exert on the joints of the grasping system, all of the passive contact forces are determinate uniquely. It is clear that rigid body models cannot be used to estimate accurately contact forces and predict force closure of passive forces. An accurate and tractable model of contact compliance is necessary for addressing these issues. In such a model of contact compliance, we assume that the object, the links of fingers, and the passive contact environment are rigid bodies, and the local contacts between them are treated as elastic contacts with friction. Under this assumption, every local contact can be replaced by a 3-D virtual spring system, namely every frictional contact has 3 springs, all going through the point of contact, and orthogonal to each other (see Fig. 7.2). In addition, every joint in the grasping system is modeled by a virtual spring which is resulted from the elastic behavior of the structure, the effects of joint servo, etc. (see Fig. 7.3).

7.3   Local Elastic Contact Model

To reflect accurately the natural compliance at contacts, we derive an effective local elastic contact model on the basis of Hertz

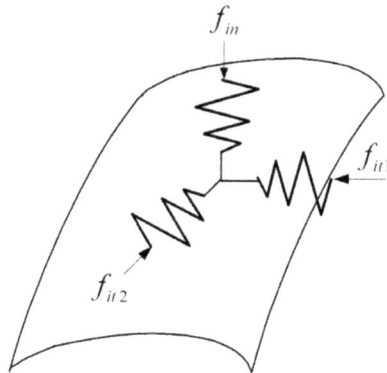

Fig. 7.2.   Compliant physical model at the ith contact.

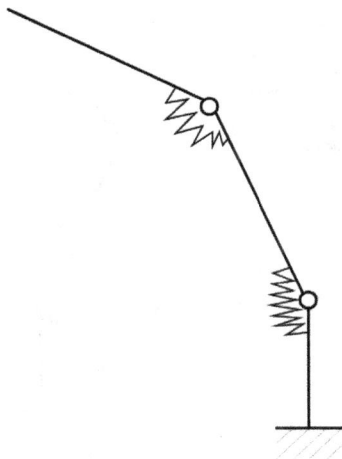

Fig. 7.3. Compliant physical models of joints in the grasping system.

elastic contact theory [34], and the discrete contact elasticity model [29].

Consider an object and the link of a finger (or environment/palm) come into contact at a point C_i, the surfaces of the object and link (or environment/palm) have mean curvature radii R_{o_i} in a neighborhood of the contact point C_i, as shown in Fig. 7.4:

$$R_{o_i} = \frac{2R_{o_{i1}} R_{o_{i2}}}{R_{o_{i1}} + R_{o_{i2}}}$$

where $R_{o_{i1}}$ and $R_{o_{i2}}$ are the two principal curvature radii of the object surface and

$$R_{l_i} = \frac{2R_{l_{i1}} R_{l_{i2}}}{R_{l_{i1}} + R_{l_{i2}}}$$

where $R_{l_{i1}}$ and $R_{l_{i2}}$ are the two principal curvature radii of the link surface.

According to the Hertz theory of elastic contact [34], the point of contact spreads into an area after the normal compressive load f_{in} is applied to the two solids, and the radius a_i of the contact area is

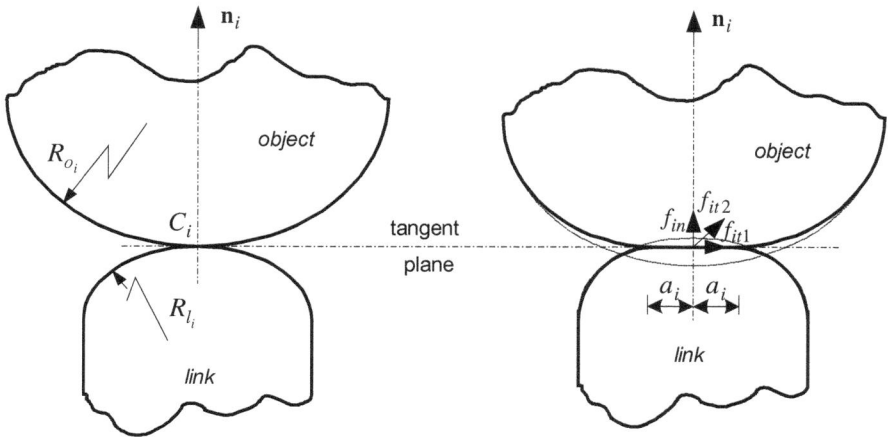

Fig. 7.4. Compression and indentation of two surfaces between the object and the link.

defined as

$$a_i = \left(\frac{3 f_{in} R_{i*}}{4 E_{i*}} \right)^{\frac{1}{3}} \tag{7.3}$$

and

$$R_{i*} = \left(\frac{1}{R_{l_i}} + \frac{1}{R_{o_i}} \right)^{-1} \tag{7.4}$$

$$E_{i*} = \left(\frac{1 - \nu_o^2}{E_o} + \frac{1 - \nu_{l_i}^2}{E_{l_i}} \right)^{-1} \tag{7.5}$$

where E_o and E_{l_i} are Young's modulus (i.e., the ratio of stress to strain on the loading plane along the loading direction) and ν_o and ν_{l_i} are Poisson's ratios (i.e., the ratio of transverse contraction strain to longitudinal extension strain in the direction of stretching force) of the object and the link at the ith contact, respectively.

The relationship between the normal deformation δd_{in} and the normal contact force component f_{in} at the ith contact can be represented as

$$\delta d_{in} = \frac{a_i^2}{R_{i*}} \tag{7.6}$$

Substituting Eq. (7.3) into Eq. (7.6) yields

$$\delta d_{in} = C_{in} f_{in}^{\frac{2}{3}} \tag{7.7}$$

where the normal deformation coefficient C_{in} is defined as

$$C_{in} = \left(\frac{9}{16 E_{i*}^2 R_{i*}} \right)^{\frac{1}{3}} \tag{7.8}$$

The tangential deformations δd_{it1} and δd_{it2} at the ith contact can be represented as

$$\delta d_{it1} = \frac{f_{it1}}{8 a_i} \left(\frac{2 - \nu_o}{G_o} + \frac{2 - \nu_{l_i}}{G_{l_i}} \right) \tag{7.9}$$

$$\delta d_{it2} = \frac{f_{it2}}{8 a_i} \left(\frac{2 - \nu_o}{G_o} + \frac{2 - \nu_{l_i}}{G_{l_i}} \right) \tag{7.10}$$

where

$$G_o = \frac{E_o}{2 (1 + \nu_o)}$$

is the shear modulus of the object and

$$G_{l_i} = \frac{E_{l_i}}{2 (1 + \nu_{l_i})}$$

is the shear modulus of the link at the ith contact, and f_{it1} and f_{it2} are the two orthogonal elements of the contact force at the ith contact.

Substituting Eq. (7.3) into Eq. (7.9) and Eq. (7.10) yields

$$\delta d_{it1} = C_{it} f_{it1} f_{in}^{-\frac{1}{3}} \tag{7.11}$$

$$\delta d_{it2} = C_{it} f_{it2} f_{in}^{-\frac{1}{3}} \tag{7.12}$$

where the tangential deformation coefficient C_{it} is defined as

$$C_{it} = \left(\frac{E_{i*}}{48 R_{i*}} \right)^{\frac{1}{3}} \left(\frac{(1 + \nu_o)(2 - \nu_o)}{E_o} + \frac{(1 + \nu_{l_i})(2 - \nu_{l_i})}{E_{l_i}} \right) \tag{7.13}$$

From Eq. (7.8) and Eq. (7.13), we can find that the deformation coefficients are related to the local geometrical and material properties of the object and links at contacts. Thus, the local elastic deformation only depends on the local geometrical and material properties of the object and links at contacts for the given contact forces. Finally, rewriting Eqs. (7.7), (7.11), and (7.12), we establish nonlinear relationships of the contact forces with the local elastic deformations as

$$f_{in} = C_{in}^{-\frac{3}{2}} (\delta d_{in})^{\frac{3}{2}} \tag{7.14}$$

$$f_{it1} = C_{it}^{-1} C_{in}^{-\frac{1}{2}} (\delta d_{in})^{\frac{1}{2}} \delta d_{it1} \tag{7.15}$$

$$f_{it2} = C_{it}^{-1} C_{in}^{-\frac{1}{2}} (\delta d_{in})^{\frac{1}{2}} \delta d_{it2} \tag{7.16}$$

These local relationships of elastic contacts establish a "constitutive" model for each individual contact, which describes the inherent relationship between the contact forces and the local elastic deformation. We call such a model the local contact compliant model.

7.4   Deformation Compatible Constraints for All Contacts

In Section 7.3, the relationship between the elastic deformations and contact forces for single contact is derived. In fact, the elastic deformations for different contacts are related to each other in the grasping/fixturing system. Here we formulate the elastic deformation compatible constraints for all contacts in the grasping system.

Consider a general object as shown in Fig. 7.5 with its reference frame $\{\mathbf{O}\}$ given in the palm frame $\{\mathbf{P}\}$. Let $\{\mathbf{F}_i\}$ be the coordinate frame of the ith link of the finger. In the palm frame, the position and orientation of the object are specified as $\mathbf{X}_o \in \Re^{3 \times 1}$ and $\mathbf{\Theta}_o \in \Re^{3 \times 1}$, while the position and orientation of the ith link are described as $\mathbf{X}_{fi} \in \Re^{3 \times 1}$ and $\mathbf{\Theta}_{fi} \in \Re^{3 \times 1}$. At this contact point, the position of the contact is represented by $\mathbf{r}_{oi} \in \Re^{3 \times 1}$ in the object frame $\{\mathbf{O}\}$ or by $\mathbf{r}_{fi} \in \Re^{3 \times 1}$ in the ith link frame $\{\mathbf{F}_i\}$.

Now consider the set of normal and tangential deformations $\mathbf{\Delta d}_i = \begin{pmatrix} \delta d_{in} & \delta d_{it1} & \delta d_{it2} \end{pmatrix}^T$ that locally occurred at this elastic contact. Under the contact forces, both the object and the links at

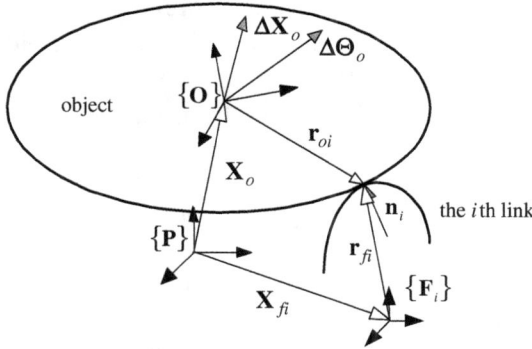

Fig. 7.5. Coordinate frames in the grasping system.

contacts are compressed elastically. The deformations will cause a change of $\mathbf{\Delta r}_{oi}$ in $\mathbf{r}_{oi}$ and a change of $\mathbf{\Delta r}_{fi}$ in $\mathbf{r}_{fi}$. The change of $\mathbf{\Delta r}_{oi}$ will be manifested in a change of

$$\mathbf{\Delta X} = \left(\mathbf{\Delta X}_o^T \;\; \mathbf{\Delta \Theta}_o^T \right)^T$$

in the position $\mathbf{X}_o$ and orientation $\mathbf{\Theta}_o$ of the object resulting from all of the deformations at all contacts. If we define a local contact frame $\{\mathbf{LC}_i\}$ of the ith contact with respect to the palm frame $\{\mathbf{P}\}$ defined by a matrix $_{lci}^{p}\mathbf{R}$ for its orientation, then the local deformations $\mathbf{\Delta d}_i$ can be related to the coordinate changes of $\mathbf{\Delta r}_{oi}$ and $\mathbf{\Delta r}_{fi}$ as

$$\mathbf{\Delta c}_i = {}_{fi}^{p}\mathbf{R}\mathbf{\Delta r}_{fi} - {}_{o}^{p}\mathbf{R}\mathbf{\Delta r}_{oi} = {}_{lci}^{p}\mathbf{R} \cdot \mathbf{\Delta d}_i \qquad (7.17)$$

where $_{o}^{p}\mathbf{R}$ and $_{fi}^{p}\mathbf{R}$ are the orientation matrices of the object frame $\{\mathbf{O}\}$ and the ith link frame $\{\mathbf{F}_i\}$ of the finger, respectively, both with respect to the palm frame $\{\mathbf{P}\}$.

Therefore, we can find a set of "compatibility" equations between the local elastic deformations and the displacement of the object as follows:

$$\mathbf{U}_{oi} \cdot \mathbf{\Delta X} = \mathbf{\Delta c}_i \qquad (7.18)$$

where $\mathbf{U}_{oi} = \left(\mathbf{I}_{3\times3} \;\vdots\; -_{o}^{p}\mathbf{R}\mathbf{r}_{oi}\times \right) \in \Re^{3\times6}$, $\mathbf{I}_{3\times3} \in \Re^{3\times3}$ is an identity matrix, and

$$\mathbf{r}_{oi}\times = \begin{pmatrix} r_x \\ r_y \\ r_z \end{pmatrix} \times = \begin{pmatrix} 0 & -r_z & r_y \\ r_z & 0 & -r_x \\ -r_y & r_x & 0 \end{pmatrix} \qquad (7.19)$$

Finally, we shall collect the compatibility equations for all l contacts containing the passive contacts. This would yield the entire $3l$ equations governing the relationships between the displacement of the object and the local elastic deformations of all contacts of the grasping system

$$\mathbf{G}^T \mathbf{\Delta X} = \mathbf{\Delta c} \tag{7.20}$$

where $\mathbf{\Delta c} = (\mathbf{\Delta c}_1^T \cdots \mathbf{\Delta c}_l^T)^T \in \Re^{3l \times 1}$ is the local elastic deformations of all contacts.

7.5   Stability of Grasping/Fixturing Systems

Grasp stability, which is the capability of returning the grasped object to its equilibrium point after it is disturbed from its equilibrium, is one of the important criteria for evaluating a grip. To investigate the stability of the compliant grasping system, we need to construct the potential energy of the system. We know the original external wrench $\mathbf{F}_e$ including the gravity force of the object can be regarded as a preload for the system; it affects the deflection of the equilibrium point only, thus the potential energy function of the system where the external wrench is not considered can be written as

$$U = \int_0^{\mathbf{\Delta c}} \mathbf{f}_c^T (\mathbf{\Delta c}) \ \mathrm{d}\mathbf{\Delta c} \tag{7.21}$$

Using Taylor expansion at the point $\mathbf{\Delta c}_0$ in Eq. (7.21), we obtain

$$\begin{aligned}
U = U\,|_{\mathbf{\Delta c}_0} &+ \nabla U^T\,|_{\mathbf{\Delta c}_0} (\mathbf{\Delta c} - \mathbf{\Delta c}_0) \\
&+ \tfrac{1}{2} (\mathbf{\Delta c} - \mathbf{\Delta c}_0)^T \nabla^2 U\,|_{\mathbf{\Delta c}_0} (\mathbf{\Delta c} - \mathbf{\Delta c}_0) \\
&+ o(\|\mathbf{\Delta c} - \mathbf{\Delta c}_0\|^2)
\end{aligned} \tag{7.22}$$

where

$$U\,|_{\mathbf{\Delta c}_0} = \int_0^{\mathbf{\Delta c}_0} \mathbf{f}_c^T (\mathbf{\Delta c}) \ \mathrm{d}\mathbf{\Delta c} \tag{7.23}$$

$$\nabla U^T\,|_{\mathbf{\Delta c}_0} (\mathbf{\Delta c} - \mathbf{\Delta c}_0) = \mathbf{f}_c^T (\mathbf{\Delta c}_0) \delta \mathbf{c} \tag{7.24}$$

Using Eq. (7.20), we can rewrite Eq. (7.24) as

$$\nabla U^T |_{\mathbf{\Delta c}_0} (\mathbf{\Delta c} - \mathbf{\Delta c}_0) = [\mathbf{Gf}_c (\mathbf{\Delta c}_0)]^T \delta \mathbf{X} = \nabla U^T |_{\mathbf{\Delta X}_0} \delta \mathbf{X} \quad (7.25)$$

where $\delta \mathbf{c} = \mathbf{\Delta c} - \mathbf{\Delta c}_0$, $\delta \mathbf{X} = \mathbf{\Delta X} - \mathbf{\Delta X}_0$, $\mathbf{\Delta X}_0$ is the displacement of the object at the point $\mathbf{\Delta c}_0$, and $\mathbf{G}^T \mathbf{\Delta X}_0 = \mathbf{\Delta c}_0$.

From Eq. (7.1), we can find when the external wrench is not considered, the equilibrium equations of the grasping system are described as

$$\mathbf{Gf}_c (\mathbf{\Delta c}_0) = \mathbf{0} \quad (7.26)$$

that is the gradient of the potential energy function U, i.e.,

$$\nabla U |_{\mathbf{\Delta X}_0} = \mathbf{0} \quad (7.27)$$

Thus, the point $\mathbf{\Delta c}_0$ (or $\mathbf{\Delta X}_0$) is an equilibrium point of the grasping system. Is it a stable equilibrium point of the compliant grasping system? To answer the problem, we give a theorem as follows:

Theorem 1. *When the compliant matrix*

$$\mathbf{G} \frac{\partial \mathbf{f}_{c(\mathbf{\Delta c})}}{\partial \mathbf{\Delta c}} \mathbf{G}^T$$

is positive definite, the compliant grasping system is stable.

Proof. The third term on the right side of Eq. (7.22) can be represented as

$$\frac{1}{2} (\mathbf{\Delta c} - \mathbf{\Delta c}_0)^T \nabla^2 U |_{\mathbf{\Delta c}_0} (\mathbf{\Delta c} - \mathbf{\Delta c}_0) = \frac{1}{2} \delta \mathbf{X}^T \nabla^2 U |_{\mathbf{\Delta X}_0} \delta \mathbf{X}$$
$$(7.28)$$

where the Hessian matrix $\nabla^2 U$ of the potential energy function U can be rewritten as

$$\nabla^2 U = \frac{\partial [\mathbf{Gf}_c (\mathbf{\Delta c})]}{\partial \mathbf{\Delta X}} = \mathbf{G} \left(\frac{\partial \mathbf{f}_c (\mathbf{\Delta c})}{\partial \mathbf{\Delta c}} \cdot \frac{\partial \mathbf{\Delta c}}{\partial \mathbf{\Delta X}} \right) \quad (7.29)$$

Using Eq. (7.20), we obtain

$$\frac{\partial \mathbf{\Delta c}}{\partial \mathbf{\Delta X}} = \mathbf{G}^T \quad (7.30)$$

Substituting Eq. (7.30) into Eq. (7.29) and yields

$$\nabla^2 U = \mathbf{G} \frac{\partial \mathbf{f}_c \left(\mathbf{\Delta c} \right)}{\partial \mathbf{\Delta c}} \mathbf{G}^T \tag{7.31}$$

From Eq. (7.22), it can be found that $U > U|_{\mathbf{\Delta c}_0}$ when the Hessian matrix $\nabla^2 U$ is positive definite, which means that the equilibrium point $\mathbf{\Delta c}_0$ (or $\mathbf{\Delta X}_0$) is the minimum potential energy point of the compliant grasping system when the matrix

$$\mathbf{G} \frac{\partial \mathbf{f}_c \left(\mathbf{\Delta c} \right)}{\partial \mathbf{\Delta c}} \mathbf{G}^T$$

in Eq. (7.31) (we call it compliant matrix) is positive definite. Thus, the compliant grasping system is stable when the compliant matrix is positive definite. □

For 3-D grasping/fixturing, the matrix

$$\frac{\partial \mathbf{f}_c \left(\mathbf{\Delta c} \right)}{\partial \mathbf{\Delta c}}$$

is represented as

$$\frac{\partial \mathbf{f}_c \left(\mathbf{\Delta c} \right)}{\partial \mathbf{\Delta c}} = \begin{bmatrix} \frac{\partial f_{1n}}{\partial \mathbf{\Delta} c_{1n}} & \frac{\partial f_{1n}}{\partial \mathbf{\Delta} c_{1t1}} & \frac{\partial f_{1n}}{\partial \mathbf{\Delta} c_{1t2}} & & \\ \frac{\partial f_{1t1}}{\partial \mathbf{\Delta} c_{1n}} & \frac{\partial f_{1t1}}{\partial \mathbf{\Delta} c_{1t1}} & \frac{\partial f_{1t1}}{\partial \mathbf{\Delta} c_{1t2}} & \mathbf{0} & \\ \frac{\partial f_{1t2}}{\partial \mathbf{\Delta} c_{1n}} & \frac{\partial f_{1t2}}{\partial \mathbf{\Delta} c_{1t1}} & \frac{\partial f_{1t2}}{\partial \mathbf{\Delta} c_{1t2}} & & \\ & & & \ddots & \\ & & & \frac{\partial f_{ln}}{\partial \mathbf{\Delta} c_{ln}} & \frac{\partial f_{ln}}{\partial \mathbf{\Delta} c_{lt1}} & \frac{\partial f_{ln}}{\partial \mathbf{\Delta} c_{lt2}} \\ & \mathbf{0} & & \frac{\partial f_{lt1}}{\partial \mathbf{\Delta} c_{ln}} & \frac{\partial f_{lt1}}{\partial \mathbf{\Delta} c_{lt1}} & \frac{\partial f_{lt1}}{\partial \mathbf{\Delta} c_{lt2}} \\ & & & \frac{\partial f_{lt2}}{\partial \mathbf{\Delta} c_{ln}} & \frac{\partial f_{lt2}}{\partial \mathbf{\Delta} c_{lt1}} & \frac{\partial f_{lt2}}{\partial \mathbf{\Delta} c_{lt2}} \end{bmatrix} \in \Re^{3l \times 3l} \tag{7.32}$$

For 2-D grasping/fixturing, the matrix

$$\frac{\partial \mathbf{f}_c \left(\mathbf{\Delta c} \right)}{\partial \mathbf{\Delta c}}$$

is represented as

$$
\frac{\partial \mathbf{f}_c\left(\mathbf{\Delta c}\right)}{\partial \mathbf{\Delta c}} =
\begin{bmatrix}
\begin{matrix}
\frac{\partial f_{1n}}{\partial \mathbf{\Delta} c_{1n}} & \frac{\partial f_{1n}}{\partial \mathbf{\Delta} c_{1t}} \\
\frac{\partial f_{1t}}{\partial \mathbf{\Delta} c_{1n}} & \frac{\partial f_{1t}}{\partial \mathbf{\Delta} c_{1t}}
\end{matrix} & & 0 \\
& \ddots & \\
0 & & \begin{matrix}
\frac{\partial f_{ln}}{\partial \mathbf{\Delta} c_{ln}} & \frac{\partial f_{ln}}{\partial \mathbf{\Delta} c_{lt}} \\
\frac{\partial f_{lt}}{\partial \mathbf{\Delta} c_{ln}} & \frac{\partial f_{lt}}{\partial \mathbf{\Delta} c_{lt}}
\end{matrix}
\end{bmatrix} \in \Re^{2l \times 2l} \qquad (7.33)
$$

Especially, if the relationship between the contact forces and the corresponding elastic deformations at contacts is linear, and the three orthogonal linear springs have the stiffness k_{in}, k_{it1}, and k_{it2}, respectively $(i = 1, \ldots, l)$, then we have

$$
\frac{\partial \mathbf{f}_c\left(\mathbf{\Delta c}\right)}{\partial \mathbf{\Delta c}} = \mathrm{diag}\begin{bmatrix} k_{1n} & k_{1t1} & k_{1t2} & \cdots & k_{ln} & k_{lt1} & k_{lt2} \end{bmatrix} \in \Re^{3l \times 3l} \qquad (7.34)
$$

The stability of the compliant grasping system is related to the local geometrical properties, such as relative curvature tensors, the material properties at contacts, and the grasp configuration [33].

7.6 Passive Force Closure Prediction

The existence of uncontrollable contact force is the main issue of the grasp force analyses in enveloping grasps [1]. To verify whether the enveloping grasp/fixturing is force closure when the object is gripped under the joint torques exerted on the joints of multifingered robot hands, we need to check whether the contact forces in the grasping system are within their corresponding friction cone FC_i $(i = 1, \ldots, l)$. First of all, we must predict the passive contact forces in the grasping system.

There exist two types of force/torque mapping relationships in the grasping system, that is,

$$
\mathbf{f}_c \xrightarrow{\mathbf{J}^T} \boldsymbol{\tau}, \quad \text{and} \quad \mathbf{f}_c \xrightarrow{\mathbf{G}} \mathbf{F}_e
$$

which can be represented as follows:

$$\mathbf{J}^T \mathbf{f}_c = \boldsymbol{\tau} \tag{7.35}$$

$$\mathbf{G}\mathbf{f}_c - \mathbf{F}_e = \mathbf{0} \tag{7.36}$$

Equations (7.35) and (7.36) can be combined into

$$\mathbf{A}\mathbf{f}_c = \mathbf{F} \tag{7.37}$$

where

$$\mathbf{A} = \begin{pmatrix} \mathbf{J}^T \\ \mathbf{G} \end{pmatrix} \in \Re^{(\sum_{i=1}^m n_i + 6) \times 3l}, \quad \mathbf{F} = \begin{pmatrix} \boldsymbol{\tau} \\ \mathbf{F}_e \end{pmatrix} \in \Re^{(\sum_{i=1}^m n_i + 6) \times 1}$$

If

$$rank\,(\mathbf{A}) = \sum_{i=1}^m n_i \quad \text{and} \quad q = 3l - \sum_{i=1}^m n_i = 0$$

the grasping system is determinate, which means we can determine the contact forces uniquely for the given joint torques and the external wrench exerted on the object.

If

$$rank\,(\mathbf{A}) = \sum_{i=1}^m n_i \quad \text{and} \quad q = 3l - \sum_{i=1}^m n_i > 0$$

the corresponding grasp system is indeterminate (q is called the degree of indeterminacy); in this case, we cannot determine the contact forces uniquely using the rigid body model Eq. (7.37), which is the intrinsic characteristic of enveloping grasp/fixturing. To obtain the solutions of contact forces accurately, we need to use the local elastic contact model Eqs. (7.14)–(7.16) and the deformation compatible constraints Eq. (7.20) and then combine them with Eq. (7.37).

Considering the dimension variation of the object to be machined for manufacturing or assembly, and the disturbance of the external wrench on the object, in order to grasp the object stably without changing the preloading torques exerted on the joints, the natural compliance of the joints in multifingered robot hands plays an important role. Especially, such a natural compliance is referred to as passive natural compliance. In this case, the model of predicting the passive contact forces in the grasping system can be described as follows:

$$
\begin{cases}
\mathbf{J}^T \cdot \mathbf{f}_c(\mathbf{\Delta c}) - (\boldsymbol{\tau}_0 + \mathbf{\Delta\tau}) = \mathbf{0} \\
\mathbf{G} \cdot \mathbf{f}_c(\mathbf{\Delta c}) - \mathbf{F}_e = \mathbf{0} \\
\mathbf{G}^T \cdot \mathbf{\Delta X} - \mathbf{\Delta c} = \mathbf{0}
\end{cases}
\tag{7.38}
$$

where

$$
\boldsymbol{\tau}_0 \in \Re^{\sum_{i=1}^{m} n_i \times 1} \quad \text{and} \quad \mathbf{\Delta\tau} \in \Re^{\sum_{i=1}^{m} n_i \times 1}
$$

are the preloading torques exerted on the joints and the passive elastically compliant joint torques, respectively.

From Eq. (7.38), we define a vector function as follows:

$$
\Im = \begin{pmatrix} \Im_1 \\ \Im_2 \\ \Im_3 \end{pmatrix} = \begin{pmatrix} \Gamma_1 \\ \vdots \\ \Gamma_L \end{pmatrix} \in \Re^{L \times 1}
\tag{7.39}
$$

where

$$
\Im_1 = \mathbf{J}^T \cdot \mathbf{f}_c(\mathbf{\Delta c}) - (\boldsymbol{\tau}_0 + \mathbf{\Delta\tau}) \in \Re^{\sum_{i=1}^{m} n_i}
$$

$$
\Im_2 = \mathbf{G} \cdot \mathbf{f}_c(\mathbf{\Delta c}) - \mathbf{F}_e \in \Re^6
$$

$$
\Im_3 = \mathbf{G}^T \cdot \mathbf{\Delta X} - \mathbf{\Delta c} \in \Re^{3l}
$$

$$
L = \sum_{i=1}^{m} n_i + 6 + 3l
$$

Thus, the problem of solving the elastic deformations using Eq. (7.38) can be transformed into the nonlinear least square [35] which can be represented as follows:

$$
\text{minimize} \left(\Gamma(\boldsymbol{\eta}) = \sum_{j=1}^{L} \Gamma_j^2(\boldsymbol{\eta}) = \Im^T(\boldsymbol{\eta})\Im(\boldsymbol{\eta}) \right)
\tag{7.40}
$$

where $\boldsymbol{\eta} = \left(\mathbf{\Delta\tau}^T \ \mathbf{\Delta X}^T \ \mathbf{\Delta c}^T \right)^T = \left(\eta_1 \cdots \eta_L \right)^T \in \Re^L$ is the design variables for the problem of the nonlinear least square.

The gradient vector $\nabla\Gamma(\boldsymbol{\eta})$ of the function $\Gamma(\boldsymbol{\eta})$ is given by

$$\nabla\Gamma(\boldsymbol{\eta}) = \begin{bmatrix} \frac{\partial\Gamma}{\partial\eta_1} \\ \vdots \\ \frac{\partial\Gamma}{\partial\eta_L} \end{bmatrix} = 2\boldsymbol{\Psi}^T(\boldsymbol{\eta})\,\Im(\boldsymbol{\eta}) \in \Re^{L\times 1} \tag{7.41}$$

where the Jacobian matrix $\boldsymbol{\Psi}$ is described by

$$\boldsymbol{\Psi} = \begin{bmatrix} \frac{\partial\Gamma_1}{\partial\eta_1} & \cdots & \frac{\partial\Gamma_1}{\partial\eta_L} \\ \vdots & & \vdots \\ \frac{\partial\Gamma_L}{\partial\eta_1} & \cdots & \frac{\partial\Gamma_L}{\partial\eta_L} \end{bmatrix} \in \Re^{L\times L} \tag{7.42}$$

The Hessian matrix $\nabla^2\Gamma(\boldsymbol{\eta})$ of the function $\Gamma(\boldsymbol{\eta})$ is represented as

$$\nabla^2\Gamma(\boldsymbol{\eta}) = \begin{bmatrix} \frac{\partial^2\Gamma}{\partial\eta_1^2} & \cdots & \frac{\partial^2\Gamma}{\partial\eta_1\partial\eta_L} \\ \vdots & & \vdots \\ \frac{\partial^2\Gamma}{\partial\eta_1\partial\eta_L} & \cdots & \frac{\partial^2\Gamma}{\partial\eta_L^2} \end{bmatrix} = 2\boldsymbol{\Psi}^T(\boldsymbol{\eta})\,\boldsymbol{\Psi}(\boldsymbol{\eta}) + 2\mathbf{S}(\boldsymbol{\eta}) \in \Re^{L\times L} \tag{7.43}$$

where

$$\mathbf{S}(\boldsymbol{\eta}) = \sum_{j=1}^{L}\Gamma_j(\boldsymbol{\eta})\cdot\nabla^2\Gamma_j(\boldsymbol{\eta})$$

From the Levenberg–Marquardt method [35], we can determine a search direction $\boldsymbol{\delta}_{(k)}$ at the kth iteration using the following equation:

$$\left(\boldsymbol{\Psi}_{(k)}^T\boldsymbol{\Psi}_{(k)} + \nu_k\mathbf{I}\right)\boldsymbol{\delta}_{(k)} = -\boldsymbol{\Psi}_{(k)}^T\Im_{(k)} \tag{7.44}$$

where $\Im_{(k)}$ and $\boldsymbol{\Psi}_{(k)}$ are the function vectors and the Jacobian matrices at the kth iteration respectively, $\nu_k \geq 0$ is a scalar and $\mathbf{I}$ is the identity matrix of order L. The corresponding recurrence formula is as follows:

$$\boldsymbol{\eta}_{(k+1)} = \boldsymbol{\eta}_{(k)} + \alpha_k\boldsymbol{\delta}_{(k)} \tag{7.45}$$

where $\alpha_k > 0$ is the step length at the kth iteration. The value of the function $\Gamma(\boldsymbol{\eta}_{(k+1)})$ at the kth iteration can be represented approximately as

$$\Gamma(\boldsymbol{\eta}_{(k+1)}) \approx \Gamma(\boldsymbol{\eta}_{(k)}) + \alpha_k \boldsymbol{\delta}_{(k)}^T \nabla \Gamma(\boldsymbol{\eta}_{(k)}) + o(\alpha_k) \qquad (7.46)$$

When $\alpha_k \to 0$, we can neglect the higher order term $o(\alpha_k)$ of α_k, thus, Eq. (7.46) can be rewritten as

$$\Gamma(\boldsymbol{\eta}_{(k+1)}) \approx \Gamma(\boldsymbol{\eta}_{(k)}) + \alpha_k \boldsymbol{\delta}_{(k)}^T \nabla \Gamma(\boldsymbol{\eta}_{(k)}) \qquad (7.47)$$

Using Eq. (7.41) and Eq. (7.44), we have the following formula:

$$\boldsymbol{\delta}_{(k)}^T \nabla \Gamma(\boldsymbol{\eta}_{(k)}) = -2 \boldsymbol{\delta}_{(k)}^T (\boldsymbol{\Psi}_{(k)}^T \boldsymbol{\Psi}_{(k)} + \nu_k \mathbf{I}) \boldsymbol{\delta}_{(k)} \qquad (7.48)$$

For a sufficiently large value of ν_k, the matrix $\boldsymbol{\Psi}_{(k)}^T \boldsymbol{\Psi}_{(k)} + \nu_k \mathbf{I}$ is positive definite which means $\boldsymbol{\delta}_{(k)}^T (\boldsymbol{\Psi}_{(k)}^T \boldsymbol{\Psi}_{(k)} + \nu_k \mathbf{I}) \boldsymbol{\delta}_{(k)} > 0$, thus $\boldsymbol{\delta}_{(k)}^T \nabla \Gamma(\boldsymbol{\eta}_{(k)}) < 0$. From Eq. (7.47), we can find

$$\Gamma(\boldsymbol{\eta}_{(k+1)}) < \Gamma(\boldsymbol{\eta}_{(k)}) \qquad (7.49)$$

Equations (7.45) and (7.49) show that $\boldsymbol{\delta}_{(k)}$ is a descent search direction at the kth iteration. Consequently, we can obtain the optimal solution using the Levenberg–Marquardt method which is globally convergent.

After obtaining the elastic deformations $\boldsymbol{\Delta}\mathbf{c}$, we can determine the contact forces $\mathbf{f}_c(\boldsymbol{\Delta}\mathbf{c})$ using the local elastic contact model Eqs. (7.14)–(7.16). Therefore, the original indeterminate problem of predicting contact forces from Eq. (7.37) is formulated as the determinate one.

It should be noted that we must be able to predict whether the slippage between the object and links (and environment) occurs besides the contact forces at contacts, that is to judge whether the enveloping grasp/fixturing is force closure under without changing the preloading torques exerted on the joints. We can imagine that it is not always true that all of the contact forces are within the corresponding friction cones for the given preloading torques exerted on the joints. We cannot choose the preloading torques which depend on the manipulating task. Thus, to plan appropriately, the preload-ing torques is necessary for guaranteeing no slippage between the object and links (and environment) so that the grasp/fixturing is force closure.

The condition of non-slippage at the ith contact can be described as

$$\varepsilon_i = \mu_i f_{in} - \sqrt{f_{it1}^2 + f_{it2}^2} \geq 0, \quad i = 1, \ldots, l \qquad (7.50)$$

where μ_i is the static friction coefficient at the ith contact. Equation (7.50) defines a friction cone

$$FC_i = \left\{ \mathbf{f}_{ci} \in \Re^{3 \times 1} \mid \mu_i f_{in} - \sqrt{f_{it1}^2 + f_{it2}^2} \geq 0 \right\}$$

If all the contact forces $\mathbf{f}_c = \left(\mathbf{f}_{c1}^T \cdots \mathbf{f}_{cl}^T \right)^T \in \Re^{3l \times 1}$ with $\mathbf{f}_{ci} = \left(f_{in} \; f_{it1} \; f_{it2} \right)^T \in \Re^{3 \times 1}$ are within their friction cones, that is,

$$\mathbf{f}_c \in FC_1 \times \cdots \times FC_l \qquad (7.51)$$

then the corresponding grasp/fixturing is called force closure. Other-wise, if any one of the contact forces is not within the corresponding friction cone, the slippage will occur at the corresponding contact, which will result in the invalidation of Coulomb's friction law describ-ing the static friction characteristics at other contacts; sequentially, further induce the slippage occurring at the other contacts; finally, the position and orientation of the object cannot be maintained, the corresponding grasp/fixturing is not force closure.

Thus, we can define ε_i as the sliding evaluation index. When $\varepsilon_i \geq 0$ for all $i = 1, \ldots, l$, then the corresponding grasp/fixturing is force closure. In converse, if one of ε_i is less than 0 ($i = 1, \ldots, l$), then the corresponding grasp/fixturing is not force closure.

7.7   Numerical Example

To validate the compliant grasping/fixturing model and the evalua-tion method of passive force closure, we give a numerical example

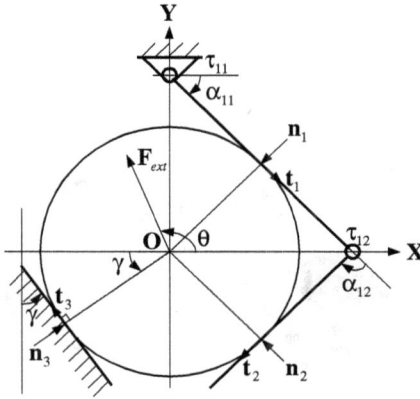

Fig. 7.6. A numerical example of enveloping grasp.

where a cylindrical object with a radius of 25 mm is grasped by a finger with two links and a passive palm (or environment), as shown in Fig. 7.6. Joint angles of the first link and the second link are α_{11} ($= 45°$) and α_{12} ($= 90°$). The position of the third contact between the object and the palm is described by the angle γ varying from $-45°$ to $45°$. The object is made from steel, and its weight is 10 N. The palm and the cylindrical links of the finger with radii of 6.35 mm are made of aluminum. We assume that both of the preloading torques of joint 11 and joint 12 are 1000 Nmm, and the external disturbance force is $\mathbf{F}_{ext} = \left(10\cos\Theta \ \ 10\sin\Theta \ \ 0 \right)^T$ (in the unit of (N)).

We use the *Levenberg–Marquardt* method in Matlab® optimization toolbox to solve the nonlinear least squares (7.40) and then predict the passive contact forces and the actual joint torques which are shown in Figs. 7.7 and 7.8 for different γ and Θ. When the static friction coefficient between the object and the links/palm is 0.35, we can obtain the corresponding sliding indices at three contacts, as shown in Fig. 7.9. From Fig. 7.9, we can find that sliding may occur in the neighborhood of the points: $\gamma = -30°$ and $\Theta = 250°$ for the 1st contact, and $\gamma = -45°$ and $\Theta = 300°$ for the 3rd contact. The compliant enveloping grasp is passive force closure except in the neighborhood mentioned above.

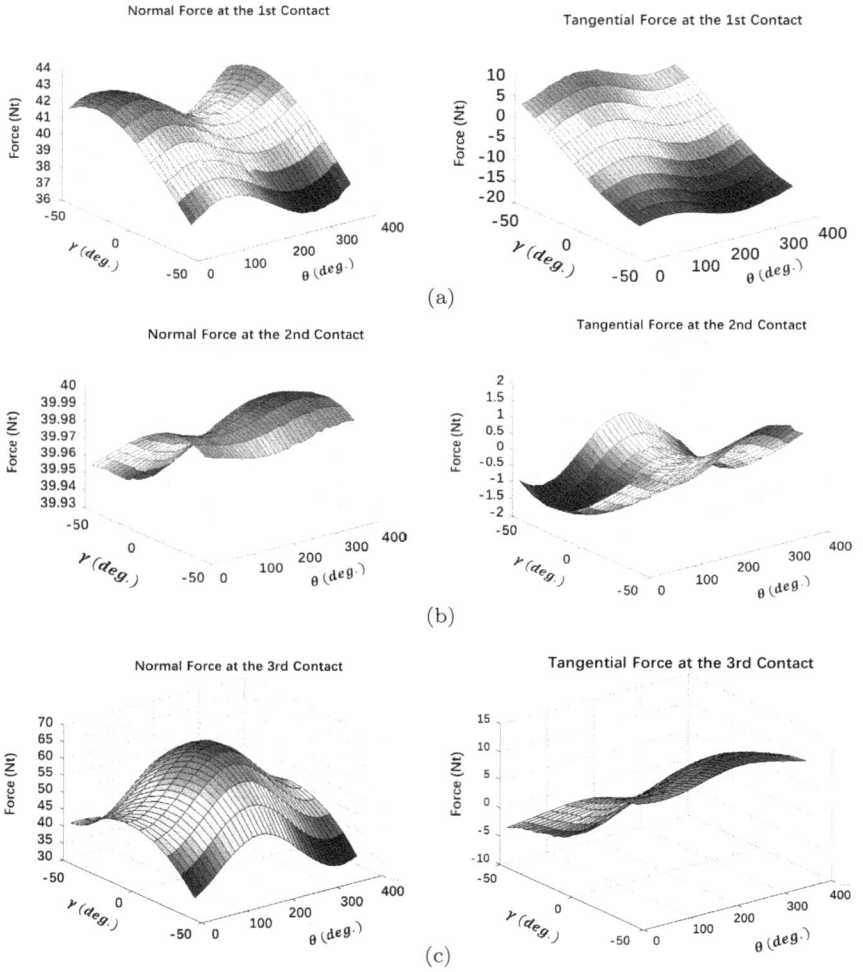

Fig. 7.7. Contact forces at three contacts. (a) 1st contact. (b) 2nd contact. (c) 3rd contact.

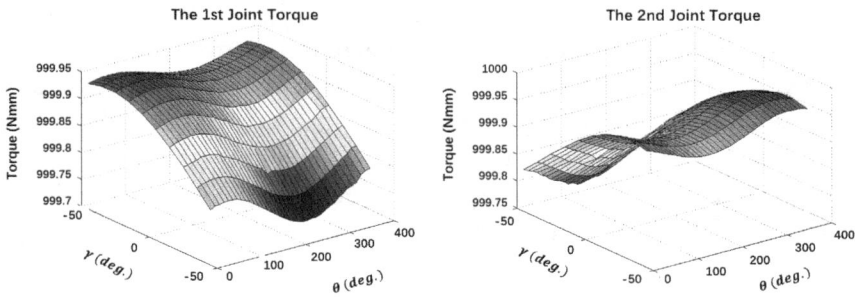

Fig. 7.8. Actual joint torques of the finger.

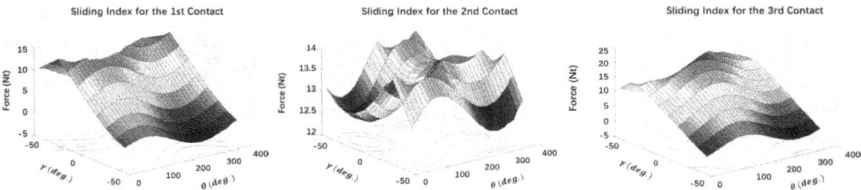

Fig. 7.9. Sliding indexes at three contacts.

7.8 Summary

A crucial problem in robot grasping is the choice of grasping forces so as to avoid or minimize the risk of slippage. The existence of uncontrollable grasp force is the main issue and characteristic of the grasp force analyses in enveloping grasp/fixturing.

This chapter focuses on the model of enveloping grasp/fixturing and the prediction method of passive force closure. Why the uncontrollable grasping forces exist is analyzed, and a physical model of compliant enveloping grasp is then formulated. A natural model of contact compliance including the normal direction and tangential direction is developed in this chapter. Using this model, nonlinear relationships of the elastic deformations with the contact forces, including normal force and tangential forces, are described. The differential motion compatibility relationships between the displacement of the object and the local elastic deformations at all contacts are derived. A natural compliance model of enveloping grasp/fixturing, where the passive compliance in joints of fingers is taken into account, is formulated. The stability of the compliant grasping system is related to the local geometrical properties such as relative curvature tensors, the material properties at contacts, and the grasp configuration. When the compliant matrix is positive definite, the compliant grasping system is stable.

The crux of judging whether the enveloping grasp is passive force closure is to predict the passive contact forces in the grasping system, which is formulated into a nonlinear least square. Consequently, the local elastic deformations are obtained by using the globally convergent Levenberg–Marquardt method. The corresponding passive contact forces can be predicted by using the natural model of contact compliance. The presented example verifies the proposed natural compliant model of enveloping grasp/fixturing and the prediction method of the passive force closure.

The proposed natural compliance model of enveloping grasp/fixturing can explain why a constraining mechanism or grasping system can maintain force closure even when an arbitrary external force is applied on the object, without changing the joint driving force/torque of the system.

References

[1] Bicchi A. and Kumar V. Robotic grasping and contact: A review. *Proceedings of IEEE International Conference on Robotics and Automation*, pp. 348–353, 2000.

[2] Asada H. and By A. B. Kinematic analysis of workpart fixturing for flexible assembly with automatically reconfigurable fixtures. *IEEE Transactions on Robotics and Automation*, 1(2), pp. 86–93, 1995.

[3] Harada K. and Kaneko M. A sufficient condition for manipulation of envelope family. *IEEE Transactions on Robotics and Automation*, 18(4), pp. 597–607, 2002.

[4] Harada K., Kaneko M., and Tsuji T. Active force closure for multiple objects. *Journal of Robotic Systems*, 19(3), pp. 133–141, 2002.

[5] Kerr J. and Roth B. Analysis of multifingered hands. *International Journal of Robotics Research*, 4(4), pp. 3–17, 1986.

[6]   Li Z. X. and Sastry S. S. Task-oriented optimal grasping by multi-fingered robot hands. *IEEE Journal of Robotics and Automation*, 4, pp. 32–43, 1988.

[7]   Liu Y. H. Qualitative test and force optimization of 3-D frictional form-closure grasps using linear programming. *IEEE Transactions on Robotics and Automation*, 15(1), pp. 163–173, 1999.

[8]   Omata T. and Nagata K. Rigid body analysis of the indeterminate grasp force in power grasps. *IEEE Transactions on Robotics and Automation*, 16(1), pp. 46–54, 2000.

[9]   Ponce J. and Faverjon B. On computing three-finger force-closure grasps of polygonal objects. *IEEE Transactions on Robotics and Automation*, 11(6), pp. 868–881, 1995.

[10] Rimon E. A. Curvature-based bound on the number of frictionless fingers required to immobilize three dimensional objects. *IEEE Transactions on Robotics and Automation*, 17(5), pp. 679–697, 2001.

[11] Wang M. Y. Passive forces in fixturing and grasping. *Proceedings of 9th IEEE Conference on Mechatronics and Machine Vision in Practice*, Chiang Mai, Thailand, pp. 2666–2671, 2002.

[12] Wang M. Y. and Pelinescu D. M. Optimizing fixture layout in a point-set domain. *IEEE Transactions on Robotics and Automation*, 17(3), pp. 312–323, 2001.

[13] Wang M. Y. and Pelinescu D. M. Contact force prediction and force closure analysis of a fixtured rigid workpiece with friction. *ASME Transactions-Journal of Manufacturing Science and Engineering*, 125, pp. 325–332, 2003.

[14] Xiong C. H., Li Y. F., Rong Y., and Xiong Y. L. Qualitative analysis and quantitative evaluation of fixturing. *Robotics and Computer Integrated Manufacture*, 18(5–6), pp. 335–342, 2002.

[15] Xiong C. H. and Xiong Y. L. Stability index and contact configuration planning for multifingered grasp. *Journal of Robotic Systems*, 15(4), pp. 183–190, 1998.

[16] Xiong Y. L., Ding H., and Wang M. Y. Quantitative analysis of inner force distribution and load capacity of grasps and fixtures. *ASME Transactions-Journal of Manufacturing Science and Engineering*, 124, pp. 444–455, 2002.

[17] Lin Q., Burdick J. W., and Rimon E. A stiffness-based quality measure for compliant grasps and fixtures. *IEEE Transactions on Robotics and Automation*, 17(5), pp. 679–697, 2000.

[18] Kao I. and Cutkosky M. R. Dextrous manipulation with compliance and sliding. *International Journal of Robotics Research*, 12(1), pp. 20–40, 1992.

[19] Xydas N. and Kao I. Modeling of contact mechanics and friction limit surface for soft fingers with experimental results. *International Journal of Robotics Research*, 18(9), pp. 941–950, 1999.

[20] Nguyen V. D. Constructing force-closure grasps. *International Journal of Robotics Research*, 7(3), pp. 3–16, 1988.

[21] Mason M. and Salisbury J. K. *Robot Hands and the Mechanics of Manipulation*. MIT Press, Cambridge, MA, USA, 1985.

[22] Murray R. M., Li Z., and Sastry S. S. *A Mathematical Introduction to Robotic Manipulation*. CRC Press, Boca Raton, FL, USA, 1994.

[23] Bicchi A. Force distribution in multiple whole-limb manipulation. *Proceedings of IEEE International Conference on Robotics and Automation*, pp. 196–201, 1993.

[24] Yoshikawa T. Passive and active closures by constraining mechanisms. *Proceedings of IEEE International Conference on Robotics and Automation*, pp. 1477–1484, 1996.

[25] Shapiro A., Rimon E., and Burdick J. W. Passive force closure and its computation in compliant-rigid grasps. *Proceedings of IEEE/RSJ International Conference on Intelligent Robots and Systems (IROS)*, pp. 1–9, 2001.

[26] Xiong C. H., Wang M. Y., Rong K. Y., and Xiong Y. L. Force closure of fixturing/grasping with passive contacts. *Proceedings of the 11th International Conference on Advanced Robotics*, Coimbra, Portugal, 3, pp. 1352–1357, 2003.

[27] Hockenberger M. J. and De Meter E. C. The application of meta functions to the quasi-static analysis of workpiece displacement within a machining fixture. *ASME Transactions-Journal of Manufacturing Science and Engineering*, 118, pp. 325–331, 1996.

[28] Li B. and Melkote S. N. An elastic contact model for the prediction of workpiece-fixture contact forces in clamping. *ASME Transactions-Journal of Manufacturing Science and Engineering*, 121, pp. 485–493, 1999.

[29] Xiong C. H., Wang M. Y., Tang Y., and Xiong Y. L. Compliant grasping with passive forces. *Journal of Robotic Systems*, 22(5), pp. 271–285, 2005.

[30] Wang Y. T. and Kumar V. Simulation of mechanical systems with multiple frictional contacts. *ASME Transactions-Journal of Mechanical Design*, 116, pp. 571–580, 1994.

[31] Howard W. S. and Kumar V. On the stability of grasped objects. *IEEE Transactions on Robotics and Automation*, 12(6), pp. 904–917, 1996.

[32] Nguyen V. D. Constructing stable grasps. *International Journal of Robotics Research*, 8(1), pp. 26–37, 1989.

[33] Xiong C. H., Li Y. F., Ding H., and Xiong Y. L. On the dynamic stability of grasping. *International Journal of Robotics Research*, 18(9), pp. 951–958, 1999.

[34] Johnson K. L. *Contact Mechanics*. Cambridge University Press, New York, USA, 1985.

[35] Scales L. E. *Introduction to Non-linear Optimization*. Macmillan, London, UK, 1985.

Chapter 8

Kinematics of Contacts and Rolling Manipulation

Kinematics of contacts is the fundamentals of fine manipulation, mobility analysis, and evaluation for multifingered grasp and workpiece-fixture systems. In this chapter, using the relationships of velocity constraint and the orientation constraint between fingertip surface and object surface, the equations of pure rolling contact over the surfaces of two contacting objects are derived. The effects of the degrees of freedom of the finger on the manipulation with rolling contact are discussed. The intrinsic characteristics of the multifingered grasp and manipulation are analyzed. We develop a direct force control method based on position control. An adjustment algorithm of the fingertip contact force is proposed for the contact transition phase. The manipulation experiment of a finger with rolling contact is set up in this chapter.

8.1 Introduction

Rolling contact represents a non-holonomic constraint between two bodies, that is, the equations relating the motion of an object to another one are expressed in terms of the velocities of these two bodies rather than their positions. Examples of such constraints can be found in wheeled vehicles, pure rolling contact gears [1], and the manipulation of multifingered robotic hands. The kinematic relationships between two rigid objects have been investigated in Refs. [2–6, 24]. Cai and Roth [2] have studied the roll-slide motions with point contact between two curves under planar motion. The velocity of the moving point and its higher order derivatives have been obtained. Montana [3,4] derived the kinematics of contact point from a geometric point of view. Kerr and Roth [5] derived the kinematic equations of pure rolling contact from the constraint that the fingertip and object velocities are equal at the point of contact. Cole *et al.* [6] derived the kinematics of rolling in $\Re^3$ using velocity constraints and normal constraints between the surfaces. However, the conditions for achieving pure rolling contact have not been given, and the kinematic relationship between two rigid objects with rolling contact has not been extended to the manipulation of multifingered hands.

In this chapter, by giving the rotational velocities of the grasped object and the fingertip in the base frame, we derive the kinematic equation of rolling using velocity constraints and the orientation constraints between the surfaces of the object and fingertip. Then, the conditions for achieving pure rolling contact are discussed.

Moreover, the contact forces between the fingers and the object change during the manipulation with multifingered hand due to the finger and object deformations, and the servo errors in the control system, which will result in the desired motion of the object, may not be satisfied; even extremely, the object may slide from the hand or be damaged. Thus, it is important for fine manipulation with multifingered hands to control appropriately the contact forces of fingers, which is another important part of this chapter.

Robot control in constrained space, such as profile tracking and deburring, is an interesting research topic. Particularly, the control of contact force between the object and the end effector of a manipulator has been one of the most attractive research areas [7–15]. Assuming that the position of the object was fixed or known, a lot of researchers studied contact control problems in the past several years [7–11]. The contact force control of multifingered hand was investigated assuming that the fingertip might be modeled as linear spring with constant stiffness [9]. However, the fact that the position and orientation of the object to be manipulated with multifingered hand depend on the positions of the fingertips is neglected in past works related to manipulation control. When a multifingered hand grasps an object, if there exist position and orientation errors for the object even though the fingertips are in their desired positions, then there are large gaps between the desired and actual contact forces at the fingertip contacts, which results in the object moving further. Thus, the problem of contact force control is not a pure force control one during the multifingered manipulation. In addition, the real time detecting of contact positions is necessary for multifingered manipulation because the contact position is one of the dominating factors influencing the grasp stability and contact forces [16, 17].

In this chapter, we discuss the coordination control strategy for the multifingered robotic hands and model the PUMA562 manipulator as a finger. By using a wrist force/torque sensor to detect the contact positions, we develop an algorithm of contact force control during the pure rolling manipulation.

8.2   Kinematics of Pure Rolling Contact

Consider a finger in contact with an object, as shown in Fig. 8.1. Choose reference frames $\{O\}$ and $\{F\}$ fixed relative to the object and the finger, respectively. Suppose $\{P\}$ be an inertial base frame, and let $S_o \subset \Re^3$ and $S_f \subset \Re^3$ be the embedding of the surfaces of the object and fingertip relative to $\{O\}$ and $\{F\}$, respectively. Let $\mathbf{n}_o$ and $\mathbf{n}_f$ be the Gauss maps (outward normal) for S_o and S_f, and (f_{oi}, U_{oi}) be an orthogonal right-handed coordinate system for S_{oi} with Gauss map $\mathbf{n}_o$. Similarly, let (f_{fj}, U_{fj}) be an orthogonal right-handed coordinate system for S_{fj} with Gauss map $\mathbf{n}_f$.

Let $\mathbf{r}_o(t) \in S_o$ and $\mathbf{r}_f(t) \in S_f$ be the positions at time t of the point of contact relative to $\{O\}$ and $\{F\}$, respectively. In general, $\mathbf{r}_o(t)$ will not remain in a single coordinate patch of the atlas $\{S_{oi}\}_{i=1}^{m_1}$ for all time, and likewise for $\mathbf{r}_f(t)$ and the atlas $\{S_{fj}\}_{j=1}^{m_2}$. Therefore, we restrict our attention to an interval I such that $\mathbf{r}_o(t) \in S_{oi}$ and $\mathbf{r}_f(t) \in S_{fj}$ for all $t \in I$ and some i and j. The coordinate systems (f_{oi}, U_{oi}) and (f_{fj}, U_{fj}) induce a normalized Gauss frame at all

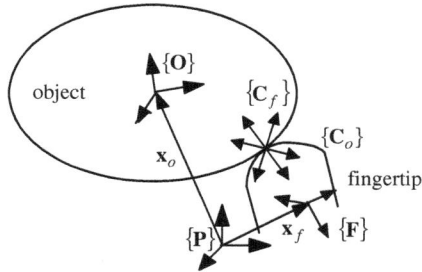

Fig. 8.1.   A finger in contact with an object.

points in S_{oi} and S_{fj}. We define the contact frames $\{\mathbf{C}_o\}$ and $\{\mathbf{C}_f\}$ as the coordinate frames that coincide with the normalized Gauss frames at $\mathbf{r}_o(t)$ and $\mathbf{r}_f(t)$, respectively, for all $t \in I$. To simplify the description, we omit the subscript i and j below.

We now define the parameters that describe the 5 degrees of freedom for the motion of the point of contact. The coordinates of the contact point relative to the coordinate systems (f_o, U_o) and (f_j, U_f) are given by

$$\mathbf{u}_o(t) = \left(u_1, u_2 \right)^T = f_o^{-1}\left(\mathbf{r}_o\left(t\right) \right) \in U_o$$

$$\mathbf{u}_f(t) = \left(v_1, v_2 \right)^T = f_f^{-1}\left(\mathbf{r}_f\left(t\right) \right) \in U_f$$

$\mathbf{u}_o(t)$ and $\mathbf{u}_f(t)$ account for 4 degrees of freedom. The final parameter is the contact angle $\varphi(t)$ which is defined as the angle between the x-axes of $\{\mathbf{C}_o\}$ and $\{\mathbf{C}_f\}$. We choose the sign of $\varphi(t)$ so that a rotation of $\{\mathbf{C}_o\}$ through angle $\varphi(t)$ around its z-axis aligns the x-axes.

The rotation matrices giving the orientations of $\{\mathbf{C}_o\}$ and $\{\mathbf{C}_f\}$ in the reference frames $\{\mathbf{O}\}$ and $\{\mathbf{F}\}$ are given as follows:

$${}^o_{co}\mathbf{R} = \left(\frac{(\mathbf{f}_o)_{u_1}}{\left\| (\mathbf{f}_o)_{u_1} \right\|} \quad \frac{(\mathbf{f}_o)_{u_2}}{\left\| (\mathbf{f}_o)_{u_2} \right\|} \quad \frac{(\mathbf{f}_o)_{u_1} \times (\mathbf{f}_o)_{u_2}}{\left\| (\mathbf{f}_o)_{u_1} \times (\mathbf{f}_o)_{u_2} \right\|} \right) \tag{8.1}$$

$${}^f_{cf}\mathbf{R} = \left(\frac{(\mathbf{f}_f)_{v_1}}{\left\| (\mathbf{f}_f)_{v_1} \right\|} \quad \frac{(\mathbf{f}_f)_{v_2}}{\left\| (\mathbf{f}_f)_{v_2} \right\|} \quad \frac{(\mathbf{f}_f)_{v_1} \times (\mathbf{f}_f)_{v_2}}{\left\| (\mathbf{f}_f)_{v_1} \times (\mathbf{f}_f)_{v_2} \right\|} \right) \tag{8.2}$$

where $(\mathbf{f}_o)_{u_1}$, $(\mathbf{f}_o)_{u_2}$ denote the partial derivatives of f_o with respect to u_1 and u_2, respectively; similarly, $(\mathbf{f}_f)_{v_1}$, $(\mathbf{f}_f)_{v_2}$ are the partial derivatives of f_f with respect to v_1 and v_2.

The curvature tensor $\mathbf{K}_o$, torsion form $\mathbf{T}_o$ and metric tensor $\mathbf{M}_o$ at the point $\mathbf{r}_o(t)$ relative to the coordinate system (f_o, U_o) are defined as

$$\mathbf{K}_o = \left(\frac{(\mathbf{f}_o)_{u_1}}{\left\|(\mathbf{f}_o)_{u_1}\right\|} \quad \frac{(\mathbf{f}_o)_{u_2}}{\left\|(\mathbf{f}_o)_{u_2}\right\|} \right)^T \left(\frac{(\mathbf{n}_o)_{u_1}}{\left\|(\mathbf{f}_o)_{u_1}\right\|} \quad \frac{(\mathbf{n}_o)_{u_2}}{\left\|(\mathbf{f}_o)_{u_2}\right\|} \right) \tag{8.3}$$

$$\mathbf{T}_o = \frac{(\mathbf{f}_o)_{u_2}^T}{\left\|(\mathbf{f}_o)_{u_2}\right\|} \left[\frac{(\mathbf{f}_o)_{u_1 u_1}}{\left\|(\mathbf{f}_o)_{u_1}\right\|^2} \quad \frac{(\mathbf{f}_o)_{u_1 u_2}}{\left\|(\mathbf{f}_o)_{u_1}\right\|\left\|(\mathbf{f}_o)_{u_2}\right\|} \right] \tag{8.4}$$

$$\mathbf{M}_o = \mathrm{diag}\left[\|(\mathbf{f}_o)_{u_1}\| \;\; \|(\mathbf{f}_o)_{u_2}\|\right] \tag{8.5}$$

where

$$(\mathbf{f}_o)_{u_1 u_1} = \partial^2 \mathbf{f}_o / \partial u_1^2, \quad (\mathbf{f}_o)_{u_1 u_2} = \partial^2 \mathbf{f}_o / \partial u_1 \partial u_2$$

and

$$\mathbf{n}_o = \frac{(\mathbf{f}_o)_{u_1} \times (\mathbf{f}_o)_{u_2}}{\left\|(\mathbf{f}_o)_{u_1} \times (\mathbf{f}_o)_{u_2}\right\|}$$

is the unit normal at the point $\mathbf{r}_o(t)$ on the surface of the object, and $(\mathbf{n}_o)_{u_1}$ and $(\mathbf{n}_o)_{u_2}$ denote the partial derivative of $\mathbf{n}_o$ with respect to u_1 and u_2.

Similarly, we can define the $\mathbf{K}_f$, $\mathbf{T}_f$, and $\mathbf{M}_f$ for the surface of fingertip.

A point on the object with coordinate $\mathbf{r}_o(\mathbf{u}_o)$ in the object frame $\{\mathbf{O}\}$ has base frame coordinates given by

$$\mathbf{x}_o + {}_o^p\mathbf{R}\mathbf{r}_o(\mathbf{u}_o) \tag{8.6}$$

and a point on the fingertip with coordinate $\mathbf{r}_f(\mathbf{u}_f)$ in the fingertip frame $\{\mathbf{F}\}$ has base frame coordinates expressed by

$$\mathbf{x}_f + {}_f^p\mathbf{R}\mathbf{r}_f(\mathbf{u}_f) \tag{8.7}$$

where $\mathbf{x}_o$, $\mathbf{x}_F \in \Re^{3\times 1}$ are the origin positions of $\{\mathbf{O}\}$ and $\{\mathbf{F}\}$ in the base frame $\{\mathbf{P}\}$, and ${}_o^p\mathbf{R}$ and ${}_f^p\mathbf{R} \in \Re^{3\times 3}$ are the orientation matrices of $\{\mathbf{O}\}$ and $\{\mathbf{F}\}$ with respect to the base frame $\{\mathbf{P}\}$, respectively.

Since there is only one contact point between the fingertip and the object, the contact point on the object is the same as that one on the fingertip. From Eqs. (8.6)and (8.7), we get

$$\mathbf{x}_o(t) + {}^p_o\mathbf{R}(t)\mathbf{r}_o(t) = \mathbf{x}_f(t) + {}^p_f\mathbf{R}(t)\mathbf{r}_f(t) \tag{8.8}$$

Differentiating Eq. (8.8) with respect to time, we have

$$\mathbf{v}_o + {}^p_o\dot{\mathbf{R}}\mathbf{r}_o + {}^p_o\mathbf{R}\dot{\mathbf{r}}_o = \mathbf{v}_f + {}^p_f\dot{\mathbf{R}}\mathbf{r}_f + {}^p_f\mathbf{R}\dot{\mathbf{r}}_f \tag{8.9}$$

where $\mathbf{v}_o = \dot{\mathbf{x}}_o(t)$ and $\mathbf{v}_f = \dot{\mathbf{x}}_f(t) \in \Re^{3\times 1}$ are the translational velocities of the origins of frames $\{\mathbf{O}\}$ and $\{\mathbf{F}\}$, and $\boldsymbol{\omega}_o$ and $\boldsymbol{\omega}_f \in \Re^{3\times 1}$ are the rotational velocities of the frames $\{\mathbf{O}\}$ and $\{\mathbf{F}\}$ relative to the base frame $\{\mathbf{P}\}$ such that

$$
{}^p_o\dot{\mathbf{R}} = (\boldsymbol{\omega}_o\times){}^p_o\mathbf{R} = \mathbf{S}\left(\boldsymbol{\omega}_o\right){}^p_o\mathbf{R} = \begin{bmatrix} 0 & -\omega_{oz} & \omega_{oy} \\ \omega_{oz} & 0 & -\omega_{ox} \\ -\omega_{oy} & \omega_{ox} & 0 \end{bmatrix} {}^p_o\mathbf{R}
$$

$$\tag{8.10}$$

$$
{}^p_f\dot{\mathbf{R}} = (\boldsymbol{\omega}_f\times){}^p_f\mathbf{R} = \mathbf{S}(\boldsymbol{\omega}_f){}^p_f\mathbf{R} = \begin{bmatrix} 0 & -\omega_{fz} & \omega_{fy} \\ \omega_{fz} & 0 & -\omega_{fx} \\ -\omega_{fy} & \omega_{fx} & 0 \end{bmatrix} {}^p_f\mathbf{R}
$$

$$\tag{8.11}$$

Since the fingertip and object keep pure rolling contact, the velocity of the contact point on the object is the same as that of the contact point on the fingertip, that is,

$$\mathbf{v}_o + \boldsymbol{\omega}_o \times {}^p_o\mathbf{R}\mathbf{r}_o = \mathbf{v}_f + \boldsymbol{\omega}_f \times {}^p_f\mathbf{R}\mathbf{r}_f \tag{8.12}$$

Equation (8.12) may be rewritten as

$$\mathbf{U}_o \begin{bmatrix} \mathbf{v}_o \\ \boldsymbol{\omega}_o \end{bmatrix} = \mathbf{U}_f \begin{bmatrix} \mathbf{v}_f \\ \boldsymbol{\omega}_f \end{bmatrix} \tag{8.13}$$

where

$$\mathbf{U}_o = \left[\mathbf{I} \vdots -({}^p_o\mathbf{R}\mathbf{r}_o\times)\right] \tag{8.14}$$

$$\mathbf{U}_f = \left[\mathbf{I} \vdots -({}^p_f\mathbf{R}\mathbf{r}_f\times)\right] \tag{8.15}$$

Here, $\mathbf{I}$ is a 3×3 identity matrix.

Subtracting Eq. (8.12) from Eq. (8.9) yields

$$
{}_o^p\mathbf{R}\dot{\mathbf{R}}_o = {}_f^p\mathbf{R}\dot{\mathbf{r}}_f \tag{8.16}
$$

Since

$$
{}_o^p\mathbf{R}\dot{\mathbf{r}}_o = {}_{co}^p\mathbf{R}{}_{co}^o\mathbf{R}^T \left[(\mathbf{f}_o)_{u_1} \ (\mathbf{f}_o)_{u_2} \right] \dot{\mathbf{u}}_o \tag{8.17}
$$

Substituting Eq. (8.1) into Eq. (8.17), we have

$$
{}_o^p\mathbf{R}\dot{\mathbf{r}}_o = {}_{co}^p\mathbf{R}
\begin{bmatrix}
\|(\mathbf{f}_o)_{u_1}\| & 0 \\
0 & \|(\mathbf{f}_o)_{u_2}\| \\
0 & 0
\end{bmatrix}
\dot{\mathbf{u}}_o \tag{8.18}
$$

Substituting Eq. (8.5) into Eq. (8.18), we get

$$
{}_o^p\mathbf{R}\dot{\mathbf{r}}_o = {}_{co}^p\mathbf{R}
\begin{bmatrix}
\mathbf{M}_o\dot{\mathbf{u}}_o \\
0
\end{bmatrix}
\tag{8.19}
$$

Similarly, we obtain

$$
{}_f^p\mathbf{R}\dot{\mathbf{r}}_f = {}_{cf}^p\mathbf{R}
\begin{bmatrix}
\mathbf{M}_f\dot{\mathbf{u}}_f \\
0
\end{bmatrix}
\tag{8.20}
$$

Thus, Eq. (8.16) can be written as

$$
{}_{co}^p\mathbf{R}
\begin{bmatrix}
\mathbf{M}_o\dot{\mathbf{u}}_o \\
0
\end{bmatrix}
= {}_{cf}^p\mathbf{R}
\begin{bmatrix}
\mathbf{M}_f\dot{\mathbf{u}}_f \\
0
\end{bmatrix}
\tag{8.21}
$$

where ${}_{co}^p\mathbf{R}$ and ${}_{cf}^p\mathbf{R} \in \Re^{3\times3}$ are the orientation matrices of $\{\mathbf{C}_o\}$ and $\{\mathbf{C}_f\}$ with respect to the base frame $\{\mathbf{P}\}$, respectively.

Equation (8.21) means that the arc length traversed by the contact point across the surface of the object is equal to that across the surface of the fingertip when the fingertip rolls on the surface of the object without slipping.

Finally, since the orientation of $\{C_o\}$ in the base frame $\{P\}$ can be expressed as $_f^p\mathbf{R}_{cf}^f\mathbf{R}_{co}^{cf}\mathbf{R}$, or $_o^p\mathbf{R}_{co}^o\mathbf{R}$, that is,

$$_f^p\mathbf{R}_{cf}^f\mathbf{R}_{co}^{cf}\mathbf{R} = _o^p\mathbf{R}_{co}^o\mathbf{R} \tag{8.22}$$

Differentiating Eq. (8.22) with respect to time yields

$$\mathbf{S}(\boldsymbol{\omega}_f)_f^p\mathbf{R}_{cf}^f\mathbf{R}_{co}^{cf}\mathbf{R} + _f^p\mathbf{R}_{cf}^f\dot{\mathbf{R}}_{co}^{cf}\mathbf{R} + _f^p\mathbf{R}_{cf}^f\mathbf{R}_{co}^{cf}\dot{\mathbf{R}}$$
$$= \mathbf{S}\left(\boldsymbol{\omega}_o\right)_o^p\mathbf{R}_{co}^o\mathbf{R} + _o^p\mathbf{R}_{co}^o\dot{\mathbf{R}} \tag{8.23}$$

i.e.,

$$\mathbf{S}(\boldsymbol{\omega}_f)_{co}^p\mathbf{R} + _{cf}^p\mathbf{R}(_{cf}^f\mathbf{R}^T{}_{cf}^f\dot{\mathbf{R}})_{co}^{cf}\mathbf{R} + _{co}^p\mathbf{R}(_{co}^{cf}\mathbf{R}^T{}_{co}^{cf}\dot{\mathbf{R}})$$
$$= \mathbf{S}(\boldsymbol{\omega}_o)_{co}^p\mathbf{R} + _{co}^p\mathbf{R}(_{co}^o\mathbf{R}^T{}_{co}^o\dot{\mathbf{R}}) \tag{8.24}$$

where

$$_{co}^{cf}\mathbf{R} = \begin{bmatrix} \mathbf{R}_\varphi & \mathbf{0} \\ \mathbf{0} & -1 \end{bmatrix} \tag{8.25}$$

with

$$\mathbf{R}_\varphi = \begin{bmatrix} c\varphi & -s\varphi \\ -s\varphi & -c\varphi \end{bmatrix} \tag{8.26}$$

and $c\varphi = \cos\varphi$, $s\varphi = \sin\varphi$.

Using Eqs. (8.1), (8.3), (8.4), and (8.5), we obtain

$$_{co}^o\mathbf{R}^T{}_{co}^o\dot{\mathbf{R}} = \begin{bmatrix} 0 & -\mathbf{T}_o\mathbf{M}_o\dot{\mathbf{u}}_o & \mathbf{K}_o\mathbf{M}_o\dot{\mathbf{u}}_o \\ \mathbf{T}_o\mathbf{M}_o\dot{\mathbf{u}}_o & 0 & \\ -(\mathbf{K}_o\mathbf{M}_o\dot{\mathbf{u}}_o)^T & 0 & \end{bmatrix} \tag{8.27}$$

Similarly, we get

$$_{cf}^f\mathbf{R}^T{}_{cf}^f\dot{\mathbf{R}} = \begin{bmatrix} 0 & -\mathbf{T}_f\mathbf{M}_f\dot{\mathbf{u}}_f & \mathbf{K}_f\mathbf{M}_f\dot{\mathbf{u}}_f \\ \mathbf{T}_f\mathbf{M}_f\dot{\mathbf{u}}_f & 0 & \\ -(\mathbf{K}_f\mathbf{M}_f\dot{\mathbf{u}}_f)^T & 0 & \end{bmatrix} \tag{8.28}$$

Substituting Eqs. (8.25), (8.27), and (8.28) into Eq. (8.24), we have

$$\mathbf{S}(\boldsymbol{\omega}_f)_{co}^p\mathbf{R} + {}_{cf}^p\mathbf{R} \begin{bmatrix} 0 & -\mathbf{T}_f\mathbf{M}_f\dot{\mathbf{u}}_f & \mathbf{K}_f\mathbf{M}_f\dot{\mathbf{u}}_f \\ \mathbf{T}_f\mathbf{M}_f\dot{\mathbf{u}}_f & 0 & \\ -(\mathbf{K}_f\mathbf{M}_f\dot{\mathbf{u}}_f)^T & 0 \end{bmatrix} {}_{co}^{cf}\mathbf{R}$$

$$+ {}_{co}^p\mathbf{R} \begin{bmatrix} 0 & -\dot{\varphi} & 0 \\ \dot{\varphi} & 0 & 0 \\ 0 & 0 & 0 \end{bmatrix} = \mathbf{S}(\boldsymbol{\omega}_o)_{co}^p\mathbf{R}$$

$$+ {}_{co}^p\mathbf{R} \begin{bmatrix} 0 & -\mathbf{T}_o\mathbf{M}_o\dot{\mathbf{u}}_o & \mathbf{K}_o\mathbf{M}_o\dot{\mathbf{u}}_o \\ \mathbf{T}_o\mathbf{M}_o\dot{\mathbf{u}}_o & 0 & \\ -(\mathbf{K}_o\mathbf{M}_o\dot{\mathbf{u}}_o)^T & 0 \end{bmatrix} \qquad (8.29)$$

Left-multiplying both sides of Eq. (8.29) with the matrix ${}_p^{co}\mathbf{R}$ yields

$$ {}_p^{co}\mathbf{R}\mathbf{S}(\boldsymbol{\omega}_f) {}_{co}^p\mathbf{R} + \begin{bmatrix} 0 & \mathbf{T}_f\mathbf{M}_f\dot{\mathbf{u}}_f & -\mathbf{R}_\varphi\mathbf{K}_f\mathbf{M}_f\dot{\mathbf{u}}_f \\ -\mathbf{T}_f\mathbf{M}_f\dot{\mathbf{u}}_f & 0 & \\ (\mathbf{R}_\varphi\mathbf{K}_f\mathbf{M}_f\dot{\mathbf{u}}_f)^T & 0 \end{bmatrix}$$

$$+ \begin{bmatrix} 0 & -\dot{\varphi} & 0 \\ \dot{\varphi} & 0 & 0 \\ 0 & 0 & 0 \end{bmatrix} = {}_p^{co}\mathbf{R}\mathbf{S}(\boldsymbol{\omega}_o) {}_{co}^p\mathbf{R}$$

$$+ \begin{bmatrix} 0 & -\mathbf{T}_o\mathbf{M}_o\dot{\mathbf{u}}_o & \mathbf{K}_o\mathbf{M}_o\dot{\mathbf{u}}_o \\ \mathbf{T}_o\mathbf{M}_o\dot{\mathbf{u}}_o & 0 & \\ -(\mathbf{K}_o\mathbf{M}_o\dot{\mathbf{u}}_o)^T & 0 \end{bmatrix} \qquad (8.30)$$

From Eq. (8.30), we obtain

$$\mathbf{K}_o\mathbf{M}_o\dot{\mathbf{u}}_o + \mathbf{R}_\varphi\mathbf{K}_f\mathbf{M}_f\dot{\mathbf{u}}_f = \begin{bmatrix} 1 & 0 & 0 \\ 0 & 1 & 0 \end{bmatrix}$$

$$\left[{}_p^{co}\mathbf{R}\left(\mathbf{S}(\boldsymbol{\omega}_f) - \mathbf{S}(\boldsymbol{\omega}_o)\right) {}_{co}^p\mathbf{R} \right] \begin{bmatrix} 0 \\ 0 \\ 1 \end{bmatrix} \qquad (8.31)$$

$$\dot{\varphi} = \mathbf{T}_o \mathbf{M}_o \dot{\mathbf{u}}_o + \mathbf{T}_f \mathbf{M}_f \dot{\mathbf{u}}_f - \begin{bmatrix} 0 & 1 & 0 \end{bmatrix}$$

$$\begin{bmatrix} {}^{co}_{p}\mathbf{R}(\mathbf{S}(\boldsymbol{\omega}_f) - \mathbf{S}(\boldsymbol{\omega}_o)){}^{p}_{co}\mathbf{R} \end{bmatrix} \begin{bmatrix} 1 \\ 0 \\ 0 \end{bmatrix} \tag{8.32}$$

Left-multiplying both sides of Eq. (8.21) with the matrix ${}^{cf}_{p}\mathbf{R}$ yields

$$\,{}^{cf}_{co}\mathbf{R} \begin{bmatrix} \mathbf{M}_o \dot{\mathbf{u}}_o \\ 0 \end{bmatrix} = \begin{bmatrix} \mathbf{M}_f \dot{\mathbf{u}}_f \\ 0 \end{bmatrix} \tag{8.33}$$

Substituting Eq. (8.25) into Eq. (8.33), we have

$$\mathbf{R}_\varphi \mathbf{M}_o \dot{\mathbf{u}}_o = \mathbf{M}_f \dot{\mathbf{u}}_f \tag{8.34}$$

From Eqs. (8.31) and (8.34), we obtain

$$\dot{\mathbf{u}}_o = \mathbf{M}_o^{-1}(\mathbf{K}_o + \widetilde{\mathbf{K}}_f)^{-1} \begin{bmatrix} 1 & 0 & 0 \\ 0 & 1 & 0 \end{bmatrix}$$

$$\begin{bmatrix} {}^{co}_{p}\mathbf{R}(\mathbf{S}(\boldsymbol{\omega}_f) - \mathbf{S}(\boldsymbol{\omega}_o)){}^{p}_{co}\mathbf{R} \end{bmatrix} \begin{bmatrix} 0 \\ 0 \\ 1 \end{bmatrix} \tag{8.35}$$

$$\dot{\mathbf{u}}_f = \mathbf{M}_f^{-1}\mathbf{R}_\varphi(\mathbf{K}_o + \widetilde{\mathbf{K}}_f)^{-1} \begin{bmatrix} 1 & 0 & 0 \\ 0 & 1 & 0 \end{bmatrix}$$

$$\begin{bmatrix} {}^{co}_{p}\mathbf{R}(\mathbf{S}(\boldsymbol{\omega}_f) - \mathbf{S}(\boldsymbol{\omega}_o)){}^{p}_{co}\mathbf{R} \end{bmatrix} \begin{bmatrix} 0 \\ 0 \\ 1 \end{bmatrix} \tag{8.36}$$

where $\widetilde{\mathbf{K}}_f = \mathbf{R}_\varphi \mathbf{K}_f \mathbf{R}_\varphi$ is the curvature of the fingertip at the contact point relative to the x- and y-axes of $\{\mathbf{C}_o\}$. Call $\mathbf{K}_o + \widetilde{\mathbf{K}}_f$ the relative curvature tensor.

Equations (8.32), (8.35), and (8.36) represent five equations which may be solved for $\dot{\mathbf{u}}_o$, $\dot{\mathbf{u}}_f$, and $\dot{\varphi}$ when given $\boldsymbol{\omega}_o$ and $\boldsymbol{\omega}_f$.

8.3 Kinematics of Manipulation with Rolling Contact

Now we discuss the problem: Given the desired motion $\left(\mathbf{v}_o^T \ \boldsymbol{\omega}_o^T\right)^T$ of the object, how to determine the necessary motion of the finger under the constraints of pure rolling contact between the fingertip and the object? First, we rewrite Eqs. (8.35) and (8.35) as follows:

$$\dot{\mathbf{u}}_o = \mathbf{M}_o^{-1}\left(\mathbf{K}_o + \tilde{\mathbf{K}}_f\right)^{-1}\Phi(\boldsymbol{\omega}_f - \boldsymbol{\omega}_o) \tag{8.37}$$

$$\dot{\mathbf{u}}_f = \mathbf{M}_f^{-1}\mathbf{R}_\varphi(\mathbf{K}_o + \tilde{\mathbf{K}}_f)^{-1}\Phi\left(\boldsymbol{\omega}_f - \boldsymbol{\omega}_o\right) \tag{8.38}$$

where $\Phi \in \Re^{2\times 3}$ is referred to as function matrix which is related to the orientation of the contact frame $\{\mathbf{C}_o\}$ with respect to the object reference frame $\{\mathbf{O}\}$.

Equations (8.37) and (8.38) can be rewritten further as

$$\dot{\mathbf{u}}_o - \mathbf{M}_o^{-1}(\mathbf{K}_o + \tilde{\mathbf{K}}_f)^{-1}\Phi\boldsymbol{\omega}_f = -\mathbf{M}_o^{-1}\left(\mathbf{K}_o + \tilde{\mathbf{K}}_f\right)^{-1}\Phi\boldsymbol{\omega}_o$$
$$\tag{8.39}$$

$$\dot{\mathbf{u}}_f - \mathbf{M}_f^{-1}\mathbf{R}_\varphi(\mathbf{K}_o + \tilde{\mathbf{K}}_f)^{-1}\Phi\boldsymbol{\omega}_f = -\mathbf{M}_f^{-1}\mathbf{R}_\varphi(\mathbf{K}_o + \tilde{\mathbf{K}}_f)^{-1}\Phi\boldsymbol{\omega}_o$$
$$\tag{8.40}$$

Since

$$\boldsymbol{\omega}_f = \mathbf{J}_{fr}\dot{\mathbf{q}} \tag{8.41}$$

where $\mathbf{J}_{fr} \in \Re^{3\times n_i}$ is the angular velocity Jacobian matrix of the finger and n_i is the degrees of freedom of the ith finger, substituting Eq. (8.41) into Eqs. (8.39) and (8.40) yields

$$\dot{\mathbf{u}}_o - \mathbf{M}_o^{-1}(\mathbf{K}_o + \tilde{\mathbf{K}}_f)^{-1}\Phi\mathbf{J}_{fr}\dot{\mathbf{q}} = -\mathbf{M}_o^{-1}(\mathbf{K}_o + \tilde{\mathbf{K}}_f)^{-1}\Phi\boldsymbol{\omega}_o$$
$$\tag{8.42}$$

$$\dot{\mathbf{u}}_f - \mathbf{M}_f^{-1}\mathbf{R}_\varphi(\mathbf{K}_o + \tilde{\mathbf{K}}_f)^{-1}\Phi\mathbf{J}_{fr}\dot{\mathbf{q}} = -\mathbf{M}_f^{-1}\mathbf{R}_\varphi(\mathbf{K}_o + \tilde{\mathbf{K}}_f)^{-1}\Phi\boldsymbol{\omega}_o$$
$$\tag{8.43}$$

From Eq. (8.13), we obtain

$$\mathbf{U}_f \begin{bmatrix} \mathbf{v}_f \\ \boldsymbol{\omega}_f \end{bmatrix} = \begin{bmatrix} \mathbf{I} \vdots - ({}_o^p\mathbf{R}\mathbf{r}_o\times) \end{bmatrix} \begin{bmatrix} \mathbf{v}_o \\ \boldsymbol{\omega}_o \end{bmatrix} \tag{8.44}$$

Since

$$\mathbf{J}_f\dot{\mathbf{q}} = \begin{pmatrix} \mathbf{v}_f \\ \boldsymbol{\omega}_f \end{pmatrix} \tag{8.45}$$

where $\mathbf{J}_f \in \Re^{6\times n_i}$ is the Jacobian matrix of the finger, substituting Eq. (8.45) into Eq. (8.44) yields

$$\mathbf{U}_f\mathbf{J}_f\dot{\mathbf{q}} = \begin{bmatrix} \mathbf{I} \vdots - ({}_o^p\mathbf{R}\mathbf{r}_o\times) \end{bmatrix} \begin{bmatrix} \mathbf{v}_o \\ \boldsymbol{\omega}_o \end{bmatrix} \tag{8.46}$$

Thus, substituting Eq. (8.41) into Eq. (8.32) yields

$$\dot{\varphi} - \mathbf{T}_o\mathbf{M}_o\dot{\mathbf{u}}_o - \mathbf{T}_f\mathbf{M}_f\dot{\mathbf{u}}_f + \psi\mathbf{J}_{fr}\dot{\mathbf{q}} = \psi\boldsymbol{\omega}_o \tag{8.47}$$

where $\psi \in \Re^{1\times 3}$ is referred to as function matrix which is related to the orientation of the contact frame $\{\mathbf{C}_o\}$ with respect to the object reference frame $\{\mathbf{O}\}$.

From Eqs. (8.42), (8.43), (8.46), and (8.47), we obtain

$$\mathbf{J}\dot{\Theta} = \Lambda\mathbf{V} \tag{8.48}$$

where

$$\mathbf{J} = \begin{bmatrix} \mathbf{I}_{2\times 2} & \mathbf{0}_{2\times 2} & \mathbf{0}_{2\times 1} & -\mathbf{M}_o^{-1}(\mathbf{K}_o + \widetilde{\mathbf{K}}_f)^{-1}\Phi\mathbf{J}_{fr} \\ \mathbf{0}_{2\times 2} & \mathbf{I}_{2\times 2} & \mathbf{0}_{2\times 1} & -\mathbf{M}_f^{-1}\mathbf{R}_\phi(\mathbf{K}_o + \widetilde{\mathbf{K}}_f)^{-1}\Phi\mathbf{J}_{fr} \\ \mathbf{0}_{3\times 2} & \mathbf{0}_{3\times 2} & \mathbf{0}_{3\times 1} & \mathbf{U}_f\mathbf{J}_f \\ -\mathbf{T}_o\mathbf{M}_o & -\mathbf{T}_f\mathbf{M}_f & 1 & \psi\mathbf{J}_{fr} \end{bmatrix}$$
$$\in \Re^{8\times(5+n_i)}$$

is the generalized Jacobian matrix which is a time-variant nonlinear function of finger configuration, object orientation, and contact positions.

$$\Theta = \begin{pmatrix} \dot{\mathbf{u}}_o^T & \dot{\mathbf{u}}_f^T & \dot{\varphi} & \dot{\mathbf{q}}^T \end{pmatrix}^T \in \Re^{(5+n_i)\times 1}$$

is the generalized manipulation velocity.

$$
\Lambda = \begin{bmatrix}
\mathbf{0}_{2\times 3} & -\mathbf{M}_o^{-1}(\mathbf{K}_o + \tilde{\mathbf{K}}_f)^{-1}\Phi \\
\mathbf{0}_{2\times 3} & -\mathbf{M}_f^{-1}\mathbf{R}_\phi(\mathbf{K}_o + \tilde{\mathbf{K}}_f)^{-1}\Phi \\
\mathbf{I}_{3\times 3} & -({}_o^p\mathbf{R}\mathbf{r}_o\times) \\
\mathbf{0}_{1\times 3} & \psi
\end{bmatrix} \in \Re^{8\times 6}
$$

is the function of object orientation, contact positions, etc. $\mathbf{V} = \left(\mathbf{v}_o^T \; \boldsymbol{\omega}_o^T\right)^T$ is the generalized velocity of an object.

Then, we discuss the effects of the degrees of freedom of the finger on the manipulation with rolling contact:

- When the number of the rows of the generalized Jacobian matrix $\mathbf{J}$ is less than that of the columns, that is, the degrees of freedom of the finger is less than 3, giving the arbitrary object motion $\mathbf{V} = \left(\mathbf{v}_o^T \; \boldsymbol{\omega}_o^T\right)^T$, the required generalized manipulation velocity $\dot{\Theta}$ cannot be determined by Eq. (8.48). This shows that manipulation with pure rolling contact is impossible. The sliding between the object and the fingertip is inevitable.

- When the generalized Jacobian matrix $\mathbf{J}$ is a non-singular square matrix, that is, the number of degrees of freedom of the finger is 3, giving the arbitrary object motion $\mathbf{V} = \left(\mathbf{v}_o^T \; \boldsymbol{\omega}_o^T\right)^T$, the required finger motion can be exactly determined by Eq. (8.48). This shows that the relative motion between the surfaces of the fingertip and the object to be manipulated is pure rolling.

- When the number of rows of the generalized Jacobian matrix $\mathbf{J}$ is more than that of the columns, that is, the degrees of freedom of the finger is more than 3, multiple solutions exist in Eq. (8.48). At this time, $\dot{\Theta} = \mathbf{J}^+\Lambda\mathbf{V} + \left(\mathbf{E} - \mathbf{J}^+\mathbf{J}\right)\Xi$, where $\mathbf{J}^+ = \mathbf{J}^T\left(\mathbf{J}\mathbf{J}^T\right)^{-1}$ is the Moore–Penrose generalized inverse matrix of the matrix $\mathbf{J}$, $\mathbf{E} \in \Re^{(5+n_i)\times(5+n_i)}$ is an identity matrix, and $\Xi \in \Re^{(5+n_i)\times 1}$ is an arbitrary vector. This implies that we must plan the motion of the finger, according to some measures, to ensure that the relative motion between the surfaces of the fingertip and the object is pure rolling.

As discussed above, it is found that each finger requires a minimum of 3 degrees of freedom in order to permit general pure rolling motion between the surface of the fingertip and the surface of the object.

To grasp objects with arbitrary shapes and manipulate the grasped object, in general, multifingered robotic hands require a minimum of two fingers. If the multifingered hand manipulates an object with m fingers, with respect to every finger, giving the arbitrary object motion, we can determine the required motion of the finger using Eq. (8.48). However, Eq. (8.48) will not have a closed-form solution. The most straightforward way to numerically integrate Eq. (8.48) is by the Runge–Kutta method. With the increase in computer speed, real-time calculation is possible.

8.4   Coordinating Manipulation of Multifingered Robotic Hands

8.4.1   Classification of Grasp Phases

A grasp task may be classified as three phases. The first phase is called pre-contact one where the task is to manipulate the fingers to approach their desired positions following the non-collision path. The second phase is called contact one. Since all of the fingertips don't contact simultaneously the object due to the position errors of fingertips during grasping, the task is to adjust the contact positions and forces of fingertips in the contact phase. The third phase is called post-contact where all the fingertips have been keeping contact states with the object. The task is to carry out contact control so that the contact positions and forces of fingertips approach their desired values as soon as possible.

It should be noted that the second phase is a transition one from pre-contact states to post-contact states and focuses on the contact transition control which is currently an active research area.

8.4.2   Coordinating Manipulation Strategy

The multifingered manipulation is different from the manipulation of a single manipulator. When a multifingered robotic hand grasps an object, multiple closed kinematic loops will be formed. Moreover, the position and orientation of the object are determined uniquely by the positions of the fingertips, and the large gaps between the actual and the desired contact forces at contacts cause the object displacement from its desired position and orientation. Thus, the force/position hybrid control is required in multifingered manipulation so that the object can be grasped successfully. Here we discuss the coordinating manipulation strategy for the multifingered robotic hands, especially, for the contact transition phase.

Assume that the contacts between the object and fingertips are hard contacts with friction. From Eq. (4.3) in Chapter 4, we have

$$\mathbf{G}\mathbf{f}_c = \mathbf{F}_e \qquad (8.49)$$

Given the positions of the fingertips, Eq. (8.49) describes the relationship between the contact forces of fingertips and the external wrench exerted on the object.

When $R(\mathbf{G}) = 6$, $dim\,\mathbf{N}(\mathbf{G}) = 3(M - 2)$ (m is the number of the fingertips used to manipulate the object, generally, $m \geq 3$), if all of the fingertips are in their desired positions, and the contact forces of $m - 2$ fingertips can track the desired contact forces by force control, then the contact forces of the other 2 fingertips approach theoretically their desired values. Thus, it is not necessary for multifingered manipulation to use force and position control for each fingertip in any case. We should develop different manipulation strategies corresponding to the different cases.

When the position of the object is not constrained, for example, if an object is on a table, then the object will not be constrained in the orthogonal directions of each other besides in the direction normal to the surface of the table. In this case, we may use position control for 2 fingertips with high load capability and position precision, and force/position hybrid control for the other $m - 2$ fingertips so that the position displacements of the object may be minimized, and the object may be grasped successfully.

On the contrary, when the position of the object is constrained, for example, if we expect to pull a peg from a hole with a multifingered robotic hand, then the position of the peg is constrained by the hole in this case. Thus, we may use force control for all of the fingertips because the adjustment of the fingertip contact forces will not change the position of the peg when the multifingered robotic hand approaches the peg so that the peg may be pulled out from the hole.

8.5   Adjustment of Fingertip Contact Forces

Generally, force control is indispensable for multifingered grasp and manipulation. Force control includes direct force control and indirect force control. The aim of direct force control is to maintain that the actual contact force follows the planned force trajectory. The direct force control may be implemented by position control where a force controller in the outer loop provides the position adjustment values for the position controller in the inner loop. The direct force control may be carried out by moment control as well, where we need to obtain the force error between the detected and the desired contact forces, and the product of the transpose of the Jacobian of the manipulator and the force error which provides a torsion correct signal for the actuator of joints.

There are no force set-points in the indirect force control. The objective of the indirect force control is to adjust adaptively the contact force between the end-effector of the manipulator and the environment so that the large contact force undulation may be avoided due to position errors. The indirect force control methods include impedance control [18–20], compliance control [8, 14, 15], etc.

The contact force of the fingertip must be able to real-time track the planned force trajectory during the multifingered grasp and manipulation. Since position control is usually used in most industrial manipulators, here we apply the direct force control method based on position control to adjust the fingertip contact force, as shown in Fig. 8.2.

In Fig. 8.2, $\mathbf{f}_{cd}$ and $\mathbf{f}_c$ are the desired and actual contact forces of fingertips, respectively, $\mathbf{X}_d$ and $\mathbf{X}$ are the desired and actual positions of fingertips, respectively, $\Delta\mathbf{X}$ is the adjustments of the positions of fingertips, and θ_d and θ are the desired and actual joint angles of fingers, respectively.

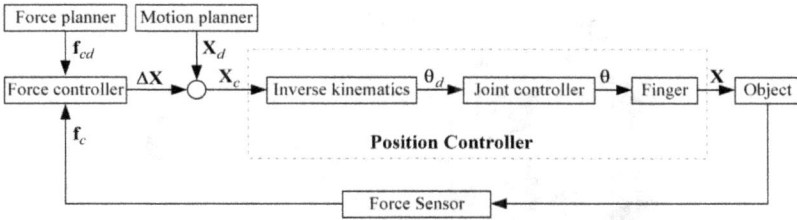

Fig. 8.2. Force control system based on the position control.

In order to adjust the positions of fingertips so that the contact forces between the object and fingertips approach the desired ones, we have to determine the relationship between the contact forces and the position displacements of fingertips (i.e., the finger stiffness). Kao and Cutkosky [21, 22] studied the grasp stiffness and showed that the grasp stiffness was related to the finger structure compliance, joint stiffness, and grasp configuration. Shimoga and Goldenberg [23] compared the material characteristics of fingertips and pointed out that soft fingertips are suitable for stable grasps where the finger stiffness mostly depends on the soft fingertip stiffness. It should be noted that the manipulation of a hard object with soft fingertips is almost the same as the manipulation of a soft object with hard fingertips.

Here we discuss the manipulation of the elastic object with hard fingertips and give an experiment to verify the adjustment algorithm of contact forces. In this experiment set-up, a PUMA562 manipulator is modeled as a finger of the multifingered robotic hand where the fingertip is a disk, and the object to be manipulated is an elastic ball, as shown in Figs. 8.3 and 8.4. The experiment reveals the relationship between the contact force and the position displacement of the fingertip, as shown in Fig. 8.5 and Table 8.1.

We use the rule-based method to construct the force controller where the adjustment of fingertip contact force is transformed into the fingertip position adjustment in the normal direction so that the normal contact force of the fingertip maintains $800g$ during the ball rolling on the work table with the fingertip along a straight line. The motion of the ball along the straight line is implemented by using the position control.

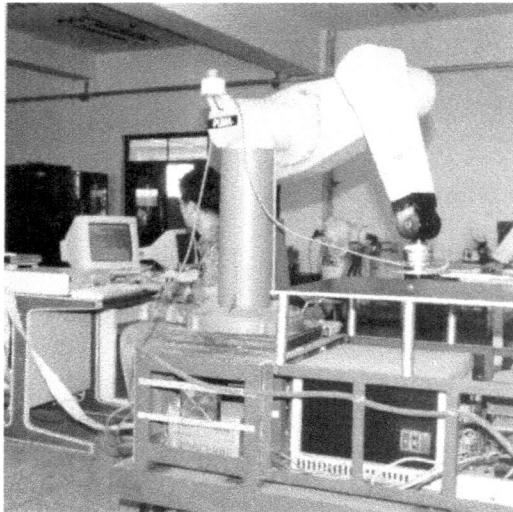

Fig. 8.3.   Experimental system.

Fig. 8.4.   Ball rolled by fingertip.

From the data in Table 8.1, the adjustment algorithm of the fingertip position is formulated as follows:

(i) *If the normal contact force of the fingertip $f_{cn} < N_0$, then the tangent increment of the displacement of the fingertip is 0, and the normal increment is given by $Z_d - Z_0$.*

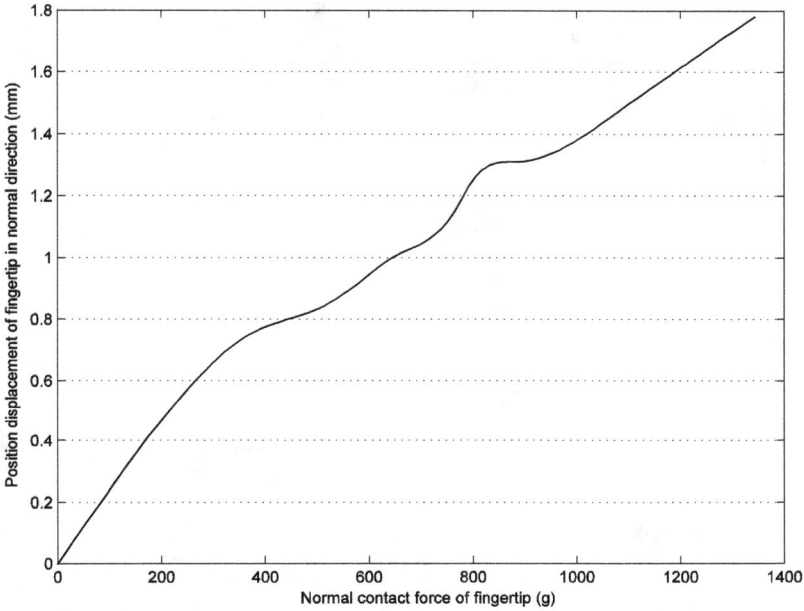

Fig. 8.5. Relationship between contact force and position displacement of fingertip.

(ii) *If the normal contact force of the fingertip $f_{cn} > N_9$, then the tangent increment of the displacement of the fingertip is $0.5\,\mathrm{mm}$, and the normal increment is given by $Z_d - Z_9$.*

(iii) *If the normal contact force of the fingertip $N_i \leq f_{cn} \leq N_{i+1}$, then the tangent increment of the displacement of the fingertip is $0.5\,\mathrm{mm}$, and the normal increment is given by Δ_i, $i = 0, 1, \ldots, 8$.*

The normal increment of the displacement of the fingertip is given by the following formula:

$$\Delta_i = Z_d - \left[\frac{(f_{cn} - N_i)\, Z_{i+1}}{N_{i+1} - N_i} + \frac{(N_{i+1} - f_{cn})\, Z_i}{N_{i+1} - N_i} \right] \qquad (8.50)$$

where Z_d is the desired normal compression of the fingertip ($Z_d = 1.25\,\mathrm{mm}$ in this experiment) and f_{cn} is the actual normal contact force of the fingertip at each sampling time.

Since the original control system of the PUMA562 manipulator is based on position control and cannot deal with the contact force

Table 8.1.   Relationship between the contact force and the position displacement of fingertip.

Sequence no.: i	Normal contact force (g): N_i	Position displacement in normal normal direction (mm): Z_i
0	370	0.75
1	510	0.84
2	590	0.93
3	645	1.00
4	765	1.15
5	800	1.25
6	890	1.31
7	1078	1.47
8	1206	1.62
9	1343	1.78

information, we rebuild the control system of the PUMA562 manipulator. Particularly, we use a personal computer (PC) as the motion planner and force planner, to produce the set points for the position controller and the desired normal contact force of the fingertip. At the same time, the calculation required during the rolling manipulation is also implemented in the PC. The results of comparing the desired normal contact force and the detecting one with the wrist force/torque sensor are used to adjust the fingertip position in the normal direction based on the above adjustment algorithm so that the constant normal contact force can be maintained during the rolling manipulation. The communication of the PC with the position controller and that of the PC with the wrist force/torque sensor are in a parallel mode through the function board $AX5214$. The PUMA562 manipulator requests the PC to send data in an interrupt mode through the function board $AX5214$. The rebuilt control system is shown in Fig. 8.6 where the rebuilt part is represented in the dashed frame.

The control strategies are classified into two phases:

(i) *Before the fingertip touches the ball, that is, the value of the wrist force/torque sensor is 0, the position control is carried out for the PUMA562 manipulator so that it approaches the ball along*

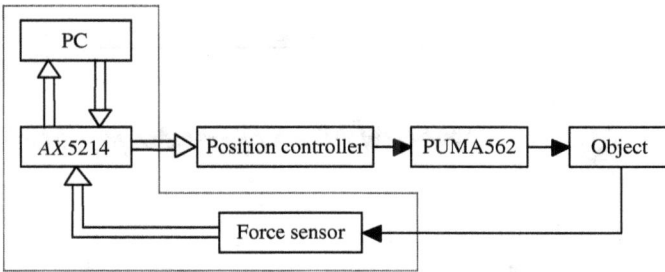

Fig. 8.6. Rebuilt control system.

the normal vector of the work table and doesn't move in the plane of the work table.

(ii) When the fingertip approaches the ball, if the wrist force/torque sensor detects the contact force which means that the fingertip has contacted the ball, then the position control should be switched to force/position hybrid control, and the fingertip motion is implemented by using the adjustment algorithm.

In a word, after the fingertip moves a step along a straight line in the plane of the work table, the position control is switched to the force control, and the normal position of the fingertip is adjusted with the adjustment algorithm so that the normal contact force approaches the desired value, then the force control is switched to the position control so that the fingertip moves a new step along a straight line in the plane of the work table. Repeat the process until the manipulation task is finished. It is clear that the force/position hybrid control is the intrinsic characteristic in the multifingered grasp and manipulation.

8.6 Experimental Results

Using the adjustment algorithm developed in Section 8.5, we obtain the response curve of the fingertip normal contact force, as shown in Fig. 8.7. It can be found from Fig. 8.7 that the rebuilt control system with the adjustment algorithm has the fast force tracking capability, and the overrun and undulation are very small.

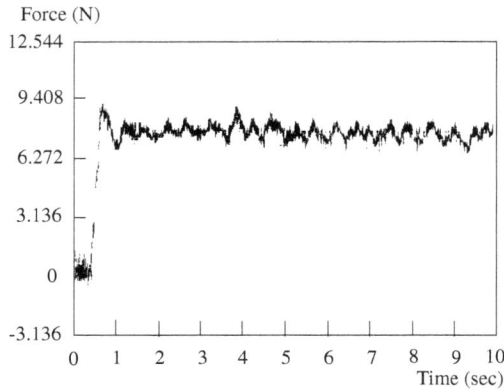

Fig. 8.7.   Response curve of the fingertip normal contact force.

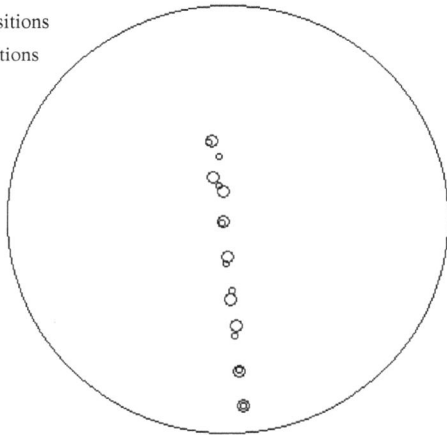

Fig. 8.8.   Contact points on the fingertip.

Using the detected information of the force/torque sensor mounted on the wrist of the PUMA562 manipulator, we can reckon the actual positions of the contact points on the fingertip. Figure 8.8 shows a series of the actual and desired contact points on the fingertip.

It can be found from Fig. 8.8 that there exists departure between the actual and desired contact points. The absolute and relative position errors between the actual and desired contact points are shown

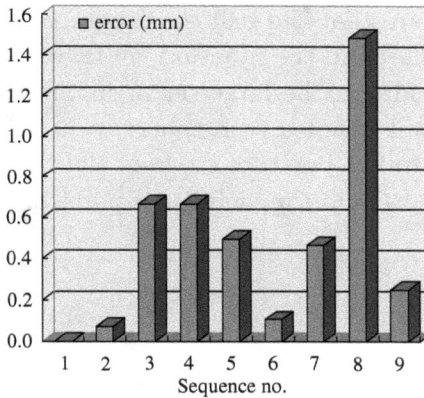

Fig. 8.9. Absolute errors of contact point positions.

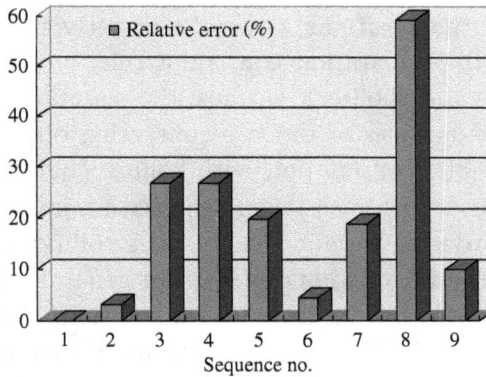

Fig. 8.10. Relative errors of contact point positions.

in Figs. 8.9 and 8.10, respectively. The desired distance between two adjacent points is 2.5 mm in Fig. 8.8. The absolute error of the distance between two adjacent points is defined as the absolute value of the difference between the actual and desired distances of the two adjacent points. The relative error is defined as the ratio of the absolute error to the desired distance. Figures 8.9 and 8.10 show that the absolute error and the relative error are small besides in the vicinities of points 4 and 8. The position errors of contact points are caused by the rotundity errors of the ball, not by the slippage during the rolling manipulation. The reasons why the errors are larger in the vicinities of points 4 and 8 are that the ball consists of two hemispheres, and there exist bumps around the juncture circle between the two hemispheres, which results in the larger error in the vicinity of point 4. In addition, the ball is not a true sphere where the radii along the two poles normal to the juncture circle of the ball are larger, which results in a larger error in the vicinity of point 8.

8.7   Summary

Rolling constraint is a classical example of a non-holonomic constraint. The equations relating the motion of the object to the motion of a fingertip are expressed in terms of the velocities of the two bodies rather than the positions of the bodies. By using the velocity constraint and the orientation constraint relationships between the fingertip surface and the body surface, the equations of pure rolling contact over the surfaces of the two contacting objects are derived. The kinematics of manipulation with rolling contact is developed. The analysis shows that each finger requires a minimum of 3 degrees of freedom in order to permit general pure rolling motion between the surfaces of the fingertip and object. In general, the equations of manipulation do not have a closed-form solution. The equation can be used not only to plan the motion of the multifingered robotic hand and analyze the dynamic stability of grasping but also to design the pure rolling contact gears.

In order to perform manipulation tasks with multifingered robotic hands, we may classify a task into three phases: pre-contact, contact transition, and post-contact phases. The contact transition phase is the key for finishing successfully a manipulation task. The force/position hybrid control is the intrinsic characteristic in the multifingered grasp and manipulation. The rolling manipulation experiment shows that the control system with the adjustment algorithm has fast force-tracking capability, and the overrun and undulation are very small. Moreover, the experiment also shows that using the detected information of the force/torque sensor mounted on the finger, the actual positions of the contact points on the fingertip may be reckoned without solving the inverse kinematics of rolling manipulation.

References

[1]   Wagner M. J., Ng W. F., and Dhande S. G. Profile synthesis and kinematic analysis of pure rolling contact gears. *Transactions of the ASME-Journal of Mechanical Design*, 114, pp. 326–333, 1992.

[2]   Cai C. S. and Roth B. On the planar motion of rigid bodies with point contact. *Mechanism and Machine Theory*, 21(6), pp. 453–466, 1986.

[3]   Montana D. J. The kinematics of contact and grasp. *International Journal of Robotics Research*, 7(3), pp. 17–31, 1988.

[4]   Montana D. J. The kinematics of multi-fingered manipulation. *IEEE Transactions on Robotics and Automation*, 11(4), pp. 491–503, 1988.

[5]   Kerr J. and Roth B. Analysis of multifingered hands. *International Journal of Robotics Research*, 4(4), pp. 3–17, 1986.

[6]   Cole A., Hauser J., and Sastry S. Kinematics and control of a multifingered robot hand with rolling contact. *IEEE Transactions on Automation Control*, 34(4), pp. 398–403, 1989.

[7]   Mandal N. and Payandeh S. Control strategies for robotic contact tasks: An experimental study. *Journal of Robotic Systems*, 12(1), pp. 67–92, 1995.

[8]   Seraji H., Lim D., and Steele R. Experiments in contact control. *Journal of Robotic Systems*, 13(2), pp. 53–73, 1996.

[9]   Payandeh S. and Saif M. Force and position control of grasp in multiple robotic mechanisms. *Journal of Robotic Systems*, 13(8), pp. 515–525, 1996.

[10] Weng S. W. and Young K. Y. An impact control scheme inspired by human reflex. *Journal of Robotic Systems*, 13(12), pp. 837–855, 1996.

[11] Tarokh M. and Bailey S. Adaptive fuzzy force control of manipulators with unknown environment parameters. *Journal of Robotic Systems*, 14(5), pp. 341–353, 1997.

[12] Whitney D. E. Historical perspective and state of the art in robot force control. *International Journal of Robotics Research*, 6(1), pp. 3–13, 1987.

[13] Raibert M. H. and Craig J. J. Hybrid position/force control of manipulators. *Transactions of the ASME-Journal of Dynamics Systems, Measurement, and Control*, 102, pp. 126–133, 1981.

[14] Schutter J. D. and Brussel H. V. Compliant robot motion I: A formalism for specifying compliant motion tasks. *International Journal of Robotics Research*, 7(4), pp. 3–17, 1988.

[15]   Schutter J. D. and Brussel H. V. Compliant robot motion II: A control approach based on external control loops. *International Journal of Robotics Research*, 7(4), pp. 18–32, 1988.

[16]   Kaneko M., Imamura N., and Honkawa K. Contact points detection for inner link based grasps. *Advanced Robotics*, 9(5), pp. 519–533, 1995.

[17]   Son J. S., Cutkosky M. R., and Howe R. D. Comparison of contact sensor localization abilities during manipulation. *Robotics and Autonomous Systems*, 17, pp. 217–233, 1996.

[18]   Seraji H. and Colbaugh R. Force tracking in impedance control. *International Journal of Robotics Research*, 16(1), pp. 97–117, 1997.

[19]   Bonitz R. G. and Hsia T. C. Internal force-based impedance control for cooperating manipulators. *IEEE Transactions on Robotics and Automation*, 12(1), pp. 78–89, 1996.

[20]   Schneider S. A. and Cannon R. H. Object impedance control for cooperative manipulation: Theory and experimental results. *IEEE Transactions on Robotics and Automation*, 8(3), pp. 383–394, 1992.

[21]   Kao I., Cutkosky M. R., and Johansson R. S. Robotic stiffness control and calibration as applied to human grasping tasks. *IEEE Transactions on Robotics and Automation*, 13(4), pp. 557–566, 1997.

[22]   Cutkosky M. R. and Kao I. Computing and controlling the compliance of robotic hand. *IEEE Transactions on Robotics and Automation*, 5(2), pp. 151–165, 1989.

[23]   Shimoga K. B. and Goldenberg A. A. Soft robotic fingertips Part I: A comparison of construction materials. *International Journal of Robotics Research*, 15(4), pp. 320–340, 1996.

[24]   Xiong C. H., Li Y. F., Xiong Y. L., and Zhang W. P. Kinematics of finger with rolling contact. *Progress in Natural Science*, 9(3), pp. 189–197, 1999.

Chapter 9

Dynamic Stability of Grasping/Fixturing

Stability is one of the important properties that a robotic grasp/ fixture must possess to be able to perform tasks similar to those performed by human hands. This chapter discusses the dynamic stability of a grasped/fixtured object. To analyze the stability of grasps, we build the model of the dynamics of the grasped/fixtured object in response to the small perturbations. Furthermore, we determine the conditions associated with dynamic stability and discuss the effects of various factors on grasp stability. A quantitative measure for evaluating grasps is then presented. Finally, the effectiveness of the proposed theory is verified via examples.

9.1 Introduction

Grasp stability is one of the important criteria for evaluating a grip. In recent years, extensive investigations related to grasp stability have been carried out. Hanafusa and Asada [3] first investigated the stability of an articulated grasp. Their quasi-static stability analysis was based on minimizing the total elastic energy of the grasp. Nguyen [8] presented a method of constructing stable planar and spatial grasps of n elastic fingers and showed that the grasp stiffness matrix must be positive definite so that the grasp is quasi-statically stable. Cutkosky and Kao [2] were the first to provide a systematic method for determining the stability of a grasp when the compliance of the finger joints and links is included. Brodsky and Shoham [1] investigated the stability of planar grasps. Howard and Kumar [4] established a framework for analyzing the stability of grasps and showed that the stability of a grasped object depends on the local curvature properties at the contacts as well as on the magnitude and arrangement of the applied forces. Xiong *et al.* [11, 13] studied the grasp capability and presented a stability index for contact configuration planning. So far, the stability analyses have largely been based on the quasi-static assumptions, and hence the results are limited to grasps with negligible dynamic forces.

To address the dynamic issues, Nakamura, Nagai, and Yoshikawa [7] proposed a measure to evaluate the dynamic contact stability. Montana [6] formulated a model of the dynamics of two-fingered grasps. He also derived a quantitative measure of contact stability. However, he did not consider the dynamic stability of the grasped object. Other researchers investigated the dynamic stability of grasping [5, 9, 10, 14] and the natural compliance in fixturing and grasping arrangements [15, 16]. Rimon and Burdick [9] analyzed the effect of curvature of both the fingertip and object at contact points on dynamic stability. Jen, Shoham, and Longman [5] suggested using the stability theory of differential equations as a means of defining stable grasps. Shimoga [10] presented the stability algorithms that aim at achieving positive definite grasp impedance matrices by solving for the required fingertip impedances. Grasp stability is achieved by controlling the fingertip contact forces or the apparent impedance of each finger [10]. Xiong *et al.* analyzed the dynamic stability of compliant grasps using a linear spring-damper model for the fingers [14].

In this chapter, we take a different approach from the above to study the dynamic stability of the grasped object. In this investigation, fingertips are considered as elastic structures and the object as a rigid body. The contact between the object and each fingertip is modeled as pure rolling contact with friction. This implies that the locations of the points of contact over the surfaces of the contacting object and fingertip are not fixed when the object is perturbed. Given the rotational velocities of the grasped object and the fingertip in the base frame, we derived independently the kinematics of rolling [12]. Using the derived kinematics, we develop the dynamic equations of the grasped object in response to small disturbances from equilibrium. Furthermore, we determine the conditions of dynamic stability of the grasped object and define a measure of grasp stability.

9.2 Dynamic Equations of Motion for a Grasped/Fixtured Object

We assume the following:

(1) The fingertips maintain their grasp of the object, that is, maintain static friction at all points of contact. Due to static friction, the object can roll but not slip at the points of contact.
(2) The fingertips are modeled as elastic structures and the object as a rigid body. Every fingertip makes a frictional point contact with the object.
(3) Disturbances from equilibrium are small enough so that all first-order approximations are valid, for example, the curvatures of the object and fingertip at a displaced point of contact are the same as those at the equilibrium point. We also ignore torques which are higher than first-order functions of the motion.
(4) The inertial force and moment of the fingers can be ignored.
(5) The object reference frame $\{O\}$ fixed at the center of mass of the object and its coordinate axes are principal axes of the object inertial. Furthermore, choose the inertial frame (base frame $\{P\}$) so that these two frames are coincident when the object is in the desired equilibrium orientation.

Since the disturbances are small enough, the orientation of the object frame approximately coincides with the orientation of the inertial frame after the object displaces from its equilibrium position, that is, $^P_O R$ can be viewed as a 3×3 identity matrix. Thus, the following analysis is performed in the object reference frame $\{O\}$ unless otherwise specified.

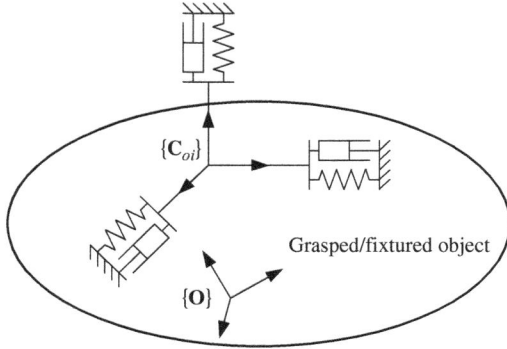

Fig. 9.1. Impedances at the ith point of contact.

According to the model of point contact with friction, every elastic fingertip can be replaced by a 3-D virtual spring-damper system, namely every frictional point contact has three linear springs and dampers all going through the point of contact (see Fig. 9.1).

Assuming that the compression of the virtual springs of the ith fingertip is $\mathbf{x}_{fi}^c$ when the grasped object is in the equilibrium state, then the torque and force equilibrium conditions can be represented as

$$\sum_{i=1}^{m} \left(\mathbf{r}_{oi} \times {}_{coi}^{o}\mathbf{R}\mathbf{f}_{ci}^e \right) = \mathbf{0} \tag{9.1}$$

$$\sum_{i=1}^{m} {}_{coi}^{o}\mathbf{R}\mathbf{f}_{ci}^e + {}^{o}\mathbf{W}_g = \mathbf{0} \tag{9.2}$$

where ${}_{coi}^{o}\mathbf{R} \in \mathbf{SO}(3)$ is the rotation matrix giving the orientation of the contact frame $\{\mathbf{C}_{oi}\}$ in $\{\mathbf{O}\}$,

$$\mathbf{f}_{ci}^e = -\mathbf{k}_{fi}\mathbf{x}_{fi}^c \in \Re^{3\times1}$$

is the equilibrium contact force of the ith fingertip, and

$$\mathbf{k}_{fi} = \mathrm{diag}(k_{fi}^x, k_{fi}^y, k_{fi}^z)$$

is the virtual fingertip stiffness of the ith finger with the stiffness components k_{fi}^x, k_{fi}^y, and k_{fi}^z along the corresponding coordinate axes of $\{\mathbf{C}_{oi}\}$.

$$^o\mathbf{W}_g \in \Re^{3 \times 1}$$

is the gravity vector of the object expressed in $\{\mathbf{O}\}$.

Let the m finger reference frames be $\{\mathbf{F}_1\}, \{\mathbf{F}_2\}, \ldots, \{\mathbf{F}_m\}$ and let the points of contact have coordinates (in each finger reference frame) $\mathbf{r}_{f1}, \mathbf{r}_{f2}, \ldots, \mathbf{r}_{fm}$, respectively. Let the corresponding contact points on the object be given by $\mathbf{r}_{o1}, \mathbf{r}_{o2}, \ldots, \mathbf{r}_{om}$ with respect to the object reference frame $\{\mathbf{O}\}$. Then, the rolling contact constraint of such a hand-object can be described as [12, 14]

$$\begin{bmatrix} \mathbf{v}_{c1} \\ \vdots \\ \mathbf{v}_{cm} \end{bmatrix} = \begin{bmatrix} \mathbf{U}_{o1} \\ \vdots \\ \mathbf{U}_{om} \end{bmatrix} \begin{bmatrix} \mathbf{v}_o \\ \boldsymbol{\omega}_o \end{bmatrix} = \mathbf{G}^T \begin{bmatrix} \mathbf{v}_o \\ \boldsymbol{\omega}_o \end{bmatrix} \tag{9.3}$$

where the matrix $\mathbf{G} = (\mathbf{U}_{o1}^T, \ldots, \mathbf{U}_{om}^T) \in \Re^{6 \times 3m}$ is referred to as grasping/fixturing matrix, $\mathbf{U}_{oi} = (\mathbf{I}_{3 \times 3} \vdots -_o^p\mathbf{R}\mathbf{r}_{oi} \times) \in \Re^{3 \times 6}$.

We assume that the grasped object experiences a small displacement

$$\left(\mathbf{\Delta x}^T, \mathbf{\Delta \theta}^T\right)^T \in \Re^{6 \times 1}$$

from its equilibrium position after a disturbance wrench is exerted on the object. At the same time, the small displacement of the ith contact point can be represented by $\mathbf{\Delta x}_{fi}^c$ in the contact frame $\{\mathbf{C}_{oi}\}$. Then, using Eq. (9.3), we obtain

$$\begin{bmatrix} \mathbf{\Delta x}_{f1}^c \\ \vdots \\ \mathbf{\Delta x}_{fm}^c \end{bmatrix} = {}^{co}_o\mathbf{R}\mathbf{G}^T \begin{bmatrix} \mathbf{\Delta x} \\ \mathbf{\Delta \theta} \end{bmatrix}, \quad \begin{bmatrix} \mathbf{\Delta \dot{x}}_{f1}^c \\ \vdots \\ \mathbf{\Delta \dot{x}}_{fm}^c \end{bmatrix} = {}^{co}_o\mathbf{R}\mathbf{G}^T \begin{bmatrix} \mathbf{\Delta \dot{x}} \\ \mathbf{\Delta \dot{\theta}} \end{bmatrix} \tag{9.4}$$

where ${}^{co}_o\mathbf{R} = \text{block diag}({}^{co1}_o\mathbf{R}, \ldots, {}^{com}_o\mathbf{R})$ with ${}^{coi}_o\mathbf{R} = {}^o_{coi}\mathbf{R}^T \in \mathbf{SO}(3)$ for $i = 1, \ldots, m$.

The ith fingertip contact force ${}^c\mathbf{f}_{ci}$ (expressed in the contact frame $\{\mathbf{C}_{oi}\}$) exerted on the object can be represented as

$$^c\mathbf{f}_{ci} = -\mathbf{k}_{fi}\left(\mathbf{x}_{fi}^c + \mathbf{\Delta x}_{fi}^c\right) - \mathbf{b}_{fi}\mathbf{\Delta \dot{x}}_{fi}^c = \mathbf{f}_{ci}^e - \mathbf{k}_{fi}\mathbf{\Delta x}_{fi}^c - \mathbf{b}_{fi}\mathbf{\Delta \dot{x}}_{fi}^c \tag{9.5}$$

where $\mathbf{b}_{fi} = \mathrm{diag}(b^x_{fi}, b^y_{fi}, b^z_{fi})$ is the virtual contact damping matrix with the damping components b^x_{fi}, b^y_{fi}, and b^z_{fi} along the corresponding coordinate axes of $\{\mathbf{C}_{oi}\}$ at the ith point of contact.

Since the grasped object is displaced from its equilibrium position, the new location of the ith contact point in the object reference frame can be written as $\mathbf{r}_{oi} + \Delta\mathbf{r}_{oi}$, where $\Delta\mathbf{r}_{oi} \in \Re^{3\times 1}$ is the small displacement of the ith contact point in the object reference frame and the torque $\Delta\boldsymbol{\tau}_i$ around the center of mass of the object in the object reference frame produced by the ith fingertip contact force $^c\mathbf{f}_{ci}$ can be represented as (neglecting the second-order terms of the motion)

$$\Delta\boldsymbol{\tau}_i = \mathbf{r}_{oi} \times {}_{coi}^{o}\mathbf{R}\mathbf{f}^e_{ci} - \mathbf{r}_{oi} \times {}_{coi}^{o}\mathbf{R}\mathbf{k}_{fi}\Delta\mathbf{x}^c_{fi} - \mathbf{r}_{oi} \times {}_{coi}^{o}\mathbf{R}\mathbf{b}_{fi}\Delta\dot{\mathbf{x}}^c_{fi}$$
$$+ \Delta\mathbf{r}_{oi} \times {}_{coi}^{o}\mathbf{R}\mathbf{f}^e_{ci} \tag{9.6}$$

Furthermore, if the ith point of contact moves a small amount $\Delta\mathbf{s}_i = [\Delta\mathbf{s}_{ix}\ \Delta\mathbf{s}_{iy}]^T$ along the object's surface, then the vector from the object's center of mass to the point of contact changes by $\Delta\mathbf{r}^c_{oi} = [\Delta\mathbf{s}_{ix}\ \Delta\mathbf{s}_{iy}\ 0]^T$ (measured relative to the object's contact frame $\{\mathbf{C}_{oi}\}$). Using Eq. (8.35), we obtain

$$\begin{bmatrix} \Delta\mathbf{s}_{ix} \\ \Delta\mathbf{s}_{iy} \end{bmatrix} = \mathbf{M}_{oi}\dot{\mathbf{u}}_{oi} = \mathbf{K}_i \begin{bmatrix} 1 & 0 & 0 \\ 0 & 1 & 0 \end{bmatrix}$$

$$\times \left[{}_{o}^{coi}\mathbf{R}\left(\mathbf{S}\left(\boldsymbol{\omega}_f\right) - \mathbf{S}\left(\boldsymbol{\omega}_o\right)\right) {}_{coi}^{o}\mathbf{R} \right] \begin{bmatrix} 0 \\ 0 \\ 1 \end{bmatrix} \tag{9.7}$$

where $\mathbf{K}_i = (\mathbf{K}_{oi} + \widetilde{\mathbf{K}}_{fi})^{-1}$, $\mathbf{K}_{oi}$ is the curvature of the object at the ith point of contact relative to the object's contact frame $\{\mathbf{C}_{oi}\}$, and $\widetilde{\mathbf{K}}_{fi}$ is the curvature of the ith fingertip at the point of contact relative to the x- and y-axes of $\{\mathbf{C}_{oi}\}$.

Assuming that the fingertips do not rotate, then Eq. (9.7) can be written as

$$\begin{bmatrix} \Delta\mathbf{s}_{ix} \\ \Delta\mathbf{s}_{iy} \end{bmatrix} = \mathbf{K}_i\Phi_i\Delta\dot{\boldsymbol{\theta}} \tag{9.8}$$

where $\mathbf{\Phi}_i \in \Re^{2 \times 3}$ is the function matrix of the orientation of the ith contact frame $\{\mathbf{C}_{oi}\}$ relative to the object reference frame $\{\mathbf{O}\}$.

Thus, the vector $\mathbf{\Delta r}_{oi}$ can be represented as

$$\mathbf{\Delta r}_{oi} = {}^{coi}_{o}\mathbf{R}^T \mathbf{\Delta r}^c_{oi} = {}^{coi}_{o}\mathbf{R} \begin{bmatrix} \mathbf{0}_{2 \times 3} & \mathbf{K}_i \mathbf{\Phi}_i \\ & \mathbf{0}_{1 \times 6} \end{bmatrix} \begin{bmatrix} \mathbf{\Delta \dot{x}} \\ \mathbf{\Delta \dot{\theta}} \end{bmatrix} = \widetilde{\mathbf{K}}_i \begin{bmatrix} \mathbf{\Delta \dot{x}} \\ \mathbf{\Delta \dot{\theta}} \end{bmatrix}$$

$$(9.9)$$

We can obtain the torque $\mathbf{\Delta \tau}$ produced by the m fingertip contact forces

$$\mathbf{\Delta \tau} = -\widetilde{\mathbf{H}}^{co}_{o}\mathbf{RG}^T \begin{bmatrix} \mathbf{\Delta \dot{x}} \\ \mathbf{\Delta \dot{\theta}} \end{bmatrix} - (\widetilde{\mathbf{D}}^{co}_{o}\mathbf{RG}^T + \widetilde{\mathbf{E}}) \begin{bmatrix} \mathbf{\Delta x} \\ \mathbf{\Delta \theta} \end{bmatrix} \qquad (9.10)$$

where

$$\widetilde{\mathbf{D}} = (\mathbf{r}_{o1} \times {}^o_{co1}\mathbf{Rk}_{f1} \cdots \mathbf{r}_{om} \times {}^o_{com}\mathbf{Rk}_{fm}) \in \Re^{3 \times 3m}$$

$$\widetilde{\mathbf{H}} = (\mathbf{r}_{o1} \times {}^o_{co1}\mathbf{Rb}_{f1} \cdots \mathbf{r}_{om} \times {}^o_{com}\mathbf{Rb}_{fm}) \in \Re^{3 \times 3m}$$

$$\widetilde{\mathbf{E}} = \left({}^o_{co1}\mathbf{Rf}^e_{c1} \times \widetilde{\mathbf{K}}_1 + \cdots + {}^o_{com}\mathbf{Rf}^e_{cm} \times \widetilde{\mathbf{K}}_m \right) \in \Re^{3 \times 6}$$

Neglecting the second-order term $\mathbf{\Delta \dot{\theta}} \times \mathbf{I}_b \mathbf{\Delta \dot{\theta}}$ ($\mathbf{I}_b \in \Re^{3 \times 3}$ the inertial matrix of the object), we can write the physical law governing the rotational motion of the object as follows:

$$\mathbf{I}_b \mathbf{\Delta \ddot{\theta}} = -\widetilde{\mathbf{H}}^{co}_{o}\mathbf{RG}^T \begin{bmatrix} \mathbf{\Delta \dot{x}} \\ \mathbf{\Delta \dot{\theta}} \end{bmatrix} - \left(\widetilde{\mathbf{D}}^{co}_{o}\mathbf{RG}^T + \widetilde{\mathbf{E}} \right) \begin{bmatrix} \mathbf{\Delta x} \\ \mathbf{\Delta \theta} \end{bmatrix}. \qquad (9.11)$$

The translational motion of the object can be written using Newton's motion law as

$$\mathbf{M}_b \mathbf{\Delta \ddot{x}} = -\mathbf{B}_b {}^{co}_{o}\mathbf{RG}^T \begin{bmatrix} \mathbf{\Delta \dot{x}} \\ \mathbf{\Delta \dot{\theta}} \end{bmatrix} - \mathbf{K}_b {}^{co}_{o}\mathbf{RG}^T \begin{bmatrix} \mathbf{\Delta x} \\ \mathbf{\Delta \theta} \end{bmatrix} \qquad (9.12)$$

where $\mathbf{M}_b \in \Re^{3 \times 3}$ is the diagonal mass matrix of the object,

$$\mathbf{B}_b = ({}^o_{co1}\mathbf{Rb}_{f1} \cdots {}^o_{com}\mathbf{Rb}_{fm}) \in \Re^{3 \times 3m},$$

$$\mathbf{K}_b = ({}^o_{co1}\mathbf{Rk}_{f1} \cdots {}^o_{com}\mathbf{Rk}_{fm}) \in \Re^{3 \times 3m}.$$

Combining Eqs. (9.11) and (9.12), we obtain the dynamic equations of motion for the object as follows:

$$\widetilde{\mathbf{M}}_b \delta \ddot{\mathbf{x}} + \widetilde{\mathbf{B}}_b \delta \dot{\mathbf{x}} + \widetilde{\mathbf{K}}_b \delta \mathbf{x} = 0 \tag{9.13}$$

where

$$\widetilde{\mathbf{M}}_b = \begin{bmatrix} \mathbf{M}_b & \mathbf{0} \\ \mathbf{0} & \mathbf{M}_b \end{bmatrix} \in \Re^{6\times6}, \quad \delta \mathbf{x} = \begin{bmatrix} \Delta \mathbf{x} \\ \Delta \boldsymbol{\theta} \end{bmatrix} \in \Re^{6\times1}$$

$$\widetilde{\mathbf{B}}_b = \begin{bmatrix} \mathbf{B}_b\,{}_o^{co}\,\mathbf{R}\mathbf{G}^T \\ \widetilde{\mathbf{H}}_o^{co}\,\mathbf{R}\mathbf{G}^T \end{bmatrix} \in \Re^{6\times6} \tag{9.14}$$

$$\widetilde{\mathbf{K}}_b = \begin{bmatrix} \mathbf{K}_b & {}_o^{co}\mathbf{R}\mathbf{G}^T \\ \widetilde{\mathbf{D}} & {}_o^{co}\mathbf{R}\mathbf{G}^T + \widetilde{\mathbf{E}} \end{bmatrix} \in \Re^{6\times6} \tag{9.15}$$

9.3   Dynamic Stability Conditions and Quality Measure

The stability criterion of Liapunov is stated as follows: A system is asymptotically stable in the vicinity of the equilibrium point at the origin if there exists a scalar function V such that

(1) $V(\mathbf{y})$ *is continuous and has continuous first partial derivatives at the origin;*
(2) $V(\mathbf{y}) > 0$ *for* $\mathbf{y} \neq \mathbf{0}$, *and* $V(\mathbf{0}) = 0$;
(3) $\dot{V}(\mathbf{y}) < 0$ *for all* $\mathbf{y} \neq \mathbf{0}$.

Now we define a state variable $\mathbf{y} = \left(\mathbf{y}_1^T\ \mathbf{y}_2^T\right)^T = (\Delta \mathbf{x}^T\ \Delta \boldsymbol{\theta}^T\ \Delta \dot{\mathbf{x}}^T\ \Delta \dot{\boldsymbol{\theta}}^T)^T$. Then the dynamic equation (9.13) of motion for the object can be written in the state equation form

$$\dot{\mathbf{y}}_1 = \mathbf{y}_2 \tag{9.16}$$

$$\dot{\mathbf{y}}_2 = -\widetilde{\mathbf{M}}_b^{-1}\widetilde{\mathbf{K}}_b\mathbf{y}_1 - \widetilde{\mathbf{M}}_b^{-1}\widetilde{\mathbf{B}}_b\mathbf{y}_2 \tag{9.17}$$

The kinetic, damping, and potential energies of the grasp system, denoted by E_k, E_d, and E_p, respectively, are given by

$$E_k = \frac{1}{2}\mathbf{y}_2^T\widetilde{\mathbf{M}}_b\mathbf{y}_2$$

$$E_d = \frac{1}{2}\mathbf{y}_2^T\widetilde{\mathbf{B}}_b\mathbf{y}_2$$

$$E_p = \frac{1}{2}\mathbf{y}_1^T\widetilde{\mathbf{K}}_b\mathbf{y}_1$$

We choose the sum of the kinetic energy and the potential energy as the Liapunov function V, that is,

$$V(\mathbf{y}) = \frac{1}{2}\mathbf{y}_2^T\widetilde{\mathbf{M}}_b\mathbf{y}_2 + \frac{1}{2}\mathbf{y}_1^T\widetilde{\mathbf{K}}_b\mathbf{y}_1 \tag{9.18}$$

It is clear that $V(\mathbf{0}) = 0$. Note that the matrix $\widetilde{\mathbf{M}}_b$ is symmetric and positive definite, and by our choice of the coordinate axes of the object frame, it is in fact diagonal. Thus, if the matrix $\widetilde{\mathbf{K}}_b$ is symmetric and positive definite, then $V(\mathbf{y})$ is positive definite. Under such conditions, differentiating Eq. (9.18) yields

$$\dot{V}(\mathbf{y}) = -\mathbf{y}_2^T\widetilde{\mathbf{B}}_b\mathbf{y}_2 \tag{9.19}$$

As can be seen, when $\mathbf{y}_1 = \mathbf{0}$, $\mathbf{y}_2 = \mathbf{0}$, $\dot{V}(\mathbf{y}) = 0$; and when $\mathbf{y}_1 \neq \mathbf{0}$, $\mathbf{y}_2 = \mathbf{0}$, $\dot{V}(\mathbf{y}) = 0$, thus $V(\mathbf{y})$ is negative semi-definite if the matrix $\widetilde{\mathbf{B}}_b$ is positive definite. For this reason, we need to find out whether $V(\mathbf{y})$ is constantly zero when $\mathbf{y}_1 \neq \mathbf{0}$, $\mathbf{y}_2 = \mathbf{0}$, for $t \geq t_0$. If $\dot{V}(\mathbf{y}) = -\mathbf{y}_2^T\widetilde{\mathbf{B}}_b\mathbf{y}_2$ is zero, then $\dot{\mathbf{y}}_2$ must be zero. From the state Eq. (9.17), we can find that $\mathbf{y}_1$ is zero. This means that $\dot{V}(\mathbf{y})$ is constantly zero if and only if $\mathbf{y}_1 = \mathbf{0}$ and $\mathbf{y}_2 = \mathbf{0}$.

Thus, the grasp system is asymptotically stable in the vicinity of the equilibrium position at the origin if its impedance matrices $\widetilde{\mathbf{M}}_b$, $\widetilde{\mathbf{B}}_b$, and $\widetilde{\mathbf{K}}_b$ are positive definite.

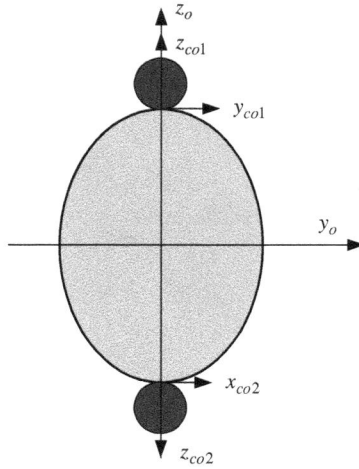

Fig. 9.2. Two-fingered grasp.

However, we need more detailed study on grasp stability than just classifying it as either stable or unstable. We note that the stiffness $\widetilde{\mathbf{K}}_b$ represents a measure of the restoring forces or torques along each axis of motion after the grasped/fixtured object is displaced from its equilibrium position. To make the grasp have similar restoring ability along each axis of motion, we define a measure of grasp stability as

$$\widetilde{w} = \lambda_1 \lambda_2 \cdots \lambda_6 \tag{9.20}$$

where λ_i $(i = 1, \ldots, 6)$ is the ith eigenvalue of the stiffness matrix $\widetilde{\mathbf{K}}_b$.

Since $det\widetilde{\mathbf{K}}_b = \lambda_1 \lambda_2 \cdots \lambda_6$, Eq. (9.20) can be represented as

$$\widetilde{w} = det\widetilde{\mathbf{K}}_b \tag{9.21}$$

Example 1. First, consider grasping/fixturing symmetrically an object with two spherical fingertips as shown in Fig. 9.2. The object coordinate frame and the contact coordinate frame are given in Fig. 9.2.

We assume that the stiffness and damping of every fingertip are the same, that is, $\mathbf{k}_{f1} = \mathbf{k}_{f2} = \mathrm{diag}(k\,k\,k)$, $\mathbf{b}_{f1} = \mathbf{b}_{f2} = \mathrm{diag}(b\,b\,b)$, and so are those in examples 2 and 3. We also assume that the equilibrium contact forces of two fingertips are $\mathbf{f}_{c1}^e = \mathbf{f}_{c2}^e = (0\,0 - f_n)^T$ (f_n is a positive constant). Let the curvature radii of the object and fingertip at the points of contact be ρ_o, ρ_f, respectively. Denote the distance from the point of contact to the center of mass of the object (i.e., the origin of the object frame) by r. Using Eqs. (9.14) and (9.15), we obtain

$$
\widetilde{\mathbf{B}}_b^1 = \begin{bmatrix} 2b & & & & & \\ & 2b & & & \mathbf{0} & \\ & & 2b & & & \\ & & & 2br^2 & & \\ & \mathbf{0} & & & 2br^2 & \\ & & & & & 0 \end{bmatrix}
$$

$$
\widetilde{\mathbf{K}}_b^1 = \begin{bmatrix} 2k & 0 & 0 & & & \\ 0 & 2k & 0 & & \mathbf{0}_{3\times 3} & \\ 0 & 0\cdot & 2k & & & \\ & & & 2kr^2 + \rho f_n & -\rho f_n & 0 \\ & \mathbf{0}_{3\times 3} & & -\rho f_n & 2kr^2 + \rho f_n & 0 \\ & & & 0 & 0 & 0 \end{bmatrix} \quad (9.22)
$$

where $\rho = \rho_o \rho_f / (\rho_o \pm \rho_f)$ is the resultant curvature radius (as is the one in examples 2 and 3), "+" is used when the grasped/fixtured object at the point of contact is convex, and "−" is used when the object is concave.

From Eq. (9.22), we can find that the damping matrix $(\widetilde{\mathbf{B}}_b^1)_{5\times 5}$ and stiffness matrix $(\widetilde{\mathbf{K}}_b^1)_{5\times 5}$ are positive definite. Thus, the grasp is stable except for the rotation around the line connecting two points of contact.

Using Eq. (9.21), we obtain

$$
det\left(\widetilde{\mathbf{K}}_b^1\right)_{5\times 5} = (2k)^3 \left(2kr^2 + 2\rho f_n\right)\left(2kr^2\right) \quad (9.23)
$$

Example 2. Next, consider grasping an object with three spherical fingertips as shown in Fig. 9.3.

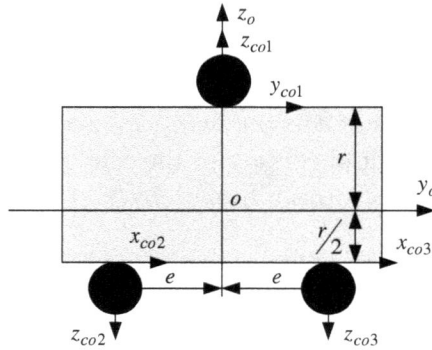

Fig. 9.3. Three-fingered grasp.

We assume that the equilibrium contact forces of the three fingertips are $\mathbf{f}_{c1}^e = (0\ 0\ -f_n)^T$ and $\mathbf{f}_{c2}^e = \mathbf{f}_{c3}^e = (0\ 0\ -0.5f_n)^T$ (f_n is a positive constant), respectively. We also assume that the curvature radii of the object and fingertip at every point of contact are the same, which are ρ_o, ρ_f, respectively. The locations of points of contact are given in Fig. 9.3. Using Eqs. (9.14) and (9.15), we obtain

$$
\widetilde{\mathbf{B}}_b^2 =
\begin{bmatrix}
3b & & & & & \\
& 3b & & & \mathbf{0} & \\
& & 3b & & & \\
& & & \frac{3r^2+4e^2}{2}b & & \\
& \mathbf{0} & & & \frac{3r^2}{2}b & \\
& & & & & 2e^2b
\end{bmatrix}
$$

$$
\widetilde{\mathbf{K}}_b^2 =
\begin{bmatrix}
3k & 0 & 0 & & & \\
0 & 3k & 0 & & \mathbf{0}_{3\times3} & \\
0 & 0 & 3k & & & \\
& & & \frac{3r^2+4e^2}{2}k + \rho f_n & -\rho f_n & 0 \\
& \mathbf{0}_{3\times3} & & -\rho f_n & \frac{3r^2}{2}k + \rho f_n & 0 \\
& & & 0 & 0 & 2e^2k
\end{bmatrix}
\tag{9.24}
$$

From Eq. (9.24), we can find that the damping matrix $\widetilde{\mathbf{B}}_b^2$ and stiffness matrix $\widetilde{\mathbf{K}}_b^2$ are positive definite. Thus, the grasp is stable.

Using Eq. (9.21), we obtain

$$det\widetilde{\mathbf{K}}_b^2 = (3k)^3 \left(\frac{(3r^2 + 4e^2)(3r^2k + 2k\rho f_n)}{4} + \frac{3r^2k}{4}\rho f_n \right) (2e^2k)$$

(9.25)

Example 3. Finally, consider grasping symmetrically an object with three spherical fingertips as shown in Fig. 9.4. The object coordinate frame and the contact coordinate frame are given in Fig. 9.4 ($\alpha = 30°$).

We assume that the equilibrium contact forces of the three fingertips are $\mathbf{f}_{c1}^e = \mathbf{f}_{c2}^e = \mathbf{f}_{c3}^e = (0\ 0\ -f_n)^T$ (f_n is a positive constant). We also assume that the curvature radii of the object and fingertip at every point of contact are the same, which are ρ_o, ρ_f, respectively. Let the distance from the center of mass to every point of contact be r. Using Eqs. (9.14) and (9.15), we obtain

$$\widetilde{\mathbf{B}}_b^3 = \begin{bmatrix} 3b & & & & & \\ & 3b & & & \mathbf{0} & \\ & & 3b & & & \\ & \mathbf{0} & & \frac{3r^2}{2}b & & \\ & & & & \frac{3r^2}{2}b & \\ & & & & & \frac{3r^2}{2}b \end{bmatrix}$$

$$\widetilde{\mathbf{K}}_b^3 = \begin{bmatrix} 3k & 0 & 0 & & & \\ 0 & 3k & 0 & & \mathbf{0}_{3\times3} & \\ 0 & 0 & 3k & & & \\ & & & \frac{3r^2}{2}k & 0 & 0 \\ & \mathbf{0}_{3\times3} & & 0 & \frac{3r^2}{2}k + \frac{3}{2}\rho f_n & 0 \\ & & & 0 & 0 & \frac{3r^2}{2}k + \frac{3}{2}\rho f_n \end{bmatrix}$$

(9.26)

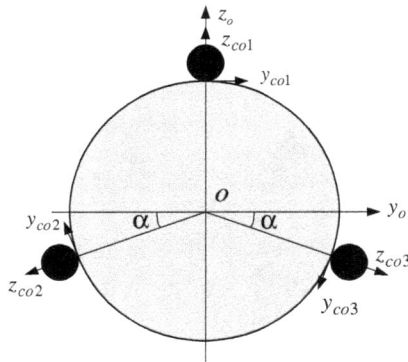

Fig. 9.4. Three-fingered symmetrically grasp.

From Eq. (9.26), we can find that the damping matrix $\widetilde{\mathbf{B}}_b^3$ and stiffness matrix $\widetilde{\mathbf{K}}_b^3$ are positive definite. Thus, the grasp is stable.

Using Eq. (9.21), we obtain

$$
det\widetilde{\mathbf{K}}_b^3 = (3k)^3 \left(\frac{3r^2}{2}k + \frac{3}{2}\rho f_n \right)^2 \left(\frac{3r^2}{2}k \right) \tag{9.27}
$$

Discussion.

From Eqs. (9.23), (9.25), and (9.27), we can see the following:

(1) When ρ, r, and e (in Example 2) are constant, if f_n increases, then the $det(\widetilde{\mathbf{K}}_b^1)_{5\times5}$, $det\,\widetilde{\mathbf{K}}_b^2$, and $det\,\widetilde{\mathbf{K}}_b^3$ also increase, which means that the grasp becomes more stable while the fingertip contact forces increase.

(2) When ρ and f_n are constants, if r or e or both of them increase(s), then the $det(\widetilde{\mathbf{K}}_b^1)_{5\times5}$, $det\,\widetilde{\mathbf{K}}_b^2$, and $det\,\widetilde{\mathbf{K}}_b^3$ also increase, which implies that the grasp becomes more stable while the distance between two points of contact (in Example 1) and area of grasp triangle formed by the three points of contact increase.

(3) When f_n, r, and e (in Example 2) are constant, if ρ increases, then the $det(\widetilde{\mathbf{K}}_b^1)_{5\times5}$, $det\,\widetilde{\mathbf{K}}_b^2$, and $det\,\widetilde{\mathbf{K}}_b^3$ also increase, which means that the grasp becomes more stable while the resultant curvature radius increases (see Fig. 9.5).

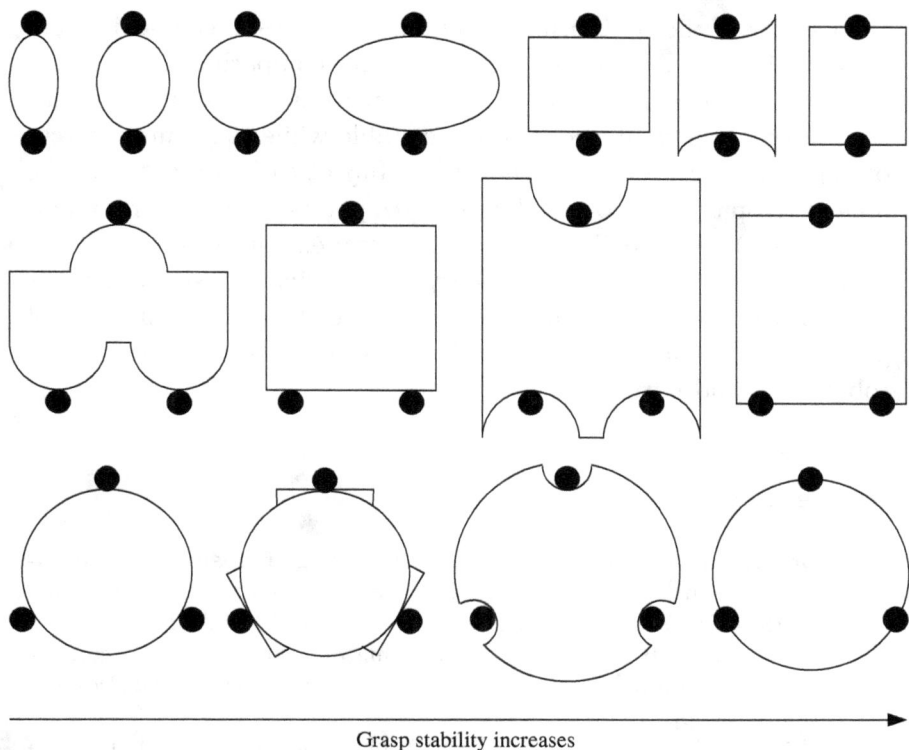

Grasp stability increases

Fig. 9.5. Effect of the resultant curvature radius.

In a word, for stable grasp, the grasp will become more stable while the contact forces of fingertips, the resultant curvature radius at each contact, and the area of grasp triangle formed by the three points of contact (for trifingered grasp) increase.

9.4 Summary

In this chapter, we have formulated a general framework for analyzing the dynamic stability of the grasped/fixtured object with multiple pure rolling contacts, where the fingertips are modeled as elastic structures and the object as a rigid body. The conditions of dynamic stability of the grasped object were achieved by using Lyapunov's direct method. That is, the grasp system is asymptotically stable in the vicinity of the equilibrium position if its impedance matrices are positive definite. A quantitative measure for comparing the dynamic stability of the grasped object has been derived. We showed that the stable grasp would become more stable while the contact forces of fingertips, the resultant curvature radius at each contact, and the area of grasp triangle formed by the three points of contact (for trifingered grasp) increase. The presented several examples show that the proposed theory is valid. The theory can provide a basis for contact configuration planning of multifingered robot hands. The method of grasp stability analysis is applicable to diverse areas such as multiple robot arms and fixtures.

References

[1]   Brodsky V. and Shoham M. On the modeling of grasps with a multi-fingered hand. In: Merlet J. P. and Ravani B. (eds.), *Computational Kinematics '95*. Kluwer Academic, Boston, MA, USA, 1995.

[2]   Cutkosky M. R. and Kao I. Computing and controlling the compliance of a robotic hand. *IEEE Transactions on Robotics and Automation*, 5(2), pp. 151–165, 1989.

[3]   Hanafusa H. and Asada H. Stable prehension by a robot hand with elastic fingers. *Proceedings of the 7th International Symposium on Industrial Robots*, Tokyo, Japan, 1977.

[4]   Howard W. S. and Kumar V. On the stability of grasped objects. *IEEE Transactions on Robotics and Automation*, 12(6), pp. 904–917, 1996.

[5]   Jen F., Shoham M., and Longman R. W. Liapunov stability of force-controlled grasps with a multifingered hand. *International Journal of Robotics Research*, 15(2), pp. 137–154, 1996.

[6]   Montana D. J. Contact stability for two-fingered grasps. *IEEE Transactions on Robotics and Automation*, 8(4), pp. 421–430, 1992.

[7]   Nakamura Y., Nagai K., and Yoshikawa T. Dynamics and stability in coordination of multiple robotic mechanisms. *International Journal of Robotics Research*, 8(2), pp. 44–61, 1989.

[8]   Nguyen V. D. Constructing stable grasps. *International Journal of Robotics Research*, 8(1), pp. 26–37, 1989.

[9]   Rimon E. and Burdick J. W. Mobility of bodies in contact. Part II: How forces are generated by curvature effects. *IEEE Transactions on Robotics and Automation*, 14(5), pp. 709–717, 1998.

[10] Shimoga K. B. Robot grasp synthesis algorithms: A survey. *International Journal of Robotics Research*, 15(3), pp. 230–266, 1996.

[11] Xiong C. H., Li Y. F., Xiong Y. L., Ding H., and Huang Q. Grasp capability analysis of multifingered robot hands. *Robotics and Autonomous Systems*, 27(4), pp. 211–224, 1999a.

[12] Xiong C. H., Li Y. F., Xiong Y. L., and Zhang W. P. Kinematics of finger with rolling contact. *Progress in Natural Science*, 9(3), pp. 189–197, 1999b.

[13] Xiong C. H. and Xiong Y. L. Stability index and contact configuration planning of multifingered grasp. *Journal of Robotic Systems*, 15(4), pp. 183–190, 1998.

[14] Xiong C. H., Li Y. F., Ding H., and Xiong Y. L. On the dynamic stability of grasping. *International Journal of Robotics Research*, 18(9), pp. 951–958, 1999.

[15] Xiong C. H., Wang M. Y., Tang Y., and Xiong Y. L. Compliant grasping with passive forces. *Journal of Robotic Systems*, 22(5), pp. 271–285, 2005.

[16] Lin Q., Burdick J. W., and Rimon E. Computation and analysis of natural compliance in fixturing and grasping arrangements. *IEEE Transactions on Robotics*, 20(4), pp. 651–667, 2004.

Chapter 10

Locating Error Analysis and Configuration Planning of Fixtures

It is one of the fundamental issues in fixture automation design and planning to evaluate and control the influence of the geometric tolerances of locators on the locating errors of the workpiece. Workpiece localization accuracy is primarily determined by the positioning accuracy of the locators and their layout in the fixture, while the choice of a set of clamps is equally important for maintaining the desired positions and orientations of workpieces during machining or manufacturing processes. In this chapter, a mapping model between the error space of locators and the workpiece locating error space is built up for 3-D workpieces. Given the geometrical tolerance specification of a key feature on a workpiece, the geometry design requirements can be determined for all of the locators by using the model. On the other hand, given the geometric tolerances of locators, the calculating methods of the locating errors of the workpiece are developed for fully constrained localization, over-constrained localization, and under-constrained localization cases by using the mapping model. In the analysis of the locator and clamp configuration characteristics, the free motion cone, which is used to judge whether the workpiece is accessible to the fixture as well as detachable from the fixture, is defined. According to the duality theory in convex analysis, the polar of the free motion cone, namely the constrained cone, is derived.

By using the constrained cone, the positions of clamps and the feasible clamping domain are determined, where the workpiece is fully constrained.

To plan the optimal fixturing configuration, three indices are defined: (1) the locating robustness index used to evaluate the configurations of locators; (2) the stability index used to evaluate the capability to withstand any external disturbance wrench for the fixturing system; (3) the fixturing resultant index used to evaluate the robustness to the position errors of locators as well as the stability under the external disturbance wrench for the fixturing system. Then three constrained nonlinear programming methods are presented to determine the optimal locator and clamp configuration. Finally, some examples are given to verify the effectiveness of the proposed fixturing analysis models and planning methods and compare the planning results of the three planning methods.

10.1 Introduction

Fixtures are used to locate and hold workpieces with locators and clamps respectively so that the desired positions and orientations of the workpieces can be maintained during machining or manufacturing processes. Fixture design involves setup planning, fixture planning, fixture structural design, and design verification [26]. The fundamental problems to be solved in automated fixture planning include the following:

(1) How to determine the locator configurations which satisfy the accessibility to the fixture and detachability from the fixture [1]?
(2) How to determine the clamping positions which make a workpiece fully constrained?
(3) How to plan an optimal locator configuration which minimizes the position and orientation errors of a workpiece?
(4) How to plan a fixturing (locating and clamping) configuration which makes the fixturing stability assured by appropriately distributing the locators and clamps on the surfaces of a workpiece?

(5) How to plan a fixturing configuration which makes the fixturing system have both properties of minimizing the position and orientation errors of a workpiece and maximizing the stability of the fixturing system?
(6) How to determine the clamping forces which make the fixturing system withstand the time- and position-variant external wrench such as cutting forces without sliding between the workpiece and fixels (locators and clamps)?

In order to provide fundamental solutions to these problems, the following modeling issues need to be studied:

(1) the mapping between locator errors and the position and orientation errors of the workpiece;
(2) the forward problem in fixture verification, that is, given the locator tolerances and their configurations, how to predict the position and orientation precision of the workpiece;
(3) the inverse problem in fixture verification, that is, given the allowable position and orientation variations of the workpiece, how to determine the tolerances of locators;
(4) the problem of finding an optimal clamping scheme without sliding between the workpiece and fixels.

The geometrical accuracy of a machined feature on a workpiece depends on, partially, the machining fixture's ability to precisely locate the workpiece, which is in fact related to the locators' configuration and the tolerance of each locator. The positions of clamps affect directly the form closure of fixturing. Consequently, it is one of the keys for fixture design automation to investigate the fundamentals of locating and clamping.

For almost a hundred years, it has been known that four and seven unilateral point contacts are the minimum numbers needed for 2-D and 3-D form closure, respectively, and the related proofs can be found in Refs. [14,21,34,35]. The analysis of form closure shows that six locators (namely, 3-2-1 locating principle, see Refs. [7,11,14,19]) and one clamp are at least needed to fully constrain a workpiece in a fixture. Necessary and sufficient conditions for the fully constrained localization of 3-2-1 locator schemes were derived [19]. More recently, Rimon [25] extended their results to a much larger class of 3-D objects and stated that only four frictionless fingers or fixels are required to immobilize a generic 3-D object when second-order geometrical effects, namely curvature effects, are taken into account. To reach form closure and force closure, modular vise algorithms for designing planar fixtures [4] and 3-D modular grippers [5] have been developed. In the algorithms, it is assumed that a part only has contact with vertical surfaces on the fingers [5]. Therefore, using the vise algorithms, we cannot always obtain the modular fixturing solutions for an arbitrary part [36]. The selection of the suitable clamping region for planar fixturing is discussed [31]. However, the proposed method cannot extend to 3-D fixturing. An algorithm to find the clamping positions on 3-D workpieces is presented [20], but how to determine the feasible clamping domain that makes the workpiece totally constrained is not given in the algorithm. The ultimate reasons why there exist essential differences between grasps and fixtures are as follows: All fingers during grasping with multifingered robotic hands can be considered as active end-effectors, but all locators during fixturing are passive while only clamps are active. When all fingers during grasping are active, the desired position and orientation of the object to be grasped can be achieved by actively controlling the multifingered robotic hands. Thus, the robotic grasping mainly concerns holding feasibility [25], compliance [17, 39], and stability [3, 32, 33]. In contrast, because the position and orientation precision of the workpiece to be fixtured depends on the passive locators' tolerances and configuration, fixturing for machining emphasizes on accurate localization of the workpiece. To meet the accuracy requirement, one might not fixture a workpiece on some surfaces [11], although they are feasible from holding point of view.

Asada and By [1] analyzed the problem of automatically locating fixture elements using robot manipulators. The kinematic problem for the fully constrained localization was characterized by analyzing the functional constraints posed by the fixtures on the surface of a rigid workpiece. Desirable fixture configuration characteristics are obtained for loading and unloading the workpiece successfully despite errors in workpiece manipulation. In the study of manufacturing processes, error sources were investigated for precision machining [11,36], and locating error analysis models for fully constrained localization were proposed [6,8,11,15,26,30]. The error sensitivity equation was formed for the fully constrained localization [6,12]. The impact of a locator tolerance scheme was modeled and analyzed on the potential datum related, geometric errors of linear, machined features for the fully constrained localization [8,26]. However, the under-constrained and over-constrained localizations are not considered in the proposed locating error analysis models. In addition, these models neglect the resultant effects of all types of errors between the workpiece and locators, which means that developing a general fixturing error model is necessary.

How to define and find an optimal fixturing configuration in the feasible configurations, which is another goal of this chapter, is one of the fundamentals to be addressed in fixture automation design and planning.

In general, locating errors always exists although it is desired to decrease these errors. Even though giving the same tolerances of locators, different locator configurations may result in different position and orientation errors of workpieces. In some cases, because of the unreasonable locator configuration, the ranges of the position and orientation variations of workpieces cannot be accepted. Thus, in the past few years, there have been a lot of researchers who have been investigating fixture planning. A practical automated fixture planning method was developed with integration the of CAD, based on predefined locating modes and operational rules [26]. A variational method for planning the locator configuration was developed [1]. The conditions for optimizing the fully constrained 3-2-1 locator scheme for low reaction forces and small locator locating errors were analyzed, and the near-optimal fully constrained 3-2-1 location scheme synthesis algorithms were presented [19], but the coupling effect of each locator error on the position and orientation errors of workpieces were neglected. Furthermore, how to determine the optimal clamp configuration was not taken into account in [1, 19].

It is known that compliance can play a significant role in fixturing [17]. A stiffness quality measure for compliant grasps or fixtures was defined [17]. During fixturing, the fixture elements have to protect the workpiece from deflecting under the load of the machining forces, the measure can indicate the number and location of fixture elements that best suit the given task. A method that minimizes workpiece location errors due to the local elastic deformation of the workpiece at the fixturing points by optimally placing the locators and clamps around the workpiece was presented [15], the similar fixture layout optimization problem that minimizes the deformation of the machined surface due to clamping and machining forces can be solved by using the genetic algorithm [13]. In fact, comparing with the errors of locators, the local elastic deformations of the workpiece at the fixturing points are not the predominant factors.

Robust fixture design is considered in Ref. [1]. If the position of a workpiece is insensitive to locator errors, it is said that the localization of the workpiece is robust locating. The sensitivity equation is the foundation for robust fixture design. The linearized sensitivity equation [1] and quadratic sensitivity equation [7] were developed for a fully constrained localization scheme. Two performance measures including maximization of the workpiece locating accuracy and minimization of the norm and dispersion of the locator contact forces were presented in Ref. [30], and the corresponding algorithms to plan the optimal fixture layout for the fully constrained localization were developed. Algorithms for computing all placements of (frictionless) point fingers that put a polygonal part in form closure and all placements of point fingers that achieve second-order immobility of a polygonal part were presented in Refs. [9,29]. An algorithm [18] for computing n-finger form-closure grasps on polygonal objects was proposed. Algorithms for synthesis of three/four-fingered force-closure grasps [23,24] for arbitrary polygonal objects were proposed. The contact stability [22] and the stability of grasped objects [10,28,32,33] were investigated to determine the optimal contact configuration. An algorithm to find the clamping positions on 3-D workpieces with planar and cylindrical faces was presented [20], but how to determine the feasible clamping domain which makes the workpiece totally constrained was not given in the algorithm.

The goal of this chapter is to develop a general method to solve the problems of locating error analysis and configuration planning, and the problem of clamping force planning will be discussed in Chapter 10. First, a fixturing error model is derived, which is applicable for fully constrained localization, under-constrained localization, and over-constrained localization. The characteristics of locator configurations are analyzed. The accessibility and detachability conditions are given. On the basis of form closure, the feasible clamping domain is determined. Moreover, the accessibility and detachability conditions are related to the locator and clamp configurations. Additionally, the locator configuration affects the workpiece localization accuracy, and the fixturing closure is related to the clamp configuration. Thus, it is necessary for the improvement of the locating precision of workpieces to plan the fixturing configuration including the locator and clamp configuration, which is investigated in this chapter. To plan the fixturing configuration, three performance indices: (1) the locating robustness index used to evaluate the configurations of locators; (2) the stability index used to evaluate the capability to withstand any external disturbance wrench for the fixturing system; (3) the fixturing resultant index used to evaluate the robustness to the errors of locators and the stability under the external disturbance wrench for the fixturing system are defined. The corresponding three nonlinear programming methods used to plan the optimal fixturing configuration are proposed.

10.2 Mapping Between Error Space of Locators and Workpiece Locating Error Space

Consider a general workpiece as shown in Fig. 10.1. Choose reference frame $\{O\}$ fixed to the workpiece. Let $\{P\}$ and $\{F_i\}$ be the global frame and the ith locator frame fixed relative to it. The position of the ith contact point between the workpiece and the ith locator can

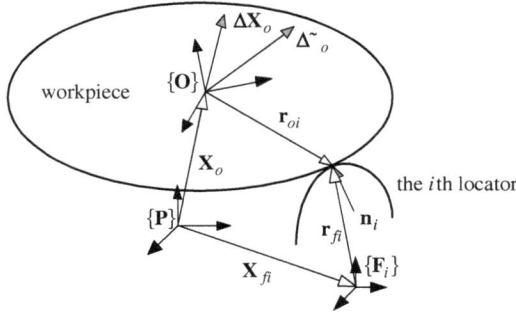

Fig. 10.1.   Fixture coordinate frames.

be described in two ways:

$$\mathbf{F}_i \left(\mathbf{X}_o, \boldsymbol{\Theta}_o, \mathbf{r}_{oi} \right) = \mathbf{X}_o + {}_o^p\mathbf{R}\mathbf{r}_{oi} \tag{10.1}$$

and

$$\mathbf{f}_i \left(\mathbf{X}_{fi}, \boldsymbol{\Theta}_{fi}, \mathbf{r}_{fi} \right) = \mathbf{X}_{fi} + {}_{fi}^p\mathbf{R}\mathbf{r}_{fi} \tag{10.2}$$

where $\mathbf{X}_o \in \Re^{3\times1}$ and $\boldsymbol{\Theta}_o \in \Re^{3\times1}$ ($\mathbf{X}_{fi} \in \Re^{3\times1}$ and $\boldsymbol{\Theta}_{fi} \in \Re^{3\times1}$) are the position and orientation of the workpiece (the ith locator) in the global frame $\{\mathbf{P}\}$, $\mathbf{r}_{oi} \in \Re^{3\times1}$ ($\mathbf{r}_{fi} \in \Re^{3\times1}$) is the position of the ith contact point between the workpiece and the ith locator in the workpiece frame $\{\mathbf{O}\}$ (the ith locator frame $\{\mathbf{F}_i\}$), ${}_o^p\mathbf{R} \in \mathbf{SO}(3)$ (${}_{fi}^p\mathbf{R} \in \mathbf{SO}(3)$) is the orientation matrix of the workpiece frame $\{\mathbf{O}\}$ (the ith locator frame $\{\mathbf{F}_i\}$) with respect to the global frame $\{\mathbf{P}\}$.

Thus, we have the following equation:

$$\mathbf{F}_i \left(\mathbf{X}_o, \boldsymbol{\Theta}_o, \mathbf{r}_{oi} \right) = \mathbf{f}_i \left(\mathbf{X}_{fi}, \boldsymbol{\Theta}_{fi}, \mathbf{r}_{fi} \right) \tag{10.3}$$

Since there may exist position errors $\boldsymbol{\Delta}\mathbf{X}_{fi}$ for the ith locator and $\boldsymbol{\Delta}\mathbf{X}_o$ for the workpiece, the contact between the workpiece and the ith locator will depart from its nominal position. Assume that $\boldsymbol{\Delta}\mathbf{X}_o \in \Re^{3\times1}$ ($\boldsymbol{\Delta}\boldsymbol{\Theta}_o \in \Re^{3\times1}$), and $\boldsymbol{\Delta}\mathbf{r}_{oi} \in \Re^{3\times1}$ are the deviations of the position $\mathbf{X}_o \in \Re^{3\times1}$ (orientation $\boldsymbol{\Theta}_o \in \Re^{3\times1}$) of the workpiece, and the position $\mathbf{r}_{oi} \in \Re^{3\times1}$ of the ith contact point, respectively. Let $\mathbf{F}_i \left(\mathbf{X}_o + \Delta\mathbf{X}_o, \boldsymbol{\Theta}_o + \Delta\boldsymbol{\Theta}_o, \mathbf{r}_{oi} + \Delta\mathbf{r}_{oi} \right)$ be the actual contact on the workpiece, then using Taylor expansion, we obtain (neglecting the higher order terms of errors)

$$
\mathbf{F}_i \left(\mathbf{X}_o + \mathbf{\Delta X}_o, \, \mathbf{\Theta}_o + \mathbf{\Delta \Theta}_o, \, \mathbf{r}_{oi} + \mathbf{\Delta r}_{oi} \right)
$$
$$
= \mathbf{F}_i \left(\mathbf{X}_o, \, \mathbf{\Theta}_o, \, \mathbf{r}_{oi} \right) + \frac{\partial \mathbf{F}_i}{\partial \mathbf{X}_o} \cdot \mathbf{\Delta X}_o + \frac{\partial \mathbf{F}_i}{\partial \mathbf{\Theta}_o} \cdot \mathbf{\Delta \Theta}_o + \frac{\partial \mathbf{F}_i}{\partial \mathbf{r}_{oi}} \cdot \mathbf{\Delta r}_{oi}
$$

$$(10.4)$$

where the 2nd term in the right side of Eq. (10.4) is the position error of the ith contact point resulting from the position error $\mathbf{\Delta X}_o$ of the workpiece, the 3rd term is the position error of the ith contact point resulting from the orientation error $\mathbf{\Delta \Theta}_o$ of the workpiece, and the 4th term is the position error of the ith contact point resulting from the workpiece geometric variation $\mathbf{\Delta r}_{oi}$ at the ith contact point.

Similarly, assume that $\mathbf{\Delta X}_{fi} \in \Re^{3 \times 1}$ ($\mathbf{\Delta \Theta}_{fi} \in \Re^{3 \times 1}$) and $\mathbf{\Delta r}_{fi} \in \Re^{3 \times 1}$ are the deviations of the position $\mathbf{X}_{fi} \in \Re^{3 \times 1}$ (orientation $\mathbf{\Theta}_{fi} \in \Re^{3 \times 1}$) of the ith locator, and the position $\mathbf{r}_{fi} \in \Re^{3 \times 1}$ of the ith contact point, respectively. Let $\mathbf{f}_i \left(\mathbf{X}_{fi} + \mathbf{\Delta X}_{fi}, \, \mathbf{\Theta}_{fi} + \mathbf{\Delta \Theta}_{fi}, \, \mathbf{r}_{fi} + \mathbf{\Delta r}_{fi} \right)$ be the contact on the ith locator, then using Taylor expansion, we obtain (neglecting the higher order terms of errors)

$$
\mathbf{f}_i \left(\mathbf{X}_{fi} + \mathbf{\Delta X}_{fi}, \quad \mathbf{\Theta}_{fi} + \mathbf{\Delta \Theta}_{fi}, \quad \mathbf{r}_{fi} + \mathbf{\Delta r}_{fi} \right)
$$
$$
= \mathbf{f}_i \left(\mathbf{X}_{fi}, \quad \mathbf{\Theta}_{fi}, \quad \mathbf{r}_{fi} \right) + \frac{\partial \mathbf{f}_i}{\partial \mathbf{X}_{fi}} \cdot \mathbf{\Delta X}_{fi} + \frac{\partial \mathbf{f}_i}{\partial \mathbf{\Theta}_{fi}}
$$
$$
\cdot \mathbf{\Delta \Theta}_{fi} + \frac{\partial \mathbf{f}_i}{\partial \mathbf{r}_{fi}} \cdot \mathbf{\Delta r}_{fi}
$$

$$(10.5)$$

where the 2nd term in the right side of Eq. (10.5) is the position error of the ith contact point resulting from the position error $\mathbf{\Delta X}_{fi}$ of the ith locator, the 3rd term is the position error of the ith contact point resulting from the orientation error $\mathbf{\Delta \Theta}_{fi}$ of the ith locator, and the 4th term is the position error of the ith contact point resulting from the locator geometric variation $\mathbf{\Delta r}_{fi}$ at the ith contact.

Although there are all kinds of errors during fixturing, the contact between the workpiece and locators must be maintained, which means we have the following equation:

$$\mathbf{F}_i \left(\mathbf{X}_o + \Delta \mathbf{X}_o, \, \mathbf{\Theta}_o + \Delta \mathbf{\Theta}_o, \, \mathbf{r}_{oi} + \Delta \mathbf{r}_{oi} \right)$$
$$= \mathbf{f}_i \left(\mathbf{X}_{fi} + \Delta \mathbf{X}_{fi}, \, \mathbf{\Theta}_{fi} + \Delta \mathbf{\Theta}_{fi}, \, \mathbf{r}_{fi} + \Delta \mathbf{r}_{fi} \right) \tag{10.6}$$

i.e.,

$$\mathbf{F}_i \left(\mathbf{X}_o, \mathbf{\Theta}_o, \mathbf{r}_{oi} \right) + \frac{\partial \mathbf{F}_i}{\partial \mathbf{X}_o} \cdot \Delta \mathbf{X}_o + \frac{\partial \mathbf{F}_i}{\partial \mathbf{\Theta}_o} \cdot \Delta \mathbf{\Theta}_o + \frac{\partial \mathbf{F}_i}{\partial \mathbf{r}_{oi}} \cdot \Delta \mathbf{r}_{oi}$$
$$= \mathbf{f}_i \left(\mathbf{X}_{fi}, \mathbf{\Theta}_{fi}, \mathbf{r}_{fi} \right) + \frac{\partial \mathbf{f}_i}{\partial \mathbf{X}_{fi}} \cdot \Delta \mathbf{X}_{fi} + \frac{\partial \mathbf{f}_i}{\partial \mathbf{\Theta}_{fi}} \cdot \Delta \mathbf{\Theta}_{fi} + \frac{\partial \mathbf{f}_i}{\partial \mathbf{r}_{fi}} \cdot \Delta \mathbf{r}_{fi} \tag{10.7}$$

Combining Eqs. (10.3) and (10.7), we obtain

$$\mathbf{U}_{oi} \cdot \Delta \mathbf{X} = \mathbf{U}_{fi} \cdot \Delta \mathbf{\Psi}_{fi} + {}^p_{fi}\mathbf{R}\Delta \mathbf{r}_{fi} - {}^p_o\mathbf{R}\Delta \mathbf{r}_{oi} \tag{10.8}$$

where

$$\mathbf{U}_{oi} = \left(\mathbf{I}_{3\times 3} \, \vdots \, -{}^p_o\mathbf{R}\mathbf{r}_{oi}\times \right) \in \Re^{3\times 6}, \Delta \mathbf{X} = \left(\Delta \mathbf{X}_o^T \, \Delta \mathbf{\Theta}_o^T \right)^T \in \Re^{6\times 1}$$

$\Delta \mathbf{X}$ is the displacement of the workpiece,

$$\mathbf{U}_{fi} = \left(\mathbf{I}_{3\times 3} \, \vdots \, -{}^p_{fi}\mathbf{R}\mathbf{r}_{fi}\times \right) \in \Re^{3\times 6}, \Delta \mathbf{\Psi}_{fi} = \left(\Delta \mathbf{X}_{fi}^T \, \Delta \mathbf{\Theta}_{fi}^T \right)^T \in \Re^{6\times 1}$$

$\Delta \mathbf{\Psi}_{fi}$ is the displacement of the ith locator frame which is fixed with respect to the ith locator, $\mathbf{I}_{3\times 3} \in \Re^{3\times 3}$ is the identity matrix.

Equation (10.8) describes the relationship between the position and orientation errors $\Delta \mathbf{X}$ of the workpiece, the position and orientation errors $\Delta \mathbf{\Psi}_{fi}$ of the ith locator, and the position errors $\Delta \mathbf{r}_{oi}$ ($\Delta \mathbf{r}_{fi}$) of the ith contact point on the workpiece (the ith locator). We call Eq. (10.8) the general locating error model in the workpiece-fixture system.

In practice, fixtures are tools to localize workpieces by using locators, thus all the locators are fixed and immovable in the workpiece-fixture systems so that the desired position and orientation of the workpiece can be obtained, which means that the position and orientation errors of locators can be neglected [19,26,37]. Thus, Eq. (10.8) can be simplified as

$$\mathbf{U}_{oi} \cdot \mathbf{\Delta X} = {}^{p}_{fi}\mathbf{R}\Delta\mathbf{r}_{fi} - {}^{p}_{o}\mathbf{R}\Delta\mathbf{r}_{oi} \tag{10.9}$$

Furthermore, it is known that the contact point $\mathbf{r}_{oi}$ is located on the datum surface of the workpiece. From Eq. (10.9), we can find two measures to improve the locating precision of fixtures, that is, increasing the manufacturing precisions of the locator and the datum surfaces, which is the reason why the data and locator surfaces are needed to be machined accurately.

Assume that the geometric position variation $\Delta\mathbf{r}_{oi}$ of the ith contact point on the workpiece is small enough so that the position errors of contact points on the datum surfaces of the workpiece can be neglected, then Eq. (10.9) can be simplified further as follows:

$$\mathbf{U}_{oi} \cdot \mathbf{\Delta X} = {}^{p}_{fi}\mathbf{R}\Delta\mathbf{r}_{fi} \tag{10.10}$$

Assuming that there exists only position error Δr_{n_i} of the ith contact point in the normal direction $\mathbf{n}_i$ for each locator, and the z-axis direction of the coordinate frame $\{\mathbf{F}_i\}$ coincides with the normal direction $\mathbf{n}_i$, i.e., ${}^{p}_{fi}\mathbf{R}\Delta\mathbf{r}_{fi} = \Delta r_{n_i} \cdot \mathbf{n}_i$.

For a locating system of m locators, we can represent the m equations in matrix form as follows:

$$\mathbf{G}_L^T\mathbf{\Delta X} = \mathbf{N}\mathbf{\Delta r} \tag{10.11}$$

where

$$\mathbf{G}_L = \begin{bmatrix} \mathbf{I}_{3\times3} & \cdots & \mathbf{I}_{3\times3} \\ \mathbf{r}_{o1}^p\times & \cdots & \mathbf{r}_{om}^p\times \end{bmatrix} \in \Re^{6\times3m}$$

$$\mathbf{\Delta X} = \begin{pmatrix} \mathbf{\Delta X}_o \\ \mathbf{\Delta\Theta}_o \end{pmatrix} \in \Re^{6\times1}$$

$$\mathbf{N} = \mathrm{diag}\begin{pmatrix} \mathbf{n}_1 & \cdots & \mathbf{n}_m \end{pmatrix} \in \Re^{3m\times m}$$

$$\mathbf{\Delta r} = \begin{pmatrix} \Delta r_{n_1} & \cdots & \Delta r_{n_m} \end{pmatrix} \in \Re^{m\times1}$$

Equation (10.11) can be rewritten as

$$\mathbf{W}_L\mathbf{\Delta X} = \mathbf{\Delta r} \tag{10.12}$$

where $\mathbf{W}_L = \mathbf{N}^T\mathbf{G}_L^T \in \Re^{m\times6}$ is referred to as the locating matrix.

Given the geometrical tolerance specification of a key machining feature on a workpiece, which can be converted into the allowed workpiece position and orientation deviation range [27], the allowable geometric variations for all of the locators can be determined by using Eq. (10.12). On the other hand, given the position tolerances of the contact points on locators, the position and orientation variation of the workpiece can be calculated as well. However, the different calculating methods need to be considered for different locating conditions of fixture, i.e., under-constrained localization, fully constrained localization, and over-constrained localization.

10.2.1 Fully Constrained Localization

When $m = 6$ and $rank\,(\mathbf{W}_L) = 6$, the fixture is referred to as fully constrained localization. Given the position errors $\Delta\mathbf{r}$ of contact points on locators, we can obtain the exact position and orientation errors $\Delta\mathbf{X}$ of the workpiece from Eq. (10.12) as follows:

$$\Delta\mathbf{X} = \mathbf{W}_L^{-1}\Delta\mathbf{r} \qquad (10.13)$$

where $\mathbf{W}_L^{-1}$ is the inverse of the locating matrix $\mathbf{W}_L$.

10.2.2 Over Constrained Localization

When $m > 6$, and $rank(\mathbf{W}_L) = 6$, the fixture is referred to as over-constrained localization. In this situation, given the position errors $\Delta\mathbf{r}$ of contact points on locators, the exact position and orientation errors $\Delta\mathbf{X}$ of the workpiece cannot be obtained from Eq. (10.12) but the least square estimation of the position and orientation errors $\Delta\mathbf{X}$ of the workpiece can be calculated as follows:

$$\Delta\mathbf{X} = \mathbf{W}_L^{+l}\Delta\mathbf{r} \qquad (10.14)$$

where $\mathbf{W}_L^{+l} = (\mathbf{W}_L^T\mathbf{W}_L)^{-1}\mathbf{W}_L^T$ is the left generalized inverse of the locating matrix $\mathbf{W}_L$.

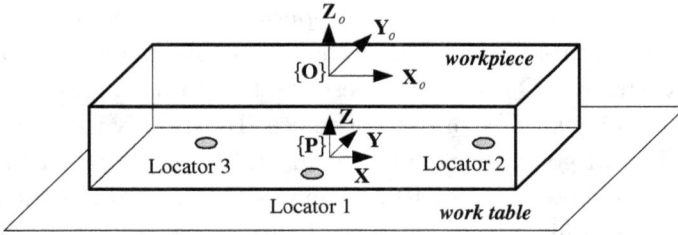

Fig. 10.2. Under-constrained localization.

10.2.3   Under-Constrained Localization

When $rank(\mathbf{W}_L) = m' < 6$, the fixture is referred to as under-constrained localization (Figure 10.2). Given the geometric errors $\mathbf{\Delta r}$ of locators, the position and orientation errors $\mathbf{\Delta X}$ of the workpiece are represented as follows:

$$\mathbf{\Delta X} = \mathbf{\Delta X}_v + \mathbf{\Delta X}_n \qquad (10.15)$$

where $\mathbf{\Delta X}_v = \mathbf{W}_L^{+r}\mathbf{\Delta r} \in \mathbf{V}$ is referred to as the minimum norm solution of the position and orientation errors $\mathbf{\Delta X}$ of the workpiece, the non-zero elements of the minimum norm solution give the estimable directions and values of the position and orientation errors $\mathbf{\Delta X}$ of the workpiece, $\mathbf{W}_L^{+r} = \mathbf{W}_L^T(\mathbf{W}_L\mathbf{W}_L^T)^{-1}$ is the right generalized inverse of the locating matrix $\mathbf{W}_L$, $\mathbf{V}$ is a subspace of Euclidean 6-D vector space $\mathbf{E}^6$, $\mathbf{\Delta X}_n = (\mathbf{I}_{6\times6} - \mathbf{W}_L^{+r}\mathbf{W}_L)\mathbf{\lambda} \in \mathbf{N}(\mathbf{W}_L)$ is referred to as the null solution of the position and orientation errors $\mathbf{\Delta X}$ of the workpiece, the non-zero elements of the null solution give the uncertain directions of the position and orientation errors $\mathbf{\Delta X}$ of the workpiece, the null space $\mathbf{N}(\mathbf{W}_L)$ of the locating matrix $\mathbf{W}_L$ is a special subspace of $\mathbf{E}^6$ (the dimension of the null space $\mathbf{N}(\mathbf{W}_L)$ is $6 - m'$), $\mathbf{N}(\mathbf{W}_L)$ and $\mathbf{V}$ are orthogonal complements of each other, we denotes this by $\mathbf{V}^\perp = \mathbf{N}(\mathbf{W}_L)$, or equivalently, $\mathbf{N}^\perp(\mathbf{W}_L) = \mathbf{V}$, $\mathbf{E}^6 = \mathbf{V} \oplus \mathbf{N}(\mathbf{W}_L)$, $\mathbf{I}_{6\times6}$ is an identity matrix, and $\mathbf{\lambda} = (\lambda_1 \cdots \lambda_6)^T \in \Re^{6\times1}$ is an arbitrary vector.

Now we give an example to explain the geometric meaning of Eq. (10.15).

Example 1. In a surface grinding operation, a workpiece is located on a machine table and in an under-constrained localization. The coordinate frame $\{O\}$ is used to describe the position and orientation of the machining surface of the workpiece. To simplify the calculation, without loss of generality, the machine table coordinate frame $\{P\}$ is defined, as shown in Fig. 10.2. The positions, errors in the normal direction, and unit normal vectors of three locators (equivalent) are as follows:

$$\mathbf{r}_{o1}^p = \begin{pmatrix} a_1 \\ b_1 \\ 0 \end{pmatrix}, \quad \mathbf{r}_{o2}^p = \begin{pmatrix} a_2 \\ b_2 \\ 0 \end{pmatrix}, \quad \mathbf{r}_{o3}^p = \begin{pmatrix} a_3 \\ b_3 \\ 0 \end{pmatrix}, \quad \Delta\mathbf{r} = \begin{pmatrix} \Delta r_{n_1} \\ \Delta r_{n_2} \\ \Delta r_{n_3} \end{pmatrix}$$

$$\mathbf{n}_1 = \begin{pmatrix} 0 \\ 0 \\ 1 \end{pmatrix}, \quad \mathbf{n}_2 = \begin{pmatrix} 0 \\ 0 \\ 1 \end{pmatrix}, \quad \mathbf{n}_3 = \begin{pmatrix} 0 \\ 0 \\ 1 \end{pmatrix}$$

where $a_1 = 0$, $a_2 = a$, $a_3 = -a$, $b_1 = -b$, $b_2 = b$, $b_3 = b$, and a and b are non-zero constants.

We obtain the locating matrix $\mathbf{W}_L$ as follows:

$$\mathbf{W}_L = \begin{bmatrix} 0 & 0 & 1 & b_1 & -a_1 & 0 \\ 0 & 0 & 1 & b_2 & -a_2 & 0 \\ 0 & 0 & 1 & b_3 & -a_3 & 0 \end{bmatrix} \qquad (10.16)$$

It is clear that $rank(\mathbf{W}_L) = m = 3$ as long as the three locators are not on the same straight line, in this situation, the right generalized reverse $\mathbf{W}_L^{+r}$ exists, and

$$\mathbf{W}_L^{+r} = \begin{bmatrix} 0 & 0 & 0 \\ 0 & 0 & 0 \\ \frac{1}{2} & \frac{1}{4} & \frac{1}{4} \\ -\frac{1}{2b} & \frac{1}{4b} & \frac{1}{4b} \\ 0 & -\frac{1}{2a} & \frac{1}{2a} \\ 0 & 0 & 0 \end{bmatrix} \qquad (10.17)$$

Further, using Eqs. (10.15) and (10.17), we obtain the locating errors of the workpiece as follows:

$$\Delta \mathbf{X} = \Delta \mathbf{X}_v + \Delta \mathbf{X}_n = \begin{pmatrix} \lambda_1 \\ \lambda_2 \\ \frac{1}{4}(2\Delta r_{n_1} + \Delta r_{n_2} + \Delta r_{n_3}) \\ \frac{1}{4b}(-2\Delta r_{n_1} + \Delta r_{n_2} + \Delta r_{n_3}) \\ \frac{1}{2a}(-\Delta r_{n_2} + \Delta r_{n_3}) \\ \lambda_6 \end{pmatrix} \tag{10.18}$$

where

$$\Delta \mathbf{X}_v = \mathbf{W}_L^{+r} \Delta \mathbf{r} = \begin{pmatrix} 0 \\ 0 \\ \frac{1}{4}(2\Delta r_{n_1} + \Delta r_{n_2} + \Delta r_{n_3}) \\ \frac{1}{4b}(-2\Delta r_{n_1} + \Delta r_{n_2} + \Delta r_{n_3}) \\ \frac{1}{2a}(-\Delta r_{n_2} + \Delta r_{n_3}) \\ 0 \end{pmatrix} \tag{10.19}$$

$$\Delta \mathbf{X}_n = (\mathbf{I}_{6\times 6} - \mathbf{W}_L^{+r} \mathbf{W}_L)\boldsymbol{\lambda} = \begin{pmatrix} \lambda_1 \\ \lambda_2 \\ 0 \\ 0 \\ 0 \\ \lambda_6 \end{pmatrix} \tag{10.20}$$

Equation (10.19) gives the translational error in **Z**-axis direction, the rotational errors around **X**- and **Y**-axis directions for the workpiece, and all of the three error elements can be determined. However, Eq. (10.20) shows the translational errors in **X**- and **Y**-axis directions, the rotational error around **Z**-axis direction for the workpiece, and all of the three error elements cannot be determined. Moreover, we can find that $\Delta \mathbf{X}_v$ and $\Delta \mathbf{X}_n$ are orthogonal to each other in Eqs. (10.19) and (10.20). All of the results are the characteristics of under-constrained localization.

Once the position and orientation precision of the workpiece is determined, the position errors of a set of critical points on the workpiece can be calculated to evaluate the variation of a feature. Assume that the position of the jth critical point is represented as $\mathbf{p}_j$, then the mapping between the position errors $\mathbf{\Delta p}_j$ of the jth critical evaluating point and the position and orientation errors $\mathbf{\Delta X}$ of the workpiece can be described as

$$\mathbf{\Delta p}_j = \left(\mathbf{I}_{3\times3} \vdots -\mathbf{p}_j \times \right) \mathbf{\Delta X} \tag{10.21}$$

In the quality evaluation of manufacturing, the position variations of critical points are often examined in a certain direction. For example, the position variations of a set of vertices in the normal direction of the workpiece surface may be considered when its flatness needs to be evaluated. Using Eq. (10.21), the mapping between the position variation v_{j_map} of the jth critical point along the direction $\mathbf{Q}_k$, and the position and orientation errors $\mathbf{\Delta X}$ of the workpiece is described as follows:

$$v_{j\_map} = \left(\mathbf{Q}_k^T \vdots -\mathbf{Q}_k^T \mathbf{p}_j \times \right) \mathbf{\Delta X} \tag{10.22}$$

Equation (10.22) can be rewritten as

$$v_{j\_map} = \boldsymbol{U} \cdot \boldsymbol{\Delta X} \tag{10.23}$$

where

$$\mathbf{U} = \left(\mathbf{Q}_k^T \vdots -\mathbf{Q}_k^T \mathbf{p}_j \times \right) = \begin{pmatrix} u_1 & u_2 & u_3 & u_4 & u_5 & u_6 \end{pmatrix}$$

$$\mathbf{\Delta X} = \left(\mathbf{\Delta X}_o^T, \mathbf{\Delta\Theta}_o^T \right)^T = \begin{pmatrix} \Delta x & \Delta y & \Delta z & \Delta\theta_x & \Delta\theta_y & \Delta\theta_z \end{pmatrix}^T$$

In general, the position and orientation errors are independent and random, thus the position variation magnitude v_j of the jth critical point along the direction $\mathbf{Q}_k$ can be described as

$$v_j = \sqrt{ \begin{array}{l} (u_1 \cdot \Delta x)^2 + (u_2 \cdot \Delta y)^2 + (u_3 \cdot \Delta z)^2 + (u_4 \cdot \Delta\theta_x)^2 \\ + (u_5 \cdot \Delta\theta_y)^2 + (u_6 \cdot \Delta\theta_z)^2 \end{array} } \tag{10.24}$$

10.3 Locator and Clamp Configuration Characteristics

10.3.1 Locator Configuration Characteristics

As mentioned in Section 10.2, the locating errors of workpieces are related to the positions and errors of locators. When the tolerance specifications of machining surfaces on a workpiece are given, the allowable geometric variations of locators can be determined. In reverse, given the position tolerances of the contact points on locators, the position and orientation errors of workpieces can be calculated. However, some locator configuration may affect workpiece accessibility to the fixture as well as detachability [1] from the fixture.

As described in Chapter 3, the motion constraint of an object with m points of frictionless contact can be represented as a convex polyhedron in space $\Re^6$. In fact, the motion constraint of the workpiece with m locators is a convex polyhedral cone which is called the free motion cone $\mathcal{K}$.

If the configuration of locators satisfies that there does not exist any non-zero feasible motion direction in the entire free motion cone $\mathcal{K}$, then the workpiece is neither accessible to the fixture nor detachable from the fixture.On the contrary, if the free motion cone $\mathcal{K}$ contains non-zero element, then the workpiece will be able to move in one or more related directions, which means the workpiece is accessible to the fixture as well as detachable from the fixture.

10.3.2 Clamp Configuration Characteristics under Accessibility and Detachability

According to the duality theory in convex analysis [40], the polar $\mathcal{K}^\circ$ of the free motion cone $\mathcal{K}$ can be described as

$$\mathcal{K}^\circ = \{\mathbf{f}_{cl} \in \Re^6 \,|\langle \mathbf{f}_{cl}, \mathbf{V}\rangle = \mathbf{f}_{cl}^T \mathbf{V} \le 0, \ \text{for all } \mathbf{V} = (\mathbf{v}_o^T \quad \boldsymbol{\omega}_o^T)^T \in \mathcal{K}\} \tag{10.25}$$

where $\mathbf{f}_{cl} \in \Re^h$ is the clamping force (wrench) ($h = 3$ for 2-D fixturing, $h = 6$ for 3-D fixturing).

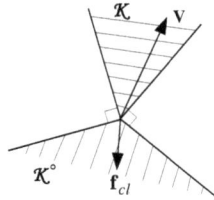

Fig. 10.3. Convex polyhedral cone and its polar.

Figure 10.3 shows the 2-D free motion cone $\mathcal{K}$ and its polar $\mathcal{K}°$. It is clear that the free motion cone $\mathcal{K}$ is a subspace of $\Re^6$ because of m locator constraints, and $\mathcal{K}°$ is the orthogonally complementary subspace. In general, for any non-empty closed convex $\mathcal{K}$, $\mathcal{K}°$ consists of all the vectors normal to $\mathcal{K}$ at $\mathbf{0}$, while $\mathcal{K}$ consists of all the vectors normal to $\mathcal{K}°$ at $\mathbf{0}$, and $(\mathcal{K}°)° = \mathcal{K}$. We call the polar $\mathcal{K}°$ the constrained cone.

The function of clamping is to balance the force system, which consists of the weight of the workpiece, cutting forces, and support forces of locators, so that the desired position and orientation of the workpiece can be maintained. To obtain the desired position and orientation, the clamping force $\mathbf{f}_{cl}$ should be inside of the constrained cone $\mathcal{K}°$. In other words, if there exist feasible motion directions of the workpiece after it is located, or if the workpiece is accessible to the fixture as well as detachable from the fixture, then the directions of the clamping force $\mathbf{f}_{cl}$ can be determined by using Eq. (10.25). Furthermore, from Eq. (10.25), it can be found that the clamp configuration is not a point but a feasible domain, which satisfies the constraints given by Eq. (10.25).

Example 2. Consider the planar positioning of a polygonal workpiece, as shown in Fig. 10.4(a). Here, three locators are used to locate the planar workpiece, the position and normal vectors of each locator are as follows:

$$\mathbf{r}_1 = \begin{pmatrix} a \\ 0 \end{pmatrix}, \quad \mathbf{r}_2 = \begin{pmatrix} 0 \\ b \end{pmatrix}, \quad \mathbf{r}_3 = \begin{pmatrix} c \\ d \end{pmatrix}, \quad \mathbf{n}_1 = \begin{pmatrix} 0 \\ 1 \end{pmatrix},$$

$$\mathbf{n}_2 = \begin{pmatrix} 1 \\ 0 \end{pmatrix}, \quad \mathbf{n}_3 = \begin{pmatrix} -\frac{\sqrt{2}}{2} \\ -\frac{\sqrt{2}}{2} \end{pmatrix}$$

Using the above position and normal vectors, and Eq. (3.4) described in Chapter 3, we can represent the motion constraint of the workpiece with three locators as

$$
\begin{bmatrix}
0 & 1 & a \\
1 & 0 & -b \\
-\frac{\sqrt{2}}{2} & -\frac{\sqrt{2}}{2} & -\frac{\sqrt{2}}{2}(d-c)
\end{bmatrix}
\begin{pmatrix}
v_x \\
v_y \\
\omega_z
\end{pmatrix} \geq 0
\tag{10.26}
$$

where a, b, c, and d are positive constants.

From Eq. (10.26), it can be found that the free motion cone $\mathcal{K}$ does not contain any other elements than $\mathbf{0}$, thus the corresponding locating configuration is neither accessible to the fixture nor detachable from the fixture when the workpiece trajectories are limited to planar motion only.

Now we change the locating configuration from Fig. 10.4(a–b), the position and normal vectors of each locator are as follows:

$$
\mathbf{r}_1 = \begin{pmatrix} a \\ 0 \end{pmatrix}, \quad
\mathbf{r}_2 = \begin{pmatrix} 0 \\ b \end{pmatrix}, \quad
\mathbf{r}_3 = \begin{pmatrix} e \\ 0 \end{pmatrix}, \quad
\mathbf{n}_1 = \begin{pmatrix} 0 \\ 1 \end{pmatrix}, \quad
\mathbf{n}_2 = \begin{pmatrix} 1 \\ 0 \end{pmatrix}
$$

$$
\mathbf{n}_3 = \begin{pmatrix} 0 \\ 1 \end{pmatrix}
$$

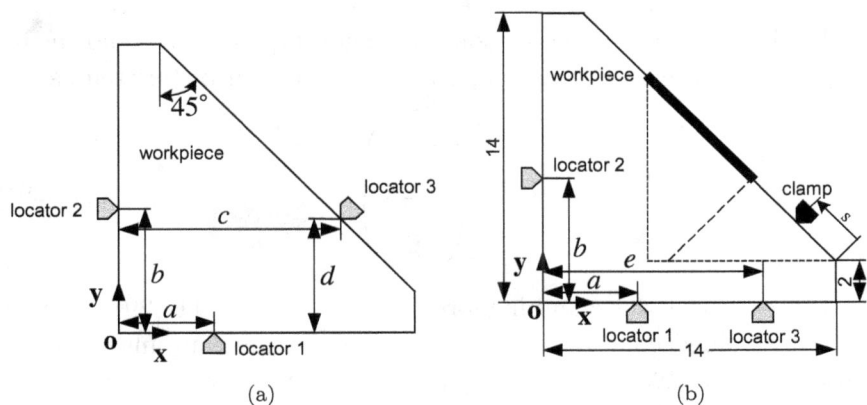

Fig. 10.4.   Planar fixturing.

Similarly, we can obtain the motion constraints of locating corresponding to Fig. 10.4(b) as follows:

$$
\begin{bmatrix} 0 & 1 & a \\ 1 & 0 & -b \\ 0 & 1 & e \end{bmatrix} \begin{pmatrix} v_x \\ v_y \\ \omega_z \end{pmatrix} \geq 0 \tag{10.27}
$$

where a, b, and e are positive constants.

From Eq. (10.27), we can find that the free motion cone $\mathcal{K}$ contains the following motion elements:

$$
\mathbf{V} = \begin{pmatrix} v_x \\ v_y \\ \omega_z \end{pmatrix} = \begin{pmatrix} 1 \\ 0 \\ 0 \end{pmatrix} \quad \text{or} \quad \mathbf{V} = \begin{pmatrix} 0 \\ 1 \\ 0 \end{pmatrix} \quad \text{or} \quad \mathbf{V} = \begin{pmatrix} u_x \\ u_y \\ 0 \end{pmatrix}
$$

where $u_x \geq 0$, $u_y \geq 0$, and $u_x^2 + u_y^2 = 1$.

The function of clamping is to resist the motions along these directions. Using Eq. (10.25), we obtain the constrained cone $\mathcal{K}^\circ$ as follows:

$$
\mathcal{K}^\circ = \left\{ \mathbf{f}_{cl} = \begin{pmatrix} f_x \\ f_y \\ m_z \end{pmatrix} \in \Re^3 \;\middle|\; \begin{cases} f_x \leq 0 \\ f_y \leq 0 \\ u_x f_x + u_y f_y \leq 0 \end{cases}, \quad \text{for all } \mathbf{V} = \begin{pmatrix} u_x \\ u_y \\ 0 \end{pmatrix} \in \mathcal{K} \right\} \tag{10.28}
$$

where f_x, f_y, and m_z are the elements of the clamping force.

If the clamp is located on the slanting edge as shown in Fig. 10.4(b), then the clamping force direction can be written as

$$
\hat{\mathbf{f}}_{cl} = \begin{pmatrix} -\frac{\sqrt{2}}{2} \\ -\frac{\sqrt{2}}{2} \\ 0 \end{pmatrix} \tag{10.29}
$$

It is clear that the clamping force described by Eq. (10.29) is inside the constrained cone $\mathcal{K}^\circ$, which implies that the clamping configuration is a feasible domain, *i.e.*, the slanting edge (see Fig. 10.4(b)), the position of clamping can be chosen at any place on the slanting edge.

Assume that $a = 2$, $b = 8$, $e = 12$, and the position of a clamp is represented by s for the fixturing configuration in Fig. 10.4(b), then the constraint matrix **GN** is written as

$$\mathbf{GN} = \begin{bmatrix} 0 & 1 & 0 & -\frac{\sqrt{2}}{2} \\ 1 & 0 & 1 & -\frac{\sqrt{2}}{2} \\ 2 & -8 & 12 & -\frac{\sqrt{2}}{2}\left(12 - \sqrt{2} \cdot s\right) \end{bmatrix} \tag{10.30}$$

According to the necessary and sufficient condition for form-closure fixturing given in Chapter 3, assuming that there exists a vector $\mathbf{y} = \left(y_1, y_2, y_3, y_4 \right)^T > 0$ such that $\mathbf{GNy} = \mathbf{0}$, then, using Eq. (10.30), we yield the position constraint of the clamp as follows:

$$4\sqrt{2} < s < 9\sqrt{2} \tag{10.31}$$

Equation (10.31) means that the position of the clamp can be chosen only in the bold solid line areas as shown in Fig. 10.4(b) to totally constrain the workpiece.

Example 3. Consider 3-D case of a polyhedral workpiece located with six locators (see Fig. 10.5(a)) (the ends of the lines coming out from the triangles represent the positions of contact points between the workpiece and locators/clamps), the workpiece is a cubic rigid

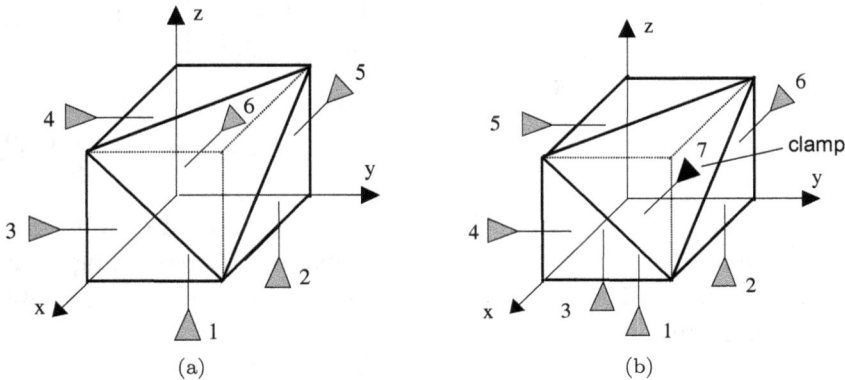

Fig. 10.5.   Space fixturing.

body with one corner cut out. The position and normal vectors of each locator are as follows:

$$\mathbf{r}_1 = \begin{pmatrix} 0.8a \\ 0.5a \\ 0 \end{pmatrix}, \quad \mathbf{r}_2 = \begin{pmatrix} 0.2a \\ 0.5a \\ 0 \end{pmatrix}, \quad \mathbf{r}_3 = \begin{pmatrix} 0.8a \\ 0 \\ 0.5a \end{pmatrix}, \quad \mathbf{r}_4 = \begin{pmatrix} 0.2a \\ 0 \\ 0.5a \end{pmatrix}$$

$$\mathbf{r}_5 = \begin{pmatrix} 0 \\ 0.5a \\ 0.5a \end{pmatrix}, \quad \mathbf{r}_6 = \begin{pmatrix} 0.8a \\ 0.8a \\ 0.4a \end{pmatrix}, \quad \mathbf{n}_1 = \begin{pmatrix} 0 \\ 0 \\ 1 \end{pmatrix}, \quad \mathbf{n}_2 = \begin{pmatrix} 0 \\ 0 \\ 1 \end{pmatrix}$$

$$\mathbf{n}_3 = \begin{pmatrix} 0 \\ 1 \\ 0 \end{pmatrix}, \quad \mathbf{n}_4 = \begin{pmatrix} 0 \\ 1 \\ 0 \end{pmatrix}, \quad \mathbf{n}_5 = \begin{pmatrix} 1 \\ 0 \\ 0 \end{pmatrix}, \quad \mathbf{n}_6 = \begin{pmatrix} -\frac{\sqrt{3}}{3} \\ -\frac{\sqrt{3}}{3} \\ -\frac{\sqrt{3}}{3} \end{pmatrix}, \quad a > 0$$

Using the above position and normal vectors, and Eq. (3.4) described in Chapter 3, we can represent the motion constraint of the workpiece with six locators as

$$\begin{bmatrix} 0 & 0 & 1 & 0.5a & -0.8a & 0 \\ 0 & 0 & 1 & 0.5a & -0.2a & 0 \\ 0 & 1 & 0 & -0.5a & 0 & 0.8a \\ 0 & 0 & 0 & -0.5a & 0 & 0.2a \\ 1 & 0 & 0 & 0 & 0.5a & -0.5a \\ -\frac{\sqrt{3}}{3} & -\frac{\sqrt{3}}{3} & -\frac{\sqrt{3}}{3} & -\frac{2\sqrt{3}}{15}a & \frac{2\sqrt{3}}{15}a & 0 \end{bmatrix} \begin{pmatrix} v_x \\ v_y \\ v_z \\ \omega_x \\ \omega_y \\ \omega_z \end{pmatrix} \geq 0$$

$$(10.32)$$

From Eq. (10.32), we can find that the free motion cone $\mathcal{K}$ does not contain any other elements than $\mathbf{0}$, thus the corresponding locating configuration is neither accessible to the fixture nor detachable from the fixture.

Now we change the locating configuration from Fig. 10.5(a–b), the position and normal vectors of each locator are as follows:

$$\mathbf{r}_1 = \begin{pmatrix} 0.8a \\ 0.5a \\ 0 \end{pmatrix}, \quad \mathbf{r}_2 = \begin{pmatrix} 0.2a \\ 0.8a \\ 0 \end{pmatrix}, \quad \mathbf{r}_3 = \begin{pmatrix} 0.2a \\ 0.2a \\ 0 \end{pmatrix}, \quad \mathbf{r}_4 = \begin{pmatrix} 0.8a \\ 0 \\ 0.5a \end{pmatrix}$$

$$\mathbf{r}_5 = \begin{pmatrix} 0.2a \\ 0 \\ 0.5a \end{pmatrix}, \quad \mathbf{r}_6 = \begin{pmatrix} 0 \\ 0.5a \\ 0.5a \end{pmatrix}, \quad \mathbf{n}_1 = \begin{pmatrix} 0 \\ 0 \\ 1 \end{pmatrix}, \quad \mathbf{n}_2 = \begin{pmatrix} 0 \\ 0 \\ 1 \end{pmatrix}$$

$$\mathbf{n}_3 = \begin{pmatrix} 0 \\ 0 \\ 1 \end{pmatrix}, \quad \mathbf{n}_4 = \begin{pmatrix} 0 \\ 1 \\ 0 \end{pmatrix}, \quad \mathbf{n}_5 = \begin{pmatrix} 0 \\ 1 \\ 0 \end{pmatrix}, \quad \mathbf{n}_6 = \begin{pmatrix} 1 \\ 0 \\ 0 \end{pmatrix}$$

The motion constraints of locating corresponding to Fig. 10.5(b) can be described as follows:

$$\begin{bmatrix} 0 & 0 & 1 & 0.5a & -0.8a & 0 \\ 0 & 0 & 1 & 0.8a & -0.2a & 0 \\ 0 & 0 & 1 & 0.2a & -0.2a & 0 \\ 0 & 1 & 0 & -0.5a & 0 & 0.8a \\ 0 & 1 & 0 & -0.5a & 0 & 0.2a \\ 1 & 0 & 0 & 0 & 0.5a & -0.5a \end{bmatrix} \begin{pmatrix} v_x \\ v_y \\ v_z \\ \omega_x \\ \omega_y \\ \omega_z \end{pmatrix} \geq 0 \qquad (10.33)$$

From Eq. (10.33), we can find that the free motion cone $\mathcal{K}$ contains the following motion elements:

$$\mathbf{V} = \begin{pmatrix} v_x & v_y & v_z & \omega_x & \omega_y & \omega_z \end{pmatrix}^T = \begin{pmatrix} 1 & 0 & \cdots & 0 \end{pmatrix}^T$$

$$\text{or} \quad \mathbf{V} = \begin{pmatrix} 0 & 1 & 0 & \cdots & 0 \end{pmatrix}^T$$

$$\text{or} \quad \mathbf{V} = \begin{pmatrix} 0 & 0 & 1 & 0 & 0 & 0 \end{pmatrix}^T$$

$$\text{or} \quad \mathbf{V} = \begin{pmatrix} u_x & u_y & u_z & 0 & 0 & 0 \end{pmatrix}^T$$

where $u_x \geq 0$, $u_y \geq 0$, $u_z \geq 0$, and $u_x^2 + u_y^2 + u_z^2 = 1$.

According to Eq. (10.25), we obtain the constrained cone $\mathcal{K}^\circ$ as follows:

$$
\mathcal{K}^\circ = \left\{ \mathbf{f}_{cl} = \begin{pmatrix} f_x \\ f_y \\ f_z \\ m_x \\ m_y \\ m_z \end{pmatrix} \in \Re^6 \;\middle|\; \begin{cases} f_x \leq 0 \\ f_y \leq 0 \\ f_z \leq 0 \\ u_x f_x + u_y f_y + u_z f_z \leq 0 \end{cases}, \quad \text{for all } \mathbf{V} = \begin{pmatrix} u_x \\ u_y \\ u_z \\ 0 \\ 0 \\ 0 \end{pmatrix} \in \mathcal{K} \right\}
$$

$$(10.34)$$

where f_x, f_y, f_z, m_x, m_y, and m_z are the elements of the clamping force.

If the clamp is located on the slanting surface as shown in Fig. 10.5(b), then the clamping force direction can be written as

$$
\hat{\mathbf{f}}_{cl} = \left(-\tfrac{\sqrt{3}}{3} \;\; -\tfrac{\sqrt{3}}{3} \;\; -\tfrac{\sqrt{3}}{3} \;\; 0 \;\; 0 \;\; 0 \right)^T \tag{10.35}
$$

It is clear that the clamping force described by Eq. (10.35) is inside the constrained cone $\mathcal{K}^\circ$, which implies that the clamping configuration is a feasible domain, i.e., the slanting surface (see Fig. 10.5(b)), the position of clamping can be chosen at any place on the slanting surface.

Assume that the position of a clamp is represented by $\mathbf{r}_{cl} = \left(x, y, 2a - x - y \right)^T$ as shown in Fig. 10.5(b), then the constraint matrix **GN** is written as

$$
\mathbf{GN} =
\begin{bmatrix}
0 & 0 & 0 & 0 & 0 & 1 & -\frac{\sqrt{3}}{3} \\
0 & 0 & 0 & 1 & 1 & 0 & -\frac{\sqrt{3}}{3} \\
1 & 1 & 1 & 0 & 0 & 0 & -\frac{\sqrt{3}}{3} \\
0.5a & 0.8a & 0.2a & 0.5a & -0.5a & 0 & \frac{\sqrt{3}}{3}(2a-x-2y) \\
-0.8a & -0.2a & -0.2a & 0 & 0 & 0.5a & -\frac{\sqrt{3}}{3}(2a-2x-y) \\
0 & 0 & 0 & 0.8a & 0.2a & -0.5a & \frac{\sqrt{3}}{3}(y-x)
\end{bmatrix}
$$

$$(10.36)$$

According to the necessary and sufficient condition for form-closure fixturing given in Chapter 3, assuming that there exists a vector $\mathbf{y} = \left(y_1, \ldots, y_7\right)^T > 0$ such that $\mathbf{GNy} = \mathbf{0}$, then, using Eq. (10.36), the position constraints of the clamp are as follows:

$$
\begin{cases}
2x + y - 1.7a > 0 \\
3y - 1.7a > 0 \\
6.3a - 4x - 5y > 0 \\
|y - x| < 0.3a \\
x + y > a
\end{cases}
\tag{10.37}
$$

The corresponding feasible clamping domain is shown in Fig. 10.6, which means that the position of the clamp can be chosen only in the closure areas inside the bold solid line (see Fig. 10.6) to totally constrain the workpiece.

10.4 Evaluation Indices of Fixturing

10.4.1 Evaluation Index of Locator Configurations

A fixture is a device to hold a workpiece in manufacturing, assembling, and inspecting. The position and orientation precision of

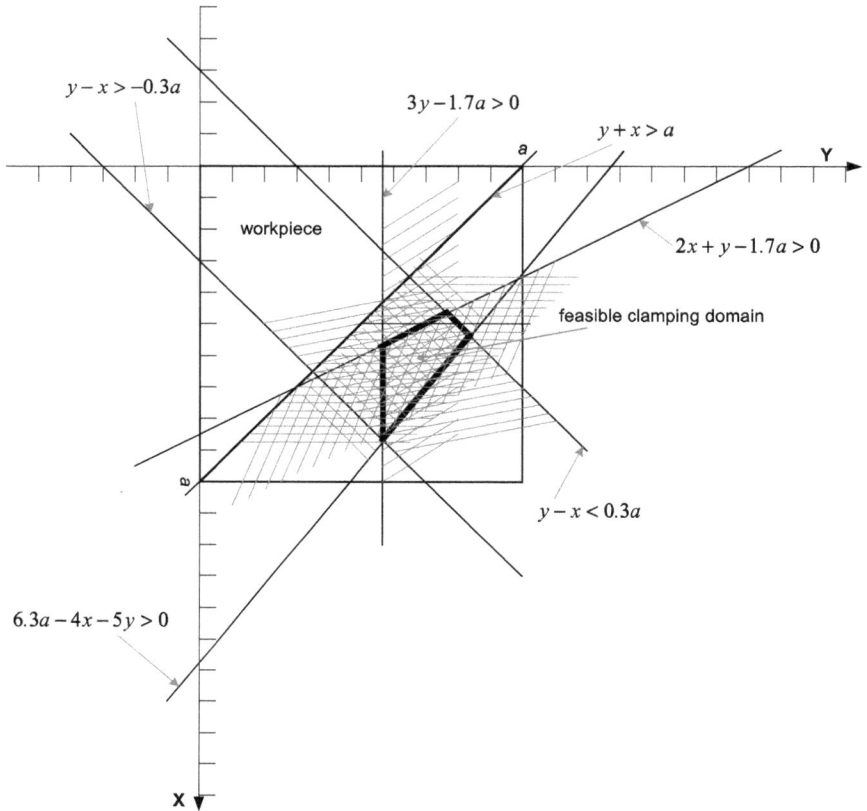

Fig. 10.6.   Planform of the workpiece.

workpieces are related to the tolerances of locators. Moreover, given the same tolerances of locators, the locating errors of the workpiece may vary with different configurations of locators. The mapping relation between the locating errors and the errors of locators can be described using Eq. (10.12). As mentioned above, the locating errors can be determined for the given errors of locators and the configuration of locators. In general, the errors of all locators are independent and random. In the viewpoint of statistics, the six elements of the locating errors for fully constrained localization can be described as

$$\Delta x = \sqrt{\left(\mathbf{W}_L^{-1}\left(1,\ 1\right)\cdot\Delta\mathbf{r}\left(1\right)\right)^2 + \cdots + \left(\mathbf{W}_L^{-1}\left(1,\ 6\right)\cdot\Delta\mathbf{r}\left(6\right)\right)^2}$$

$$\Delta y = \sqrt{\left(\mathbf{W}_L^{-1}\left(2,\ 1\right)\cdot\Delta\mathbf{r}\left(1\right)\right)^2 + \cdots + \left(\mathbf{W}_L^{-1}\left(2,\ 6\right)\cdot\Delta\mathbf{r}\left(6\right)\right)^2}$$

$$\Delta z = \sqrt{\left(\mathbf{W}_L^{-1}\left(3,1\right)\cdot\Delta\mathbf{r}\left(1\right)\right)^2 + \cdots + \left(\mathbf{W}_L^{-1}\left(3,6\right)\cdot\Delta\mathbf{r}\left(6\right)\right)^2}$$

$$\Delta\theta_x = \sqrt{\left(\mathbf{W}_L^{-1}\left(4,\ 1\right)\cdot\Delta\mathbf{r}\left(1\right)\right)^2 + \cdots + \left(\mathbf{W}_L^{-1}\left(4,\ 6\right)\cdot\Delta\mathbf{r}\left(6\right)\right)^2}$$

$$\Delta\theta_y = \sqrt{\left(\mathbf{W}_L^{-1}\left(5,\ 1\right)\cdot\Delta\mathbf{r}\left(1\right)\right)^2 + \cdots + \left(\mathbf{W}_L^{-1}\left(5,\ 6\right)\cdot\Delta\mathbf{r}\left(6\right)\right)^2}$$

$$\Delta\theta_z = \sqrt{\left(\mathbf{W}_L^{-1}\left(6,\ 1\right)\cdot\Delta\mathbf{r}\left(1\right)\right)^2 + \cdots + \left(\mathbf{W}_L^{-1}\left(6,\ 6\right)\cdot\Delta\mathbf{r}\left(6\right)\right)^2}$$

$$(10.38)$$

where $\mathbf{W}_L^{-1}(i,j)$ is the ith row, jth column element of the inverse matrix $\mathbf{W}_L^{-1}$ of the locating matrix $\mathbf{W}_L$, and $\Delta\mathbf{r}(k)$ is the error of the kth locator.

Similarly, for the over-constrained localization, the six elements of the locating errors can be written as

$$\Delta x = \sqrt{\left(\mathbf{W}_L^{+l}\left(1,\ 1\right)\cdot\Delta\mathbf{r}\left(1\right)\right)^2 + \cdots + \left(\mathbf{W}_L^{+l}\left(1,\ m\right)\cdot\Delta\mathbf{r}\left(m\right)\right)^2}$$

$$\Delta y = \sqrt{\left(\mathbf{W}_L^{+l}\left(2,\ 1\right)\cdot\Delta\mathbf{r}\left(1\right)\right)^2 + \cdots + \left(\mathbf{W}_L^{+l}\left(2,\ m\right)\cdot\Delta\mathbf{r}\left(m\right)\right)^2}$$

$$\Delta z = \sqrt{\left(\mathbf{W}_L^{+l}\left(3,\ 1\right)\cdot\Delta\mathbf{r}\left(1\right)\right)^2 + \cdots + \left(\mathbf{W}_L^{+l}\left(3,\ m\right)\cdot\Delta\mathbf{r}\left(m\right)\right)^2}$$

$$\Delta\theta_x = \sqrt{\left(\mathbf{W}_L^{+l}\left(4,\ 1\right)\cdot\Delta\mathbf{r}\left(1\right)\right)^2 + \cdots + \left(\mathbf{W}_L^{+l}\left(4,\ m\right)\cdot\Delta\mathbf{r}\left(m\right)\right)^2}$$

$$\Delta\theta_y = \sqrt{\left(\mathbf{W}_L^{+l}\left(5,\ 1\right)\cdot\Delta\mathbf{r}\left(1\right)\right)^2 + \cdots + \left(\mathbf{W}_L^{+l}\left(5,\ m\right)\cdot\Delta\mathbf{r}\left(m\right)\right)^2}$$

$$\Delta\theta_z = \sqrt{\left(\mathbf{W}_L^{+l}\left(6,\ 1\right)\cdot\Delta\mathbf{r}\left(1\right)\right)^2 + \cdots + \left(\mathbf{W}_L^{+l}\left(6,\ m\right)\cdot\Delta\mathbf{r}\left(m\right)\right)^2}$$

$$(10.39)$$

where $\mathbf{W}_L^{+l}(i, j)$ is the ith row, jth column element of the left generalized inverse matrix $\mathbf{W}_L^{+l} = \left(\mathbf{W}_L^T\mathbf{W}_L\right)^{-1}\mathbf{W}_L^T \in \Re^{6\times m}$ of the locating matrix $\mathbf{W}_L$.

However, because the eliminated degrees of freedom of the workpiece is less than six for the under-constrained localization, only partial elements of the locating errors can be determined corresponding to the special constrained directions. In Example 1, the related elements of the locating errors can be represented as

$$\Delta x = \sqrt{\left(\mathbf{W}_L^{+r}(3,\ 1)\cdot\Delta\mathbf{r}(1)\right)^2 + \cdots + \left(\mathbf{W}_L^{+r}(3, m)\cdot\Delta\mathbf{r}(m)\right)^2}$$

$$\Delta\theta_x = \sqrt{\left(\mathbf{W}_L^{+r}(4,\ 1)\cdot\Delta\mathbf{r}(1)\right)^2 + \cdots + \left(\mathbf{W}_L^{+r}(4, m)\cdot\Delta\mathbf{r}(m)\right)^2}$$

$$\Delta\theta_y = \sqrt{\left(\mathbf{W}_L^{+r}(5,\ 1)\cdot\Delta\mathbf{r}(1)\right)^2 + \cdots + \left(\mathbf{W}_L^{+r}(5, m)\cdot\Delta\mathbf{r}(m)\right)^2}$$

$$(10.40)$$

where $\mathbf{W}_L^{+r}(i, j)$ is the ith row, jth column element of the right generalized inverse matrix $\mathbf{W}_L^{+r} = \mathbf{W}_L^T\left(\mathbf{W}_L\mathbf{W}_L^T\right)^{-1} \in \Re^{6\times m}$ of the locating matrix $\mathbf{W}_L$ ($m = 3$ in Example 1 of this chapter). The remaining elements of the locating errors, which cannot be determined, are not related to the requirement of machining.

Equations (10.38), (10.39), and (10.40) show that elements of the locating errors of the workpiece depend on the tolerances and configurations of locators.

It is well known that locators are passive elements in the workpiece-fixture system. The functions of locators are to locate workpieces. However, clamps are active elements, and their functions are to maintain the obtained position and orientation of workpieces [26]. They cannot be applied to locate the workpiece. In order to obtain the exact position and orientation of a workpiece, on the one hand, the errors of locators should be as small as possible. Generally, the errors of locators cannot be fully eliminated because decreasing the errors means increasing the cost of manufacturing. On the other hand, the locating errors of workpieces are related to the locator configurations as well. To plan an optimal locator configuration so that the locating errors of workpieces can be minimized, we define the norm of the locating errors of the workpiece as the evaluation index of the locator configurations, that is,

$$\Omega_R = \|\Delta\mathbf{X}\|^2 \tag{10.41}$$

Although there may exist errors for each locator, their influence on the locating errors of the workpiece can be minimized by using the index Ω_R to plan the optimal locator configuration. The index Ω_R is called the locating robustness index of fixtures.

10.4.2   Stability Index of Fixturing

Once a workpiece is fully located by passive locators and clamped in a fixture by active clamps, that is, the fixturing is form closure. From the viewpoint of force, the form closure fixturing means that the fixture can balance any external wrench exerted on the workpiece by using a set of positive contact forces of locators and clamps.

However, the capability of withstanding the external wrench varies for different contact configurations [10, 16–18, 25, 28, 32, 33]. Obviously, in order to obtain a more stable fixturing, finding the optimal fixture contact configuration is very important. In Chapter 4, we defined a stability index described in Eq. (4.11), here we rewrite it as

$$\Omega_S = \sqrt{det\left(\mathbf{G}\mathbf{G}^T\right)} \tag{10.42}$$

The index Ω_S is called the stability index of fixturing.

10.4.3   Fixturing Resultant Index

There may be more than one index for evaluating the fixturing quality of a workpiece from different viewpoints. In the configuration planning of fixtures, we hope to achieve the best overall effect of small localization errors and great fixturing stability. Since both objectives of minimizing localization errors and maximizing fixturing stability are not reached simultaneously, a compromise is required. Therefore, to reach the best overall effect, we define the corresponding index as follows:

$$\Omega = \frac{w\Omega_R}{\Omega_S} \tag{10.43}$$

In general, the value of the index Ω_R is much smaller than the value of the stability index Ω_S. To increase the comparability of the locating robustness index Ω_R and the stability index Ω_S, a weight factor w is used in Eq. (10.43).

10.5   Configuration Planning of Fixturing

10.5.1   Constraints

No matter which index is applied in the configuration planning of fixtures, the planning must satisfy a series of constraints that can be described as follows:

(1) *Rank of Locating Matrix*: On the requirement of machining, we must constrain partially or fully the motion of a workpiece by properly arranging locators to locate it. Degrees of constraints of a workpiece depend not only on the number of locators but also on the configurations of locators, that is, the degrees of constraints of a workpiece are related to the rank of the locating matrix $\mathbf{W}_L$. Thus, the degrees of constraints can be written as

$$rank(\mathbf{W}_L) = r. \tag{10.44}$$

When $r = m < 6$, the workpiece is under-constrained localization; when $r = m = 6$, the workpiece is fully constrained localization; when $r = 6$ and $m > 6$, the workpiece is over-constrained localization.

The constraint Eq. (10.44) means that we cannot choose randomly the positions of the locators. For example, we can choose the layout of locators as shown in Fig. 10.2 for the plane grinding operation where $rank(\mathbf{W}_L) = 3$. However, if we hope to drill a hole at the desired position in the workpiece, then the chosen layout of locators must guarantee the rank of the locating matrix is full (i.e., $rank(\mathbf{W}_L) = 6$). It is clear that the layout of locators in Fig. 10.2 is not the expected one. "3-2-1" locator schemes [26] which guarantee the rank of the locating matrix is full are good guidelines for choosing the layout of locators for the workpiece with regular shapes such as rectangular objects.

(2) *Locating Surfaces*: Theoretically, locators can be arranged randomly on the surfaces of the workpiece as long as Eq. (10.44) is satisfied. However, because of different machining requirements, the positions of locators cannot be selected at an arbitrary place on the surfaces of the workpiece. For example, the contacts of locators cannot be chosen on the machining surface. The locating surface constraint for the ith locator $\mathbf{r}_i$ is written as

$$\mathbf{r}_i = \left\{ (x_i, y_i, z_i)^T \mid S_j(x_i, y_i, z_i) = 0, \ i = 1, \ldots, m; \ j = 1, \ldots, J \right\}$$
(10.45)

where $S_j(x_i, y_i, z_i) = 0$ means that the ith locator $\mathbf{r}_i$ is on the jth surface of the workpiece.

(3) *Clamping Domain*: After the passive locators locate a workpiece, the active clamps will maintain the obtained position and orientation of the workpiece by exerting clamping forces at the clamping points. If the locators and clamps can constrain totally the workpiece, then the fixturing is form-closure. It is clear that there is more than one clamping position which satisfy form closure [2]. In other words, there exists a clamping domain satisfying form closure [2, 38]. The feasible clamping domain should be determined using the principle of form-closure fixturing [2, 34, 38]. The feasible clamping domain constraints, which are derived from the necessary and sufficient condition, are written as

$$\begin{cases} \mathbf{r}_h = \{(x_h, y_h, z_h)^T \mid SC_v(x_h, y_h, z_h) = 0, \ h = m + 1, \ldots, q; \\ v = 1, \ldots, V\} \\ g_k(\mathbf{r}_1, \ldots, \mathbf{r}_q) > 0, \quad k = 1, \ldots, K \end{cases}$$
(10.46)

where $SC_v(x_h, y_h, z_h) = 0$ means that the hth clamp $\mathbf{r}_h$ is on the vth surface of the workpiece and k inequalities $g_k(\mathbf{r}_1, \ldots, \mathbf{r}_q) > 0$ form the feasible clamping domains.

It should be noted that the workpiece accessible to a fixture and the detachable from a fixture must be taken into account so that the locating surfaces and the clamping domain can be determined. In addition, the goal of clamping is to oppose the remaining motion possibility of the workpiece after it is located, thus, clamping directions must be determined appropriately, so that the corresponding fixturing is form closure.

10.5.2   Planning Methods

According to the different objective functions and constraints used in configuration planning, the methods of configuration planning are divided into the following three types:

(1) *Minimizing Locating Error Planning Method*: In this method, the objective function is the locating robustness index Ω_R. The corresponding planning method is described as

$$\underset{\substack{\mathbf{r}_i \in \mathcal{D} \\ (t=1,\dots,m)}}{\text{minimize}} \ \Omega_R \left(\mathbf{r}_1, \cdots, \mathbf{r}_m \right)$$

subject to ($\mathcal{D}$)

$$\begin{cases} rank\left(\mathbf{W}_L \right) = r \\ \mathbf{r}_i = \{ (x_i, y_i, z_i)^T \mid S_j \left(x_i, y_i, z_i \right) = 0, i = 1, \cdots, m; j = 1, \cdots, J \} \end{cases}$$

$$(10.47)$$

Thus, the problem of finding the optimal locator configuration is expressed as constrained nonlinear programming. By using this method, an optimal locator configuration can be found to minimize the locating errors of the workpiece under the constraints.

(2) *Stability-based Planning Method*: In this method, the objective function is the stability index Ω_S. The corresponding planning method is represented as follows:

$$\underset{\substack{\mathbf{r}_i \in \mathcal{D} \\ (t=1,\dots,q)}}{\text{maximize}} \ \Omega_S(\mathbf{r}_1, \dots, \mathbf{r}_q)$$

subject to($\mathcal{D}$)

$$\begin{cases} rank\left(\mathbf{W}_L \right) = r \\ \mathbf{r}_i = \{ (x_i, y_i, z_i)^T \mid S_j(x_i, y_i, z_i) = 0, i = 1, \dots, m; \\ \qquad j = 1, \dots, J \} \\ \mathbf{r}_h = \{ (x_h, y_h, z_h)^T \mid SC_v(x_h, y_h, z_h) = 0, h = m+1, \dots, q; \\ \qquad v = 1, \dots, V \} \\ g_k \left(\mathbf{r}_1, \dots, \mathbf{r}_q \right) > 0; k = 1, \dots, K \end{cases} \quad (10.48)$$

Here, the problem of the configuration planning is also changed into constrained nonlinear programming where the constraints include the clamping domain constraints besides the constraints mentioned in Eq. (10.47). By using this method, an optimal fixturing configuration can be found with the strongest capability to withstand the external disturbance wrench exerted on the workpiece under the constraints.

(3) *Fixturing Resultant Index-based Planning Method*: In this method, the objective function is the fixturing resultant evaluation index Ω. The corresponding planning method is stated as follows:

$$\underset{\substack{\mathbf{r}_i \in \mathcal{D} \\ (t=1,\ldots,q)}}{\text{minimize}} \left(\Omega \left(\mathbf{r}_1, \ldots, \mathbf{r}_q \right) = \frac{w\Omega_R \left(\mathbf{r}_1, \ldots, \mathbf{r}_m \right)}{\Omega_S \left(\mathbf{r}_1, \ldots, \mathbf{r}_q \right)} \right)$$

subject to $(\mathcal{D})$

$$\begin{cases} rank\left(\mathbf{W}_L\right) = r \\ \mathbf{r}_i = \{(x_i, y_i, z_i)^T \mid S_j(x_i, y_i, z_i) = 0, i = 1, \ldots, m; j = 1, \ldots, J\} \\ \mathbf{r}_h = \{(x_h, y_h, z_h)^T \mid SC_v(x_h, y_h, z_h) = 0, h = m+1, \ldots, q, \\ \qquad v = 1, \ldots, V\} \\ g_k\left(\mathbf{r}_1, \ldots, \mathbf{r}_q\right) > 0, k = 1, \ldots, K \end{cases}$$

$$(10.49)$$

Similarly, the third planning method is still constrained nonlinear programming with the same constraints as the second planning method. By using this method, an optimal fixturing configuration can be found to achieve the best overall effect of small localization errors and great fixturing stability under the constraints.

Example 4. Consider a workpiece fixtured with six locators and one clamp, the locator and clamp configuration is shown in Fig. 10.5(b). Assume that the side length of the cubic workpiece (a cubic rigid body with one corner cut out) is 500 (units), and all of the seven locators have the same position errors in the normal directions. Now we discuss the configuration planning of fixturing for the workpiece.

We use the function "CONSTR" of Matlab optimization toolbox version 5.0 to solve the constrained nonlinear programming problems in this example. The function "CONSTR" is an effective

Table 10.1. Optimal locator configuration (using locating robustness index as objective function).

No.	Locators' coordinates	Notation
1	(300, 250, 0); (100, 300, 0); (100, 100, 0); (300, 0, 250); (100, 0, 250); (0, 250, 250)	Initial Configuration
2	(317.8, 250.6, 0); (84.4, 323.7, 0); (92.9, 65.2, 0); (324.9, 0, 262.3); (63.1, 0, 222.7); (0, 234.9, 237.8)	Some Midway Configurations
3	(329.8, 250.9, 0); (73.9, 339.7, 0); (88.1, 41.7, 0); (341.7, 0, 270.5); (38.1, 0, 204.2); (0, 224.7, 229.5)	
4	(348.1, 251.5, 0); (57.8, 364.2, 0); (80.8, 5.7, 0); (367.4, 0, 283.2); (0, 0, 176.0); (0, 209.2, 216.8)	
5	(351.5, 252.0, 0); (55.2, 367.3, 0); (78.5, 0, 0); (371.0, 0, 284.6); (0, 0, 171.9); (0, 206.3, 214.0)	
6	(405.9, 262.9, 0); (18.3, 400.0, 0); (32.0, 0, 0); (416.3, 0, 297.1); (0, 0, 118.7); (0, 160.3, 164.2)	
7	(432.9, 268.4, 0); (0, 416.2, 0); (8.9, 0, 0); (438.8, 0, 303.3); (0, 0, 92.4); (0, 137.5, 139.5)	
8	**(442.5, 270.4, 0); (0, 421.8, 0); (0, 0, 0); (446.8, 0, 305.2); (0, 0, 82.7); (0, 128.3, 130.1)**	**Optimal Configuration**

tool to find the constrained minimum of a function of several variables (see Optimization Toolbox User's Guide, MathWorks). First, by using the planning method described in Eq. (10.47), the optimal locator configuration is obtained with some midway configurations during the process of iteration, as shown in Table 10.1 and Fig. 10.7. By using the robustness index as the objective function, the changes of the position and orientation errors of the workpiece during the process of iteration are shown in Figs. 10.8 and 10.9. From Figs. 10.8 and 10.9, it can be found that the position and orientation errors of the workpiece always show a decreasing trend during the iteration process and reach their minima at the optimal locator configuration.

However, we cannot determine the optimal clamp position using the locating robustness index Ω_R because the robustness index is not related to the clamp positions.

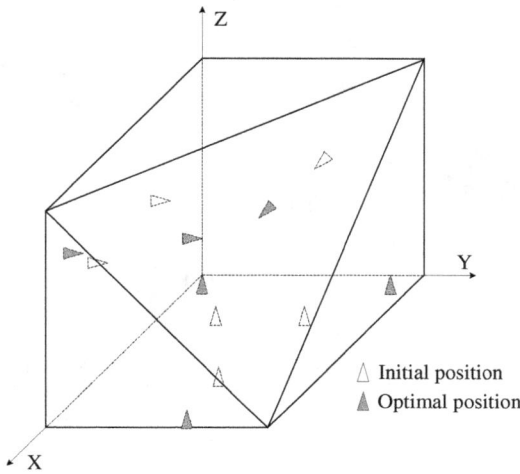

Fig. 10.7.   Robust locator configuration.

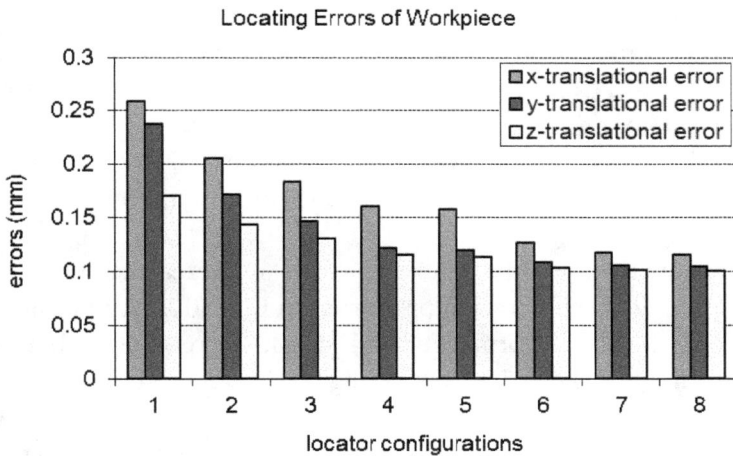

Fig. 10.8.   The position errors of the workpiece.

Now we use the second planning method described in Eq. (10.48) to plan the optimal fixture configuration under the constraints expressed in Eq. (10.48); the optimal fixture configuration is obtained with some midway configurations during the process of iteration, as shown in Table 10.2 and Fig. 10.10. By using the stability index as the objective function, the changes of the position and orientation

Locating Errors of Workpiece

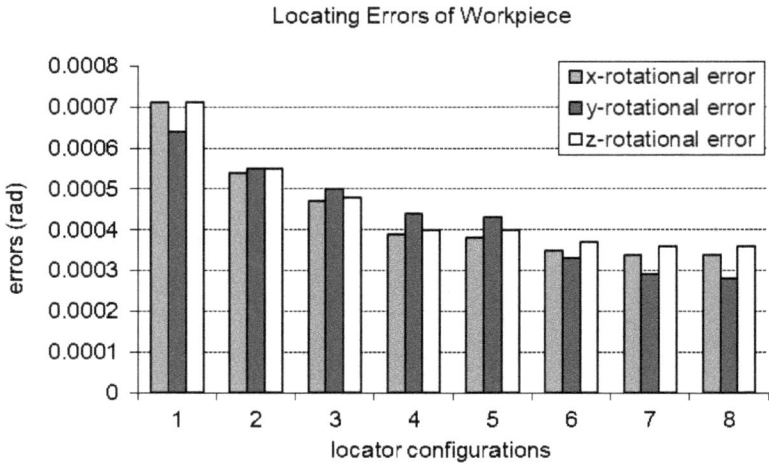

Fig. 10.9. The orientation errors of the workpiece.

errors of the workpiece, and the stability index Ω_S during the process of iteration are shown in Figs. 10.11, 10.12, and 10.13, respectively. From Figs. 10.11 and 10.12, it can be found that the changes of the position and orientation errors of the workpiece are not related to the iteration process. In other words, the position and orientation errors of the workpiece do not reach their minima at the optimal locator configuration.

However, the stability index Ω_S shows an increasing trend during the beginning of the iteration process from Fig. 10.13 (the value of the stability index Ω_S is normalized in Fig. 10.13), reaches its maximum in the vicinity of the 3rd fixture configuration, then slightly decreases along with the iteration, but is still close to the maximum, which means that the 3rd fixture configuration is optimal with respect to the stability index.

The reason why the phenomenon as shown in Figs. 10.11, 10.12, and 10.13 occur is that the index Ω_S is used to evaluate the fixturing stability of a workpiece, but not used to evaluate the locating errors of workpieces.

Finally, we use the third planning method with Eq. (10.49) where the weight factor is chosen as $w = 10^8$ to plan the fixture configuration of the workpiece. The optimal fixture configuration and

Table 10.2. Optimal fixture configuration (using the fixturing stability index as the objective function).

No.	Locators' coordinates	Clamp's coordinates	Notation
1	(300, 250, 0); (100, 300, 0); (100, 100, 0); (300, 0, 250); (100, 0, 250); (0, 250, 250)	(300, 300, 400)	Initial Configuration
2	(500, 244.2, 0); (0, 500, 0); (0, 0, 0); (500, 0, 269.2); (0, 0, 256.4); (0, 333.1, 206.9)	(320.0, 228.1, 451.9)	
3	**(500, 109.6, 0); (0, 500, 0); (0, 0, 0); (500, 0, 238.6); (0, 0, 319.5); (0, 373.9, 146.7)**	**(326.5, 200.4, 473.2)**	**Optimal Configuration**
4	(500, 53.1, 0); (0, 500, 0); (0, 0, 0); (500, 0, 225.4); (0, 0, 352.3); (0, 397.3, 122.7)	(326.7, 224.0, 449.4)	
5	(500, 46.4, 0); (0, 500, 0); (0, 0, 0); (500, 0, 225.4); (0, 0, 345.8); (0, 397.3, 122.7)	(326.7, 224.0, 449.4)	
6	(500, 50.4, 0); (0, 500, 0); (0, 0, 0); (500, 0, 225.3); (0, 0, 350.3); (0, 397.3, 122.6)	(326.7, 224.0, 449.3)	
7	(500, 173.5, 0); (0, 500, 0); (0, 0, 0); (500, 0, 220.1); (0, 0, 500); (0, 398.9, 119.0)	(327.4, 226.3, 446.3)	
8	(500, 156.8, 0); (0, 500, 0); (0, 0, 0); (500, 0, 218.1); (0, 0, 486.5); (0, 400.8, 118.9)	(326.8, 227.6, 445.7)	
9	(500, 0.0, 0); (0, 500, 0); (0, 0, 0); (500, 0, 199.4); (0, 0, 360); (0, 419.1, 118.4)	(320.8, 239.9, 439.3)	

some midway configurations during the process of iteration are shown in Table 10.3 and Fig. 10.14. By using the nonlinear programming (10.49), the changes of the position and orientation errors of the workpiece, and the stability index Ω_S during the process of iteration are shown in Figs. 10.15, 10.16, and 10.17, respectively. From Figs. 10.15 and 10.16, it can be seen that the changes of the position and orientation errors of the workpiece always show a decreasing

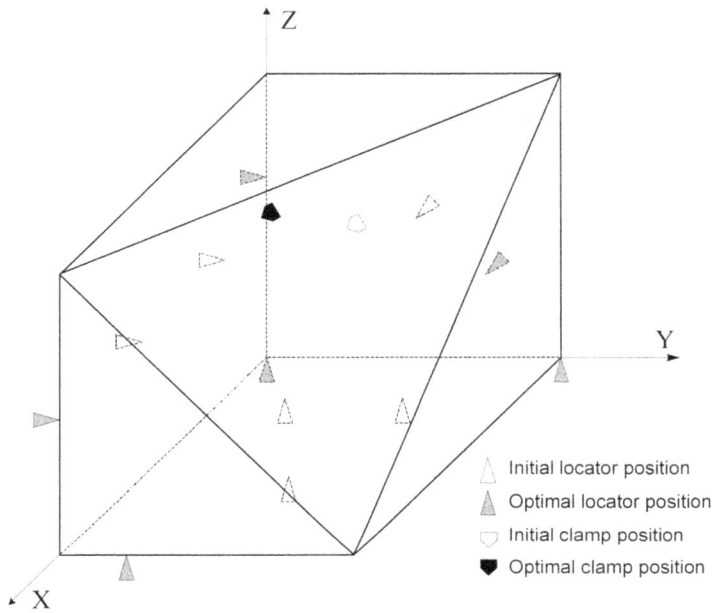

Fig. 10.10.   Stable fixture configuration.

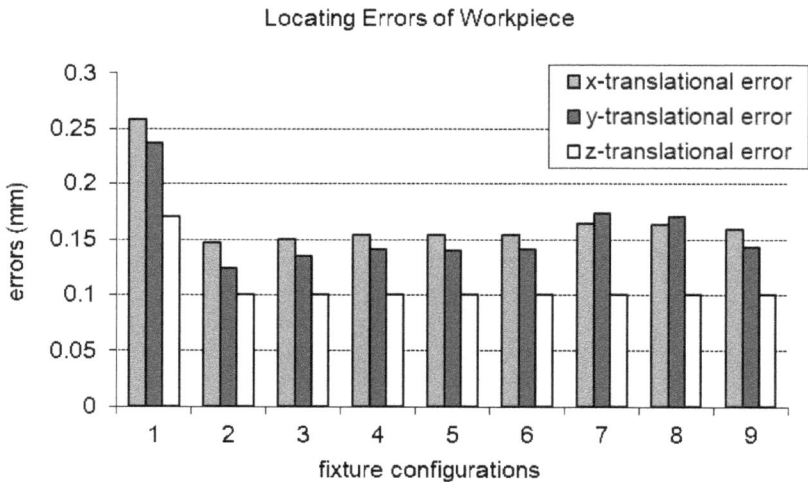

Fig. 10.11.   Changes of the position errors of the workpiece.

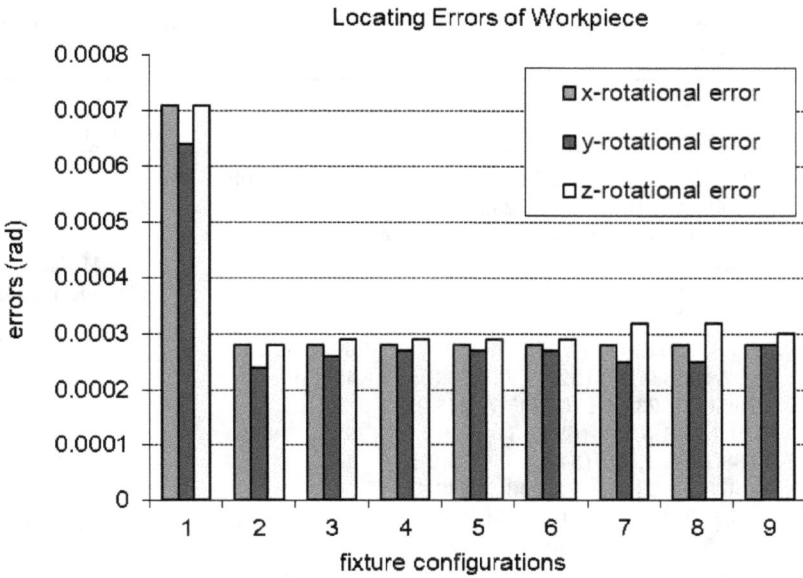

Fig. 10.12. Changes of the orientation errors of the workpiece.

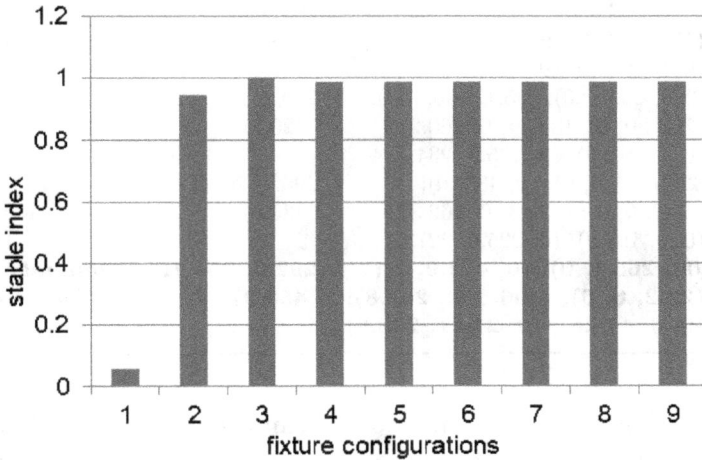

Fig. 10.13. Stability index during the process of iteration.

Table 10.3. Optimal fixture configuration (obtained using the nonlinear programming (10.49)).

No.	Locators' coordinates	Clamp's coordinates	Notation
1	(300, 250, 0); (100, 300, 0); (100, 100, 0); (300, 0, 250); (100, 0, 250); (0, 250, 250)	(300, 300, 400)	*Initial Configuration*
2	(315.7, 250.5, 0); (87.7, 317.2, 0); (95.4, 81.1, 0); (316.6, 0, 252.0); (79.3, 0, 242.9); (247.4, 247.1)	(299.9, 296.2, 403.9)	*Some Midway Configurations*
3	(325.8, 250.8, 0); (79.7, 328.3, 0); (92.4, 68.9, 0); (327.3, 0, 253.3); (66.0, 0, 238.4); (0, 245.7, 245.2)	(299.8, 293.8, 406.4)	
4	(340.4, 251.3, 0); (68.2, 344.5, 0); (88.1, 51.1, 0); (342.9, 0, 255.2); (46.6, 0, 231.7); (0, 243.2, 242.4)	(299.7, 290.3, 410.1)	
5	(355.7, 251.7, 0); (56.2, 361.2, 0); (83.6, 32.7, 0); (359.0, 0, 257.1); (26.5, 0, 224.9); (0, 240.7, 239.5)	(299.5, 286.6, 413.9)	
6	(373.5, 252.3, 0); (42.2, 380.9, 0); (78.3, 11.1, 0); (378.0, 0, 259.4); (2.9, 0, 216.8); (0, 237.7, 236.1)	(299.4, 282.4, 418.3)	
7	(375.7, 252.3, 0); (40.4, 383.3, 0); (77.7, 8.4, 0); (380.3, 0, 259.7); (0, 0, 215.8); (0, 237.3, 235.7)	(299.3, 281.8, 418.8)	
8	(383.0, 252.8, 0); (35.1, 390.5, 0); (75.1, 0, 0); (387.4, 0, 260.2); (0, 0, 212.9); (0, 235.9, 234.1)	(299.1, 280.2, 420.7)	
9	(432.3, 257.6, 0); (0, 435, 0); (54.8, 0, 0); (433.3, 0, 262.3); (0, 0, 194.9); (0, 225.6, 221.5)	(296.7, 269.4, 434.0)	
10	**(500, 265.1, 0); (0, 494.9, 0); (25.2, 0, 0); (496.3, 0, 263.8); (0, 0, 170.8); (0, 210.1, 203.5)**	**(292.0, 254.1, 453.9)**	*Optimal Configuration*

trend during the iteration process, and reach their minima at the optimal fixture configuration. At the same time, the stability index Ω_S shows an increasing trend during the iteration process from Fig. 10.17 and reaches its maximum at the optimal fixture configuration.

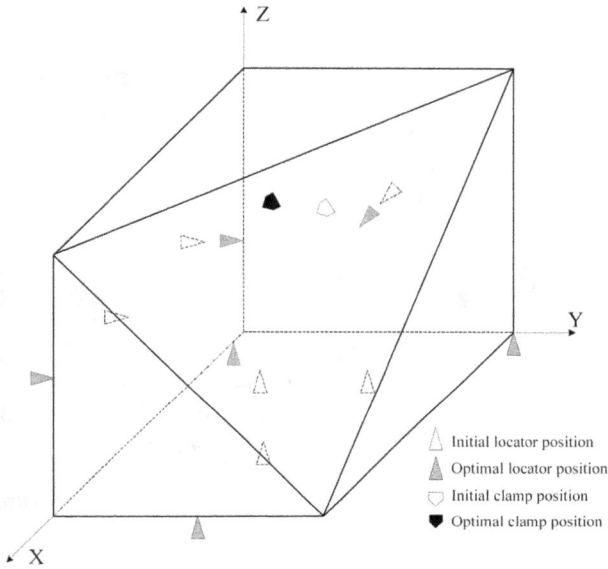

Fig. 10.14.   Optimal fixture configuration.

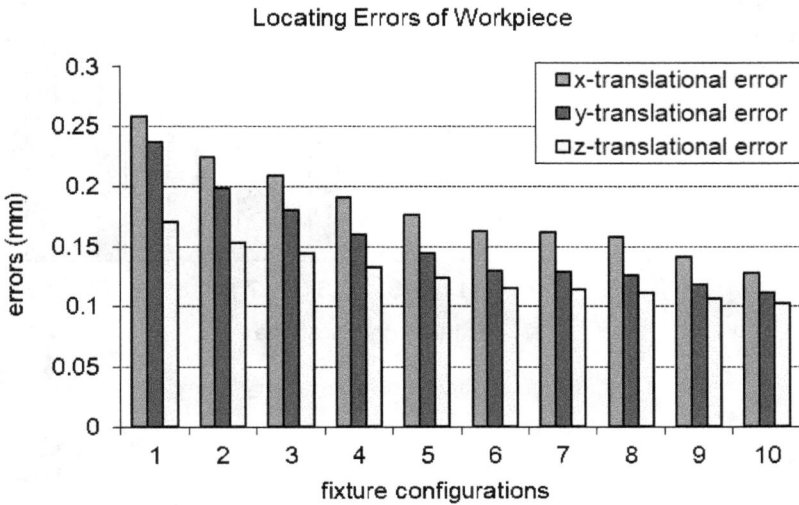

Fig. 10.15.   Changes of the position errors of the workpiece.

Locating Errors of Workpiece

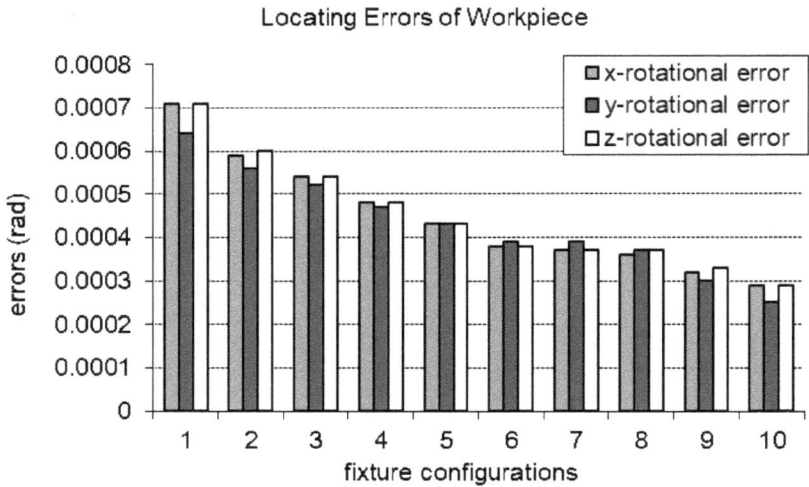

Fig. 10.16. Changes of the orientation errors of the workpiece.

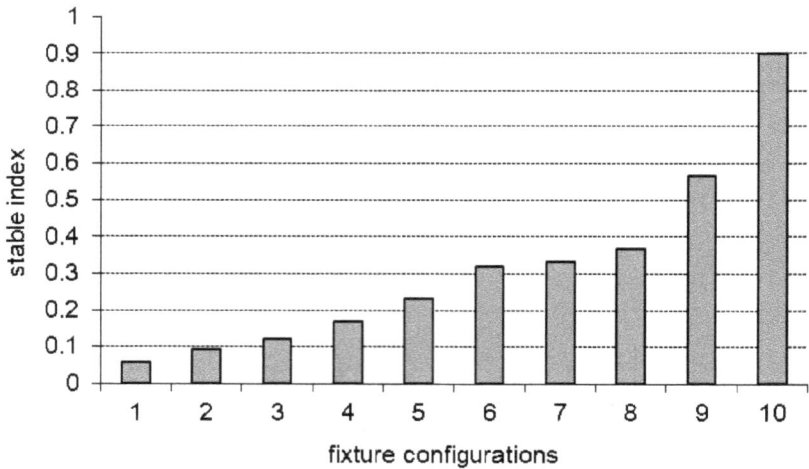

Fig. 10.17. Change of the stability index.

Comparing of three planned results mentioned above shows that the constrained nonlinear programming with Eq. (10.49) is a more effective method to find the optimal fixture configuration which can achieve the best overall effect of small localization error and great fixturing stability, though both objectives of minimizing localization errors and maximizing fixturing stability are not reached simultaneously.

10.6   Summary

There inherently exists position error for every locator. The position errors of locators will affect the position and orientation precision of the workpiece. How to evaluate and control such error influence is one of the fundamental problems in fixture automation design and planning.

This chapter first derives a general fixturing error mapping model, which is applicable for fully constrained localization, under-constrained localization, and over-constrained localization. Given the locator configuration and tolerances, we can use the model to determine the position and orientation variations of the workpiece. On the other hand, the tolerances of locators can be designed for the given tolerances of the position and orientation variations of the workpiece. Moreover, the position variations of the critical evaluation points on the workpiece can be calculated using the model, which is important in the quality evaluation of manufacturing. Based on the analysis of the free motion cone which is defined in this chapter, the accessible condition to the fixture and the detachable condition from the fixture are developed. By using the polar of the free motion cone, namely the constrained cone, the directions of clamping forces are found. The method of determining the feasible clamping domain where the fixturing of the workpiece is form closure is presented. When the configurations of locators and clamps are not reasonable, the position and orientation errors of the workpiece and the stability of fixturing would be influenced more considerably. Then this chapter defines three performance indices, namely the locating robustness index, the stability index of fixturing, and the fixturing resultant index. The locating robustness index is used to evaluate the configurations of locators. Minimizing the locating robustness index means that the position and orientation of the workpiece are most insensitive to locator errors. The stability index is used to evaluate the capability to withstand any external disturbance wrench for the fixturing system. The fixturing resultant index is used to evaluate the overall effect of localization error and fixturing stability on the fixture quality. Minimizing the fixturing resultant index implies that the best overall effect of small localization error and great fixturing stability can be achieved. Finally, three constrained nonlinear programming methods for planning the optimal fixturing configuration are formulated. Comparison of the planning results of the three methods shows that the third programming method where the objective is to minimize the fixturing resultant index is the most effective and reasonable planning method.

References

[1]   Asada H. and By A. B. Kinematic analysis of workpart fixturing for flexible assembly with automatically reconfigurable fixtures. *IEEE Journal of Robotics and Automation*, 1(2), pp. 86–93, 1985.

[2]   Bicchi A. On the closure properties of robotic grasping. *International Journal of Robotics Research*, 14(4), pp. 319–334, 1995.

[3]   Borst C., Fischer M., and Hirzinger G. A fast and robust grasp planner for arbitrary 3D objects. *Proceedings of IEEE International Conference on Robotics and Automation*, pp. 1890–1896, 1999.

[4]   Brost R. C. and Goldberg K. Y. A complete algorithm for designing planar fixtures using modular components. *IEEE Transactions on Robotics and Automation*, 12(1), pp. 31–46, 1996.

[5]   Brown R. G. and Brost R. C. A 3-D modular gripper design tool. *IEEE Transactions on Robotics and Automation*, 15(1), pp. 174–186, 1999.

[6]   Cai W., Hu S. J., and Yuan J. X. A variational method of robust fixture configuration design for 3-D workpieces. *Transactions of the ASME-Journal of Manufacturing Science and Engineering*, 119, pp. 593–602, 1997.

[7]   Carlson J. S. Quadratic sensitivity analysis of fixtures and locating schemes for rigid parts. *Transactions of the ASME-Journal of Manufacturing Science and Engineering*, 123, pp. 462–472, 2001.

[8]   Choudhuri S. A. and De Meter E. C. Tolerance analysis of machining fixture locators. *Transactions of the ASME-Journal of Manufacturing Science and Engineering*, 121, pp. 273–281, 1999.

[9] Cheong J. S., Haverkort H. J., and van der Stappen A. F. On computing all immobilizing grasps of a simple polygon with few contacts. *Proceedings of 14th International Symposium on Algorithms and Computation*, Kyoto, Japan, pp. 260–269, 2003.

[10] Howard W. S. and Kumar V. On the stability of grasped objects. *IEEE Transactions on Robotics and Automation*, 12(6), pp. 904–917, 1996.

[11] Huang X. and Gu P. Tolerance analysis in setup and fixture planning for precision machining. *Proceedings of the Fourth International Conference on Computer Integrated Manufacturing and Automation Technology*, pp. 298–305, 1994.

[12] Kang Y. Computer-aided fixture design verification. Ph.D. dissertation, Worcester Polytechnic Institute, MA, USA, 2001.

[13] Krishnakumar K. and Melkote S. N. Machining fixture layout optimization using the genetic algorithm. *International Journal of Machine Tools & Manufacture*, 40, pp. 579–598, 2000.

[14] Lakshminarayana K. Mechanics of form closure. *New York: American Society of Mechanical Engineering*, Paper No. 78-DET-32, pp. 2–8, 1978.

[15] Li B. and Melkote S. N. Improved workpiece location accuracy through fixture layout optimization. *International Journal of Machine Tools & Manufacture*, 39, pp. 871–883, 1999.

[16] Li Z. and Sastry S. S. Task-oriented optimal grasping by multifingered robot hands. *IEEE Journal of Robotics and Automation*, 4(1), pp. 32–44, 1988.

[17] Lin Q. and Burdick J. W. A task-dependent approach to minimum-deflection fixtures. *Proceedings of IEEE International Conference on Robotics and Automation*, pp. 1562–1569, 1999.

[18] Liu Y. H. Computing n-finger form-closure grasps on polygonal objects. *International Journal of Robotics Research*, 19(2), pp. 149–158, 2000.

[19] Marin R. A. and Ferreira P. M. Kinematic analysis and synthesis of deterministic 3-2-1 locator schemes for machining fixtures. *Transactions of the ASME-Journal of Manufacturing Science and Engineering*, 123, pp. 708–719, 2001.

[20] Marin R. A. and Ferreira P. M. Optimal placement of fixture clamps. *Proceedings of IEEE/ASME International Conference on Advanced Intelligent Mechatronics*, pp. 314–319, 2001.

[21] Mishra B. Grasp metrics: Optimality and complexity. In: Goldberg K. *et al.* (eds.), *Algorithmic Foundations of Robotics*, pp. 137–165, 1995.

[22]   Montana D. J. Contact stability for two-fingered grasps. *IEEE Transactions on Robotics and Automation*, 8(4), pp. 421–430, 1992.

[23]   Ponce J. and Faverjon B. On computing three-finger force-closure grasps of polygonal objects. *IEEE Transactions on Robotics and Automation*, 11(6), pp. 868–881, 1995.

[24]   Ponce J., Sullivan S., Sudsang A., Boissonnat J. D., and Merlet J. P. On computing four-finger equilibrium and force-closure grasps of polyhedral objects. *International Journal of Robotics Research*, 16(1), pp. 11–35, 1997.

[25]   Rimon E. A curvature-based bound on the number of frictionless fingers required to immobilize three-dimensional objects. *IEEE Transactions on Robotics and Automation*, 17(5), pp. 679–697, 2000.

[26]   Rong Y. and Zhu Y. *Computer-aided Fixture Design*. Marcel Dekker, Inc., New York, USA, 1999.

[27]   Xiong C. H., Rong Y., Koganti R. P., Zaluzec M. J., and Wang N. Geometric variation prediction of automotive assembly. *Assembly Automation*, 22(3), pp. 260–269, 2002.

[28]   Trinkle J. C. On the stability and instantaneous velocity of grasped frictionless objects. *IEEE Transactions on Robotics and Automation*, 8(5), pp. 560–572, 1992.

[29]   van der Stappen A. F., Wentink C., and Overmars M. H. Computing immobilizing grasps of polygonal parts. *International Journal of Robotics Research*, 19(5), pp. 467–479, 2000.

[30]   Wang M. Y. and Pelinescu D. M. Optimizing fixture layout in a point set domain. *IEEE Transactions on Robotics and Automation*, 17(3), pp. 312–323, 2001.

[31]   Wu Y., Rong Y., Ma W., and LeClair S. Automated modular fixture design: Geometric analysis. *Robotics and Computer-Integrated Manufacturing*, 14(1), pp. 1–15, 1998.

[32]   Xiong C. H. and Xiong Y. L. Stability index and contact configuration planning for multifingered grasp. *Journal of Robotic Systems*, 15(4), pp. 183–190, 1998.

[33]   Xiong C. H., Li Y. F., Ding H., and Xiong Y. L. On the dynamic stability. *International Journal of Robotics Research*, 18(9), pp. 951–958, 1999.

[34]   Xiong C. H., Li Y. F., Rong Y. K., and Xiong Y. L. Qualitative analysis and quantitative evaluation of fixturing. *Robotics and Computer Integrated Manufacturing*, 18(5-6), pp. 335–342, 2002.

[35]   Xiong Y. L. The theory and methodology for concurrent design and planning of reconfiguration fixture. *Proceedings of IEEE International Conference on Robotics and Automation*, pp. 305–311, 1993.

[36] Zhang Y., Hu W., Kang Y., Rong Y., and Yen D. W. Locating error analysis and tolerance assignment for computer-aided fixture design. *International Journal of Production Research*, 39(15), pp. 3529–3545, 2001.

[37] Gopalakrishnan K., Goldberg K., Bone G. M., Zaluzec M. J., Koganti R., Pearson R., and Deneszczuk P. A. Unilateral fixtures for sheet-metal parts with holes. *IEEE Transactions on Automation Science and Engineering*, 1(2), pp. 110–120, 2004.

[38] Xiong C. H., Rong Y., Tang Y., and Xiong Y. L. Fixturing model and analysis. *International Journal of Computer Applications in Technology*, 28(1), pp. 34–45, 2007.

[39] Xiong C. H., Wang M. Y., Tang Y., and Xiong Y. L. Compliant grasping with passive forces. *Journal of Robotic Systems*, 22(5), pp. 271–285, 2005.

[40] Rockafellar R. T. *Convex analysis*. Princeton University Press, Princeton, USA, 1970.

Chapter 11

Clamping Planning in Workpiece-Fixture Systems

Deformation at contacts between the workpiece and locators/clamps resulting from large contact forces causes overall workpiece displacement and affects the localization accuracy of the workpiece. An important characteristic of a workpiece-fixture system is that locators are passive elements and can only react to clamping forces and external loads, whereas clamps are active elements and apply a predetermined normal load to the surface of the workpiece to prevent it from losing contact with the locators. Clamping forces play an important role in determining the final workpiece quality. In order to obtain the higher localization accuracy for the workpiece, we need to plan the clamping forces including their magnitudes and positions. This chapter defines the minimum norm of the elastic deformation at contacts as the objective function and formulates the problem of determining the optimal clamping forces as a constrained nonlinear programming which guarantees that the fixturing of the workpiece is force closure. The proposed planning method of optimal clamping forces, which may also have an application to other passive, indeterminate problems such as power grasps in robotics, is illustrated with a numerical example.

221

11.1   Introduction

The workpiece localization accuracy is primarily determined by the positioning accuracy of the locators and their layout in the fixture [1, 3, 23, 24, 26–29, 31–35], while the choice of a set of clamps is equally important for guaranteeing the workpiece to maintain the desired position and orientation with a suitable set of contact forces during machining [2, 4]. Under the frictional contact condition, this requirement means that the contact forces should be always within the corresponding friction cones even when the external wrench may change during the course of a manufacturing process. In this chapter, we address the problem of finding an optimal clamping scheme.

The simplest approach to fixture design is to assume that the contacts between the workpiece and the fixture elements are frictionless [1, 17, 23, 24]. In this case, the analysis of contact forces is simple since under this assumption the contact forces are determinate in a static equilibrium with six locators for a general 3-D workpiece. There exist methods for computing optimal positions of fixture clamps on 3-D parts with force closure [17, 24]. However, frictional forces are important in practical cases to help prevent workpiece from slipping [9, 13, 28, 29]. In reality, the influence of frictional forces cannot be simply neglected. When frictional forces are taken into account, one of the fundamental problems is that the equilibrium equation of the fixtured workpiece cannot determine the contact forces uniquely in general, which means that the system is indeterminate in the rigid body framework.

Several researchers have addressed fixture clamping force optimization based on the rigid-body model with friction [13, 18, 22, 31]. Goyal developed the concept of limit friction surface [6]. Tao *et al.* [22] proposed an analytical approach for automatic fixture configuration analysis to determine whether a particular fixturing configuration is valid with respect to the machining process. A nonlinear programming was incorporated into the analysis to obtain the minimum clamping forces needed to counterbalance the dynamic cutting forces. Marin and Ferreira [18] presented a linear programming method to compute the optimal positions of clamps and clamping forces for 3-D parts with planar surfaces. The optimization goal is to minimize the maximum normal component of the clamping forces.

A primary limitation of the rigid-body analysis is its inadequacy to deal with passive forces. It has been recognized that the contact forces in the workpiece-fixture system include both active and passive contact forces [16, 23, 28, 34], as compared to active forces in a multi-fingered robotic hand [33]. Fixture locators are passive elements, whereas clamps are active loading elements and only they can be considered to be active [23, 24, 28, 34]. When the workpiece is clamped by a clamping force, contact forces at the locators are "generated" as a reaction. Similarly, any external force such as gravity or cutting force will also cause reactive forces "distributed" among the passive locators. In a workpiece-fixture system, force closure depends on the magnitude and position of the clamping force(s). Variation of the magnitude and/or the position of a clamping force results in changes in the contact forces between the workpiece and the locators as well. Thus, how to plan the clamping forces and their positions is one of the fundamental problems in fixture design.

In Refs. [10–12], the genetic algorithms were used to determine the most statically stable fixture layout, however, the effect of clamping forces was not included in their work. Kulankara *et al.* [10] used the genetic algorithm to optimize fixture layout and clamping forces for a compliant workpiece. The fitness value is the maximum workpiece elastic deformation. The input of the layout/clamping force algorithm is the stiffness matrix which is extracted from a finite element model. The basic issue involves an appropriate model for the determination of the contact forces [15, 16, 23, 25, 28]. A comprehensive approach is to consider the workpiece-fixture system as an elastic system with friction which can be analyzed with a finite element model [3, 4, 35]. Such a finite element approach often results in a large size model and requires high computational effort. The model is also sensitive to the boundary conditions [10]. Another approach is to use a discrete contact linear elasticity model to represent unidirectional contacts [25]. By applying the principle of minimum total complementary energy, this model yields a constrained quadratic program for predicting the contact forces [14]. However, the discrete contact elasticity model requires prior knowledge of the contact state of each passive contact. One may have to first guess whether a particular contact is in a state of lift-off, stick, or slip. Subsequently, the general model must be assembled and solved numerically. Afterward, the inequality constraints associated with the contact states must be verified. If any of the inequality constraints is violated, a new assumption must be made and the procedure is repeated until all inequality constraints are satisfied. Furthermore, each contact is modeled with an elastic deformation region [25], which increases the modeling and computational complexity considerably.

In this chapter, the contacts between the workpiece and locators/clamps are modeled by local elastic contacts with friction, while the workpiece and fixture elements are otherwise sturdy, and can be treated as rigid bodies. In the model the relationship of the contact force with the local elastic deformation is nonlinear. Within this framework, a general method for optimal planning of the magnitudes and the positions of the clamping forces is presented. It is known that the passive contact forces between the workpiece and the locators under a given clamping force depend also on the locators' configuration [27]. The design of a fixture is usually carried out in two stages. The first stage is to determine an appropriate scheme for locating the part which has been discussed in Chapter 10, generally based on the required localization accuracy [24]. Then, the choice of a set of clamps is developed. This chapter deals with the second design stage and is concerned with excessive contact forces that might induce unacceptable elastic deformations at contacts, which will adversely affect the workpiece localization accuracy and, in turn, the final workpiece quality.

We present a technique to compute optimal clamping schemes for a frictional workpiece-fixture system. A clamping scheme consists of the positions of the clamps on the surface of the workpiece and the magnitudes of the clamping forces to counteract the external forces. The design objective is to minimize the norm of the elastic deformations at contacts so that their influence on the localization accuracy of the workpiece is minimized. In this work, the positions of the locators are assumed to have been already defined in the first design stage. For each clamp, a feasible region on the workpiece surface is specified. Previous methods aimed at solving the same problem are with more restrictions. For example, there are methods capable of computing either the optimal clamping force [15] or the optimal clamp location [23]. Another method is applicable only to planar workpiece surfaces [17, 18]. The technique presented here computes the optimal magnitude and the position of a clamping force for a general workpiece of arbitrary geometry and with friction. The problem is formulated here as a constrained nonlinear programming and is solved using the Levenberg-Marquardt method which is globally convergent.

11.2 Planning of Magnitudes and Positions of Clamping Forces

The task of clamp planning is to determine the best clamping scheme to maintain a specified position and orientation of the workpiece in the presence of external disturbance forces such as cutting forces/torques and clamping forces. In the workpiece-fixture system, locators are passive elements and can only react to external loads, whereas clamps are active elements and must exert suitable forces on the surface of workpiece to prevent it from losing contact with the locators. However, large clamping forces will cause large elastic deformations at these contacts, which will result in a large disturbance in the position and orientation of the workpiece. On the other hand, insufficient clamping forces may not maintain permanent contacts between the workpiece and locators, which means that localization accuracy can not be guaranteed due to slide at contacts. In the following, we address the problem of finding an optimal clamping scheme.

11.2.1 Objective Function

The active clamping forces and the positions of clamps (clamp configuration), and the configuration of the passive locators on a fixture are fixed during the period of machining. However, the elastic deformation at each contact between the workpiece and fixture is usually time variant due to the time-variant cutting forces/torques. Generally, we can not change the cutting loads during machining. Given the cutting loads and the locator configuration on a fixture, the elastic deformations at contacts depend on the magnitudes and positions of clamping forces exerted on the workpiece. In order to obtain high-quality machined workpiece, we have to plan the clamp configuration and clamping forces so that the large local contact elastic deformations can be avoided; consequently, the displacement of a workpiece due to local contact elastic deformations can be minimized.

The objective of clamp planning is to find the optimal clamp configuration and clamping forces. Here we define the minimum norm of the elastic deformations at contacts during the period of machining as the objective function:

$$\text{minimize} \sum_{i=1}^{N} (\Delta \mathbf{d}_i^T \Delta \mathbf{d}_i) \tag{11.1}$$

where N is the number of contacts, $N = m + n$ for m locators and n clamps, and $\boldsymbol{\Delta}\mathbf{d}_i = \left(\delta d_{in} \ \ \delta d_{it1} \ \ \delta d_{it2} \right)^T \in \Re^{3\times1}$ is the elastic deformation at the ith contact as described in Section 7.3.

11.2.2 Kinematic Conditions

The workpiece-fixture system under investigation must satisfy a series of kinematic conditions as described as follows:

(1) *Static Equilibrium*: Generally, a workpiece-fixture system consists of a workpiece, m locators, and n clamps of the fixture. A given external wrench $\mathbf{F}_e \in \Re^{6\times1}$ (including applied forces and weight of the workpiece) is exerted on the workpiece. The contacts between the workpiece and the locators and the clamps are considered as point contacts with friction.

Let $\mathbf{n}_i^l \in \Re^{3\times1}$ ($\mathbf{n}_j^c \in \Re^{3\times1}$) be the unit inner normal vector of the workpiece at the contact position $\mathbf{r}_i^l \in \Re^{3\times1}$ ($\mathbf{r}_j^c \in \Re^{3\times1}$) of the ith locator (the jth clamp). Moreover, let $\mathbf{t}_{i1}^l \in \Re^{3\times1}$ and $\mathbf{t}_{i2}^l \in \Re^{3\times1}$ ($\mathbf{t}_{j1}^c \in \Re^{3\times1}$ and $\mathbf{t}_{j2}^c \in \Re^{3\times1}$) be the two orthogonal unit tangential vectors of the workpiece at the ith locator (the jth clamp) contact, respectively. For the ith locator, we denote by

$$\mathbf{f}_i^l = \left(f_{in}^l \ f_{it1}^l \ f_{it2}^l \right)^T \in \Re^{3 \times 1}$$

the three elements of the contact force $\mathbf{f}_i^l$ along the unit normal vector $\mathbf{n}_i^l$ and the unit tangential vectors $\mathbf{t}_{i1}^l$ and $\mathbf{t}_{i2}^l$, respectively. Similarly, for the jth clamp, we denote by

$$\mathbf{f}_j^c = \left(f_{jn}^c \ f_{jt1}^c \ f_{jt2}^c \right)^T \in \Re^{3 \times 1}$$

the three elements of the contact force $\mathbf{f}_j^c$ along the unit normal vector $\mathbf{n}_j^c$ and the unit tangential vectors $\mathbf{t}_{j1}^c$ and $\mathbf{t}_{j2}^c$ respectively. Since a clamp force is usually provided by a hydraulic actuator, the normal clamping force f_{jn}^c may be assumed to be active and prescribed. Therefore,

$$f_{in}^l, \quad f_{it1}^l, \quad f_{it2}^l \quad \text{and} \quad f_{jt1}^c, \quad f_{jt2}^c$$

are considered to be passive and unknown. Here, $i = 1 \cdots m$ and $j = 1 \cdots n$.

Thus, the force equilibrium of the workpiece-fixture system is described as

$$\mathbf{G}_l \mathbf{F}_l + \mathbf{G}_{ct} \mathbf{F}_{ct} = -\mathbf{G}_{cn} \mathbf{F}_{cn} + \mathbf{F}_e \tag{11.2}$$

where

$$\mathbf{G}_l = \begin{bmatrix} \mathbf{n}_1^l & \mathbf{t}_{11}^l & \mathbf{t}_{12}^l & \cdots & \mathbf{n}_m^l & \mathbf{t}_{m1}^l & \mathbf{t}_{m2}^l \\ \mathbf{r}_1^l \times \mathbf{n}_1^l & \mathbf{r}_1^l \times \mathbf{t}_{11}^l & \mathbf{r}_1^l \times \mathbf{t}_{12}^l & \cdots & \mathbf{r}_m^l \times \mathbf{n}_m^l & \mathbf{r}_m^l \times \mathbf{t}_{m1}^l & \mathbf{r}_m^l \times \mathbf{t}_{m2}^l \end{bmatrix} \in \Re^{6 \times 3m}$$

$$\mathbf{G}_{ct} = \begin{bmatrix} \mathbf{t}_{11}^c & \mathbf{t}_{12}^c & \cdots & \mathbf{t}_{n1}^c & \mathbf{t}_{n2}^c \\ \mathbf{r}_1^c \times \mathbf{t}_{11}^c & \mathbf{r}_1^c \times \mathbf{t}_{12}^c & \cdots & \mathbf{r}_n^c \times \mathbf{t}_{n1}^c & \mathbf{r}_n^c \times \mathbf{t}_{n2}^c \end{bmatrix} \in \Re^{6 \times 2n}$$

$$\mathbf{G}_{cn} = \begin{bmatrix} \mathbf{n}_1^c & \cdots & \mathbf{n}_n^c \\ \mathbf{r}_1^c \times \mathbf{n}_1^c & \cdots & \mathbf{r}_n^c \times \mathbf{n}_n^c \end{bmatrix} \in \Re^{6 \times n}$$

$$\mathbf{F}_l = \left((\mathbf{f}_1^l)^T \ \cdots \ (\mathbf{f}_m^l)^T \right)^T \in \Re^{3m \times 1}$$

$$\mathbf{F}_{ct} = \left(f_{1t1}^c \ f_{1t2}^c \ \cdots \ f_{nt1}^c \ f_{nt2}^c \right)^T \in \Re^{2n \times 1}$$

$$\mathbf{F}_{cn} = \left(f_{1n}^c \ \cdots \ f_{nn}^c \right)^T \in \Re^{n \times 1}$$

and $\mathbf{G}_l$ is often referred to as the locating matrix.

In a more compact form, Eq. (11.2) can be rewritten as

$$\mathbf{G}_{lct} \mathbf{F}_{lct} = -\mathbf{G}_{cn} \mathbf{F}_{cn} + \mathbf{F}_e \tag{11.3}$$

where

$$\mathbf{G}_{lct} = [\mathbf{G}_l \vdots \mathbf{G}_{ct}] \in \Re^{6 \times (3m+2n)}$$

$$\mathbf{F}_{lct} = (\mathbf{F}_l^T \vdots \mathbf{F}_{ct}^T)^T \in \Re^{(3m+2n) \times 1}$$

and $\mathbf{G}_{lct}$ is referred to as the configuration matrix.

Substitute Eqs. (7.14–7.16) into Eq. (11.3), we can rewrite six force equilibrium equations as

$$\mathbf{\Phi}\left(\mathbf{\Delta c}_1 \cdots \mathbf{\Delta c}_{m+n}\right) + \mathbf{G}_{cn}\mathbf{F}_{cn} - \mathbf{F}_e = 0 \tag{11.4}$$

where

$$\mathbf{\Phi}\left(\mathbf{\Delta c}_1 \quad \cdots \quad \mathbf{\Delta c}_{m+n}\right) = \begin{pmatrix} \varphi_1\left(\mathbf{\Delta c}_1 \cdots \mathbf{\Delta c}_{m+n}\right) \\ \vdots \\ \varphi_6\left(\mathbf{\Delta c}_1 \cdots \mathbf{\Delta c}_{m+n}\right) \end{pmatrix} \in \Re^{6 \times 1}$$

are the nonlinear functions of the elastic deformation vectors $\mathbf{\Delta c}_1, \ldots,$ and $\mathbf{\Delta c}_{m+n}$ which are described in Chapter 7.

(2) *Conditions of Compatible Deformations*: As mentioned in Chapter 7, the elastic deformations for different contacts are related to each other in the grasping/fixturing. The $3N$ "compatibility" equations in matrix form for the workpiece-fixture system with $N = m + n$ contacts can be described as follows:

$$\mathbf{G}^T \mathbf{\Delta X} - \mathbf{\Delta c} = 0 \tag{11.5}$$

(3) *Friction Model*: Furthermore, we have to guarantee that the condition of non-slippage at each contact is satisfied during the planning of clamping. Using Coulomb's model, we can represent the non-slippage constraints as follows:

$$\sqrt{f_{it1}^2 + f_{it2}^2} \leq \mu_i f_{in}, \quad i = 1, \ldots, m+n \tag{11.6}$$

where μ_i is the static friction coefficient at the ith contact. Equation (11.6) describes a friction cone

$$FC_i = \left\{ \mathbf{f}_{ci} \in \Re^{3 \times 1} | \mu_i f_{in} - \sqrt{f_{it1}^2 + f_{it2}^2} \geq 0 \right\}$$

We define $\varepsilon_i = \mu_i f_{in} - \sqrt{f_{it1}^2 + f_{it2}^2}$ as the sliding index.

If all the contact forces $\mathbf{f}_c = (\,\mathbf{f}_{c1}^T \cdots \mathbf{f}_{c(m+n)}^T\,)^T \in \Re^{3(m+n)\times 1}$ with $\mathbf{f}_{ci} = (\,f_{in}\ f_{it1}\ f_{it2}\,)^T \in \Re^{3\times 1}$ are within their friction cones, that is,

$$\mathbf{f}_c \in FC_1 \times \cdots \times FC_{m+n} \qquad (11.7)$$

then no slippage of the workpiece at any locator/clamp contact is guaranteed, which means that all of the sliding index $\varepsilon_i \geq 0$ $(i = 1, \ldots, m+n)$.

(4) *Feasible Clamping Domain*: Generally speaking, there always exist regions on the workpiece surface, such as the datum surfaces and interior surfaces of holes, at which clamps cannot be positioned. Thus, we define the feasible clamping domain on the workpiece surface as follows:

$$\begin{cases} \mathbf{r}_{oi} = \{(\,x_i,\,y_i,\,z_i\,)^T \,|\, S_h(\,x_i,\,y_i,\,z_i\,) = 0, \quad h = 1, \ldots, H\} \\ l_i\,(\,x_i,\,y_i,\,z_i\,) \leq 0, \quad i = m+1, \ldots, m+n \end{cases} \qquad (11.8)$$

where $S_h(\,x_i,\,y_i,\,z_i\,) = 0$ defines the ith clamp $\mathbf{r}_{oi}$ on the hth surface of the workpiece and the n inequalities $l_i(\,x_i,\,y_i,\,z_i\,) \leq 0$ $(i = m+1, \ldots, m+n)$ form the feasible clamping domain.

11.2.3 Solution of Optimal Clamping Forces

Using the objective function and the constraints described above, we can formulate the planning of optimal clamping forces as the following constrained nonlinear programming problem:

$$\underset{\eta \in D}{\text{minimize}} \sum_{i=1}^{m+n} (\Delta\mathbf{d}_i^T \Delta\mathbf{d}_i)$$

subject to $\qquad\qquad\qquad\qquad\qquad\qquad\qquad (11.9)$

$$D: \begin{cases} g_i(\eta) \leq 0, & i = 1, \ldots, m + 2n \\ q_j(\eta) = 0, & j = 1, \ldots, 6 + 3(m+n) \end{cases}$$

where

$$\eta = (\,\Delta\mathbf{c}^T\ \Delta\mathbf{X}^T\ \mathbf{X}_{cp}^T\,)^T = (\,\eta_1 \cdots \eta_{3m+5n+6}\,)^T \in \Re^{(3m+5n+6)\times 1}$$

denotes the design variables,

$$\mathbf{X}_{cp} = \begin{pmatrix} x_{c1} & y_{c1} & \cdots & x_{cn} & y_{cn} \end{pmatrix}^{T} \in \Re^{2n \times 1}$$

defines the coordinates of clamp positions on the surface of the workpiece, $g_i(\boldsymbol{\eta}) \leq 0$ $(i = 1, \ldots, m + 2n)$ represents all the inequality constraints of Eqs. (11.6) and (11.8), and $q_j(\boldsymbol{\eta}) = 0$ $(j = 1, \ldots, 6 + 3(m + n))$ describes all the equality constraints defined in Eqs. (11.4) and (11.5).

Using the exterior penalty function method [5], the constrained nonlinear programming problem (11.9) can be transformed into an unconstrained nonlinear programming problem as follows:

$$\text{minimize} \quad \left\{ \sum_{i=1}^{m+n} \left(\Delta \mathbf{d}_i^T \Delta \mathbf{d}_i \right) \right.$$

$$\left. + r \left[\sum_{i=1}^{m+2n} \left(g_i(\eta) \right)^2 + \sum_{j=1}^{6+3(m+n)} \left(q_j(\eta) \right)^2 \right] \right\} \quad (11.10)$$

where r is a positive penalty parameter. The role of the penalty parameter is obvious: As r increases, so does the penalty associated with a given choice of η that violates one or more of the constraints $g_i(\eta) \leq 0$ $(i = 1, \ldots, m + 2n)$ and $q_j(\eta) = 0$ $(j = 1, \ldots, 6 + 3(m + n))$. For the exterior penalty function method to work, the penalty parameter r must be very large. Theoretically, the minimum of the problem (11.10) corresponds to the solution of the original problem (11.9) only as $r \rightarrow \infty$.

In fact, Eq. (11.10) represents the non-linear least square [21] which can be rewritten as

$$\text{minimize} \quad \left[\Gamma(\eta) = \sum_{i=1}^{5m+6n+6} \Gamma_i^2(\eta) = \Im^T(\eta) \Im(\eta) \right] \quad (11.11)$$

Consequently, as described in Chapter 6, we can obtain the optimal solution for the nonlinear least square problem (11.11) using the Levenberg–Marquardt method which is globally convergent.

11.3 Verification of Force Constraints

In the process of optimization, we need to compute the passive contact forces in workpiece-fixture system and to check whether the contact forces are within their corresponding friction cone FC_i $(i = 1, \ldots, m + n)$. It can be found that Eqs. (11.4) and (11.5) provide $6 + 3(m + n)$ equality constraints. Given n normal elements $\mathbf{F}_{cn}$ of n clamping forces and the external wrench $\mathbf{F}_e$, we need to calculate all $3(m + n)$ elements of the elastic deformations $\Delta\mathbf{c}_1, \ldots$, and $\Delta\mathbf{c}_{m+n}$ as well as the six elements of the workpiece displacement $\Delta\mathbf{X}$.

In the $6 + 3(m + n)$ system equations, the workpiece displacement vector $\Delta\mathbf{X}$ appears linearly. Thus, it may be eliminated first from the system equations to reduce the size of the nonlinear system. This can be accomplished by utilizing the fact that the motion of the workpiece may be determined by three locator contacts modeled using the locally elastic contact law [8, 20]. This fact is carried out as follows.

If we choose three locator contacts arbitrarily, say $i = 1, 2, 3$. Then, from Eq. (7.18), we have the following equations:

$$\left(\mathbf{I}_{3\times3} \quad \vdots \quad -{}_o^p\mathbf{R}\mathbf{r}_{o1\times}\right)\Delta\mathbf{X} = \Delta\mathbf{c}_1 \tag{11.12}$$

$$\left(\mathbf{I}_{3\times3} \quad \vdots \quad -{}_o^p\mathbf{R}\mathbf{r}_{o2\times}\right)\Delta\mathbf{X} = \Delta\mathbf{c}_2 \tag{11.13}$$

$$\left(\mathbf{I}_{3\times3} \quad \vdots \quad -{}_o^p\mathbf{R}\mathbf{r}_{o3\times}\right)\Delta\mathbf{X} = \Delta\mathbf{c}_3 \tag{11.14}$$

Let denote $\hat{\mathbf{G}} = (\mathbf{U}_{o1}^T \ \mathbf{U}_{o2}^T \ \mathbf{U}_{o3}^T)$ with $\mathbf{U}_{oi} = (\mathbf{I}_{3\times3} \vdots -{}_o^p\mathbf{R}\mathbf{r}_{oi\times}) \in \Re^{3\times6}$ $(i = 1, 2, 3)$. Then, if the rank of $\hat{\mathbf{G}}$ is equal to 6, the displacement $\Delta\mathbf{X}$ of the workpiece can be determined uniquely by the local deformation vectors $\Delta\mathbf{c}_1, \ldots$, and $\Delta\mathbf{c}_{m+n}$, such that

$$\Delta\mathbf{X} = \hat{\mathbf{G}}^{+l}\Delta\hat{\mathbf{c}} \tag{11.15}$$

where

$$\hat{\mathbf{G}}^{+l} = (\hat{\mathbf{G}}\hat{\mathbf{G}}^T)^{-1}\hat{\mathbf{G}}$$

is the left general inverse matrix of the matrix $\hat{\mathbf{G}}$ and

$$\Delta\hat{\mathbf{c}} = (\,\Delta\mathbf{c}_1^T \quad \Delta\mathbf{c}_2^T \quad \Delta\mathbf{c}_3^T\,)^T$$

It should be pointed out that two locator contacts are not sufficient to establish this relationship. One can only determine five independent motion elements of the workpiece for the elastic deformations of any two locators, but a rotational motion around the line connecting the two contact points is left undetermined. Mathematically, this fact can be explained as follows: matrix

$$\mathbf{U}_{oij} = \begin{bmatrix} \mathbf{U}_{oi} \\ \mathbf{U}_{oj} \end{bmatrix}$$

can be transformed to an echelon form matrix

$$\begin{bmatrix} \mathbf{I}_{3\times3} & \vdots & -\mathbf{r}_{oi}\times \\ \mathbf{0}_{3\times3} & \vdots & (\mathbf{r}_{oi} - \mathbf{r}_{oj})\times \end{bmatrix}$$

by elementary row operations. Thus,

$$rank\,(\mathbf{U}_{oij}) = rank\,(\mathbf{I}_{3\times3}) + rank\,[(\mathbf{r}_{oi} - \mathbf{r}_{oj})\times] = 3 + 2 = 5$$

In order to maintain contacts between the workpiece and locators/clamps during fixturing, the following "compatibility" equations must be satisfied:

$$\tilde{\mathbf{G}}^T\Delta\mathbf{X} = \Delta\tilde{\mathbf{c}} \tag{11.16}$$

where

$$\tilde{\mathbf{G}} = (\,\mathbf{U}_{o3}^T \cdots \mathbf{U}_{o(m+n)}^T\,) \in \Re^{6\times3(m+n-2)}$$

and

$$\Delta\tilde{\mathbf{c}} = (\,\Delta\mathbf{c}_3^T \cdots \Delta\mathbf{c}_{m+n}^T\,) \in \Re^{3(m+n-2)\times1}$$

Substituting Eq. (11.15) into Eq. (11.16), we obtain the compatible elastic deformation conditions in a more compact form involving the local deformations only:

$$\tilde{\mathbf{G}}^T\hat{\mathbf{G}}^{+l}\Delta\hat{\mathbf{c}} = \Delta\tilde{\mathbf{c}} \tag{11.17}$$

Now the $3(m + n)$ nonlinear system Eqs. (11.4) and (11.17) involve $3(m + n)$ unknown variables of the local elastic deformations at all contacts, and the $3(m + n)$ elements of the elastic deformation vectors $\Delta c_1, \ldots,$ and Δc_{m+n} can be determined uniquely by solving the nonlinear system with a nonlinear programming method.

From Eq. (11.4), we define a vector function as follows:

$$\varphi = \left(\gamma_1 \cdots \gamma_6 \right)^T = \Phi \left(\Delta c_1 \cdots \Delta c_{m+n} \right) + G_{cn} F_{cn} - F_e \quad (11.18)$$

Similarly, from Eq. (11.17), we define another vector function as follows:

$$\tilde{\varphi} = \left(\gamma_7 \cdots \gamma_{3(m+n)} \right)^T = \tilde{G}^T \hat{G}^{+l} \Delta \hat{c} - \Delta \tilde{c} \quad (11.19)$$

Thus, the problem of solving the elastic deformations using Eqs. (11.4) and (11.17) can be transformed into the following unconstrained nonlinear programming problem:

$$\min_{\Delta c_i} \left(\xi = \sum_{i=1}^{3(m+n)} \gamma_i^2 \right) \quad (11.20)$$

After obtaining the elastic deformations by solving the nonlinear programming problem (11.20), we can determine the passive contact forces $F_{lct} \in \Re^{(3m+2n) \times 1}$ using Eqs. (7.14–7.16).

Using the sliding index ε_i defined in Eq. (11.6), we can further verify the conditions of frictional constraints. When $\varepsilon_i \geq 0$ for all $i = 1, \ldots, m + n$, then all of the frictional forces satisfy Coulomb's friction law, thus corresponding to a fully constrained workpiece-fixture system. In converse, if one of ε_i $(i = 1, \ldots, m+n)$ is less than 0, then the workpiece will not be constrained totally by the fixture.

11.4 Numerical Example

In this section, an example is presented to illustrate the method of optimal planning of clamping forces. From the processes of deriving the local contact compliant model in Chapter 7 and the clamping force planning method mentioned above, it can be found that the

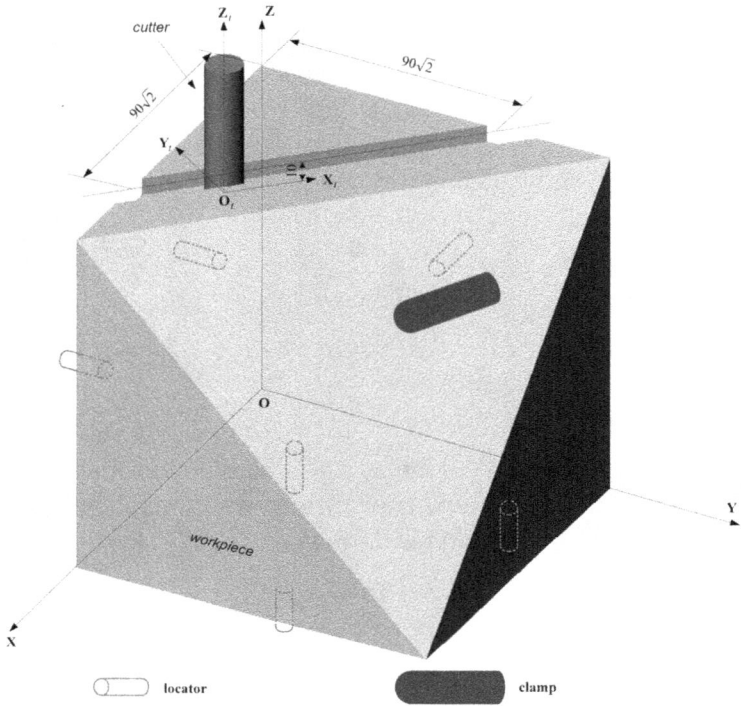

Fig. 11.1.   A workpiece under milling.

planning method of the optimal clamping forces proposed in this chapter can be applicable for any complex workpieces and is suitable for fixtures with more than one clamp. In order to simplify the complexity, without loss generality, we assume that a workpiece is located by six spherically tipped locating pins according to the 3-2-1 locating principle and is to be clamped by one clamp, as shown in Fig. 11.1. The workpiece is in fact a cubic rigid body with one corner cut out, and its side length is 200 mm. The coordinates of the six locators and the normal and tangential unit vectors at the corresponding contacts are shown in Table 11.1. The workpiece is assumed to be made of 7075-T6 aluminum with $E_o = 70.3$ GPa and $\nu_o = 0.354$. Its weight is 50N. The radii of the spherical tips of the locators and the clamp are 6.35 mm and 9.53 mm, respectively. Both the locator and clamp contact elements are made of tool steel with $E_l = 207$ GPa

Table 11.1. Position and orientation of 6 locators.

Coordinates	Unit normal vectors	Unit tangential vectors	
(160, 100, 0)	(0, 0, 1)	(1, 0 ,0)	(0, 1, 0)
(40, 160, 0)	(0, 0, 1)	(1, 0 ,0)	(0, 1, 0)
(40, 40, 0)	(0, 0 ,1)	(1, 0, 0)	(0, 1, 0)
(160, 0, 100)	(0, 1, 0)	(1, 0, 0)	(0, 0, 1)
(40, 0, 100)	(0, 1, 0)	(1, 0, 0)	(0, 0, 1)
(0, 100, 100)	(1, 0, 0)	(0, 1, 0)	(0, 0, 1)

and $\nu_l = 0.292$. The static friction coefficient between the workpiece and the locators/clamp is assumed to be 0.25.

A slot milling operation is going to be performed on the workpiece to produce a through slot with a feed rate of $5\sqrt{2}$ mm/s. The instantaneous milling forces (f_{tx}, f_{ty} and f_{tz}) (N) and couple $m_{tz}(Nmm)$ described in the local tool frame $\mathbf{O}_t \ \mathbf{X}_t \ \mathbf{Y}_t \ \mathbf{Z}_t$ are applied on the workpiece:

$$f_{tx} = 30$$
$$f_{ty} = 25 \sin (\pi t/4)$$
$$f_{tz} = -20$$
$$m_{tz} = 800$$

Note that the cutting force component f_{ty} is time-variant.

A clamp is to be applied on the inclined top surface of the workpiece. A hydraulic cylinder with a clamping force of magnitude f_{in}^c actuates the clamp. The clamp coordinates are $\left(x_{c1}, y_{c1}, 400 - x_{c1} - y_{c1} \right)$. The goal of the clamp planning is to determine the force magnitude f_{1n}^c and the clamp position x_{c1} and y_{c1}.

In our numerical implementation, we use the Levenberg–Marquardt method of the Matlab optimization toolbox to solve the nonlinear least squares problem (11.11) on a PC with Pentium 2.4 GHz processor. The convergence process of the optimization is shown in Figs. 11.2 and 11.3. The step length is large at the beginning of the iteration process, and it becomes smaller as the iteration converges. The optimal clamping force $f_{1n}^c = 81.6$ N is obtained at the position $x_{c1} = 122.1$ mm and

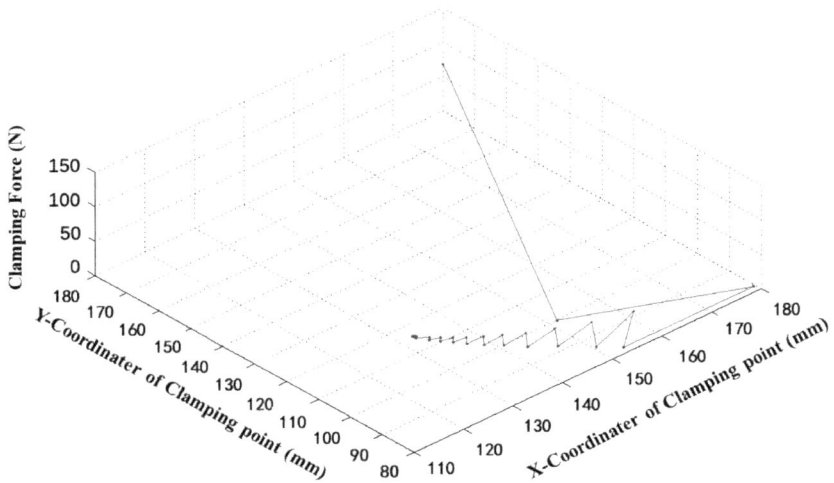

Fig. 11.2.   Clamping force magnitude and position during the iteration process.

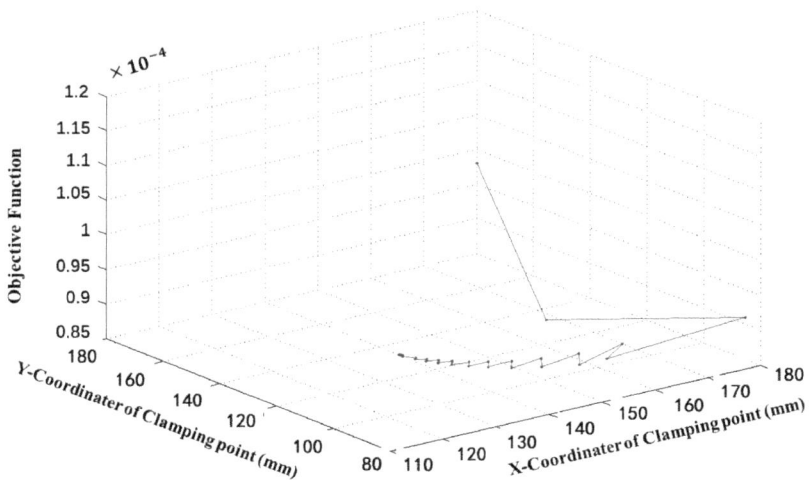

Fig. 11.3.   Objective function changes with the position of clamping during iterations.

$y_{c1} = 99.1$ mm, as shown in Fig. 11.2 (the penalty parameters are finally chosen as 10^5 corresponding to the inequality constraints (11.6) and (11.8), 10^7, 10^7, 10^7, 10^5, 10^5, and 10^5 corresponding to the equality constraints (11.4), 10^8 corresponding to the equality

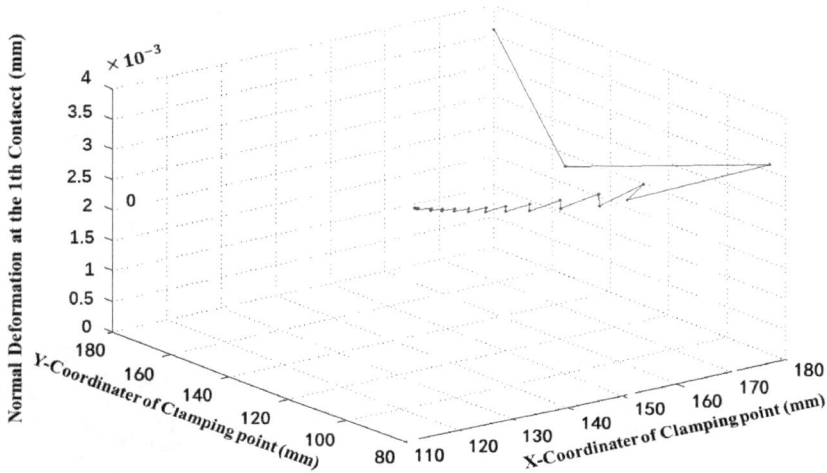

(1a) Normal deformation change with clamping positions at 1st contact.

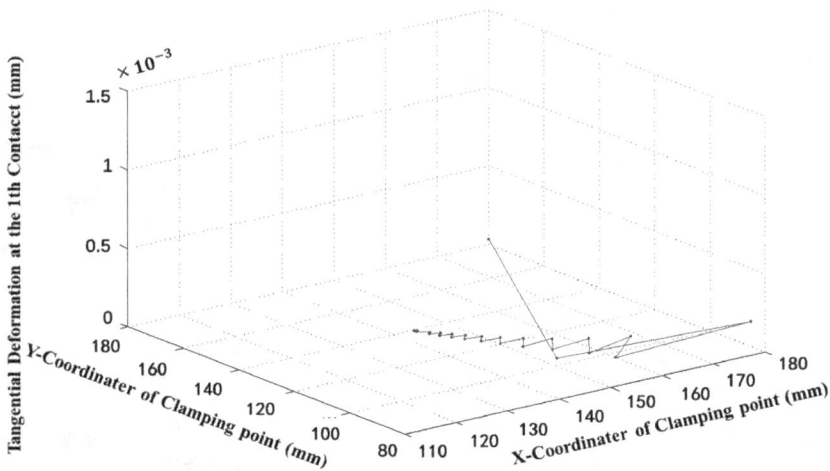

(1b) Tangential deformation change with clamping positions at 1st contact.

Fig. 11.4. Elastic deformation changes with the position of clamping during iterations.

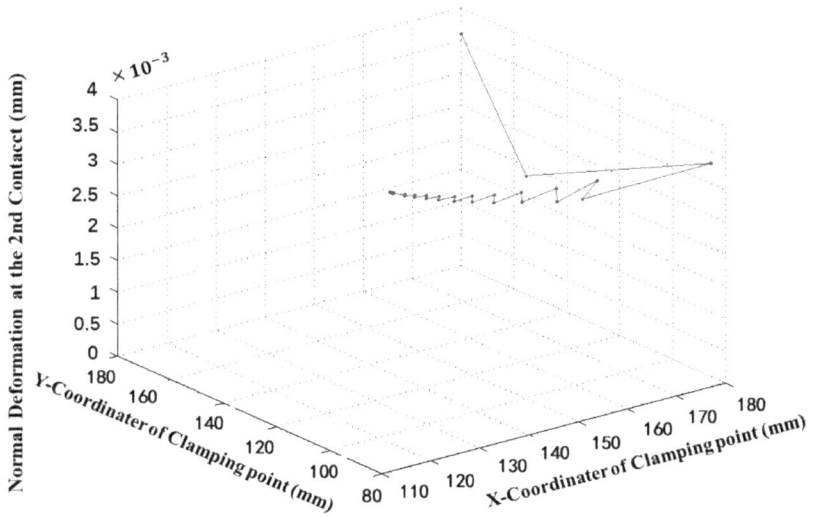

(2a) Normal deformation change with clamping positions at 2nd contact.

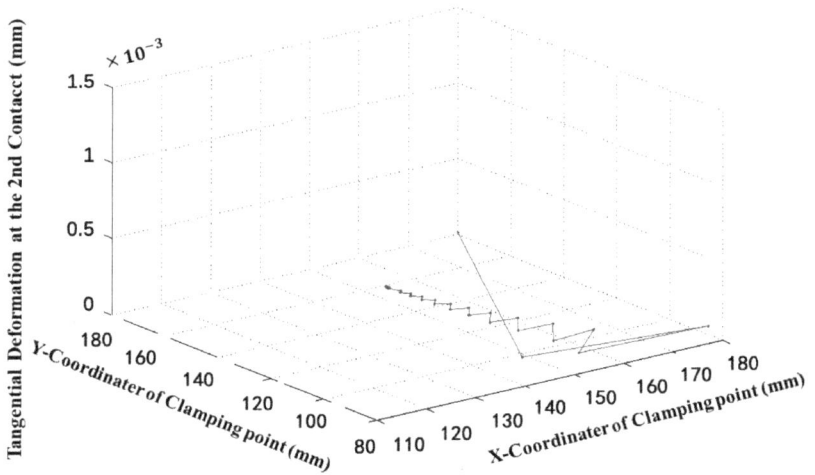

(2b) Tangential deformation change with clamping positions at 2nd contact.

Fig. 11.4. (*Continued*)

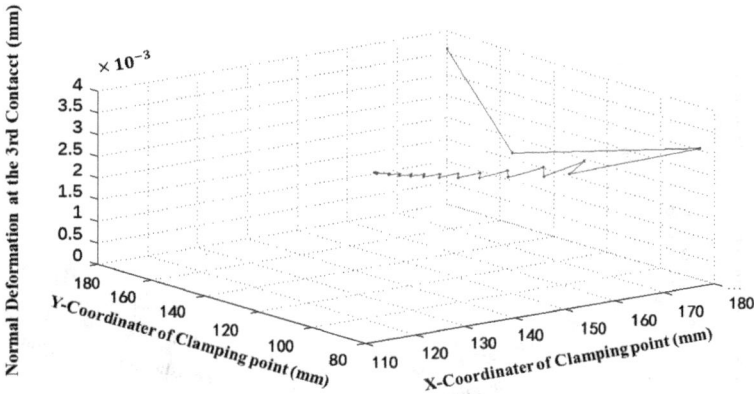

(3a) Normal deformation change with clamping positions at 3rd contact.

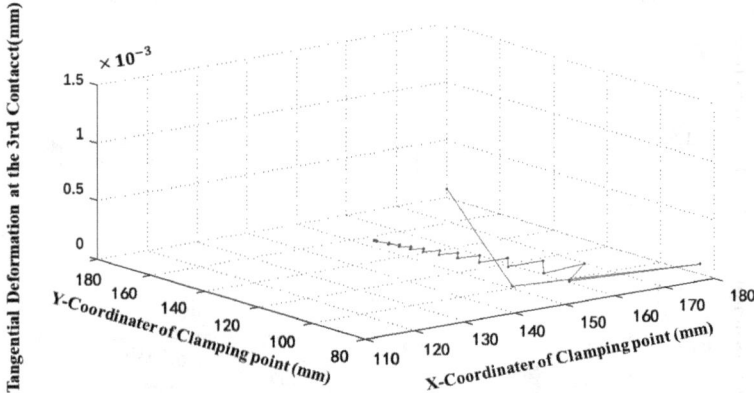

(3b) Tangential deformation change with clamping positions at 3rd contact.

Fig. 11.4. (*Continued*)

constraints (11.5), respectively). The computation costs less than half an hour. The objective function

$$\sum_{i=1}^{7} (\Delta \mathbf{d}_i^T \Delta \mathbf{d}_i)$$

decreases in the process of optimization as shown in Fig. 11.3 with respect to the position of clamping force. The corresponding elastic deformations at contacts during iterations are shown in Fig. 11.4.

With the obtained optimal clamping force and its optimal location, we can predict the contact forces between the workpiece and

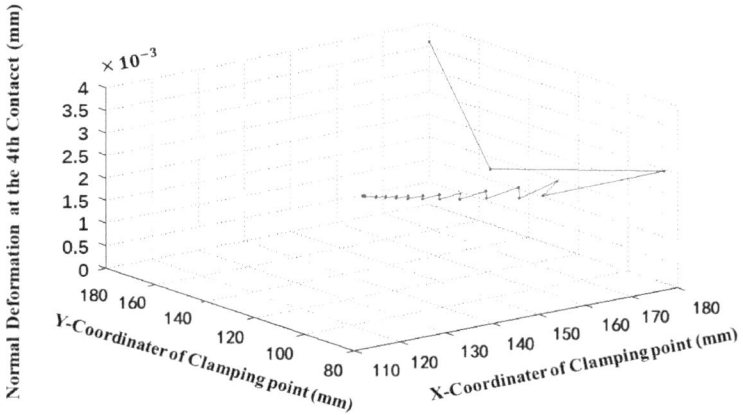

(4a) Normal deformation change with clamping positions at 4th contact.

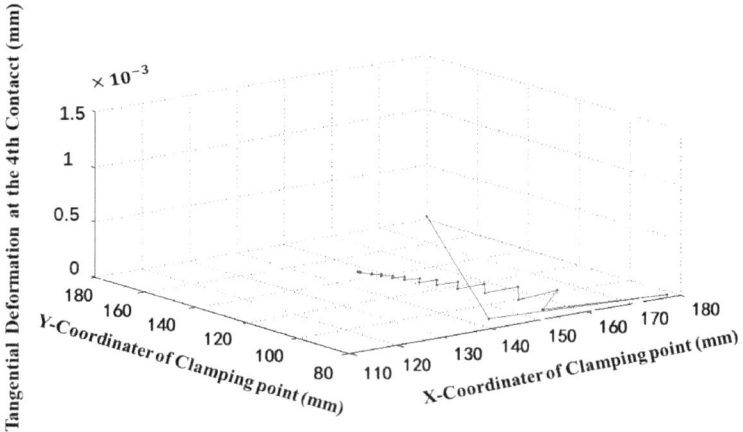

(4b) Tangential deformation change with clamping positions at 4th contact.

Fig. 11.4. (*Continued*)

locator/clamp contacts during the entire machining process. The normal, tangential contact forces and sliding indices for seven contacts between the workpiece, six locators, and one clamp are shown in Figs. 11.5–11.11. It can be found that no slippage occurs at any contact during the whole machining process, which means that all of the frictional constraints (11.6) are satisfied, that is, the clamping force can maintain the workpiece to be constrained totally by the fixture.

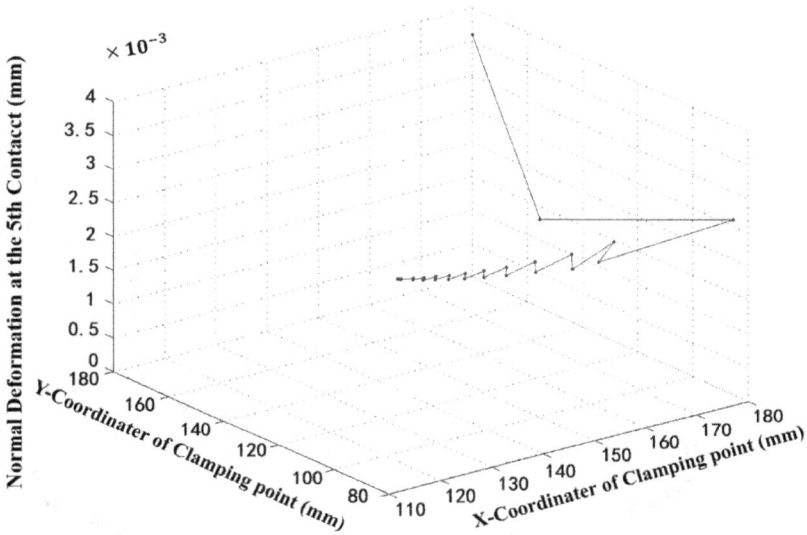

(5a) Normal deformation change with clamping positions at 5th contact.

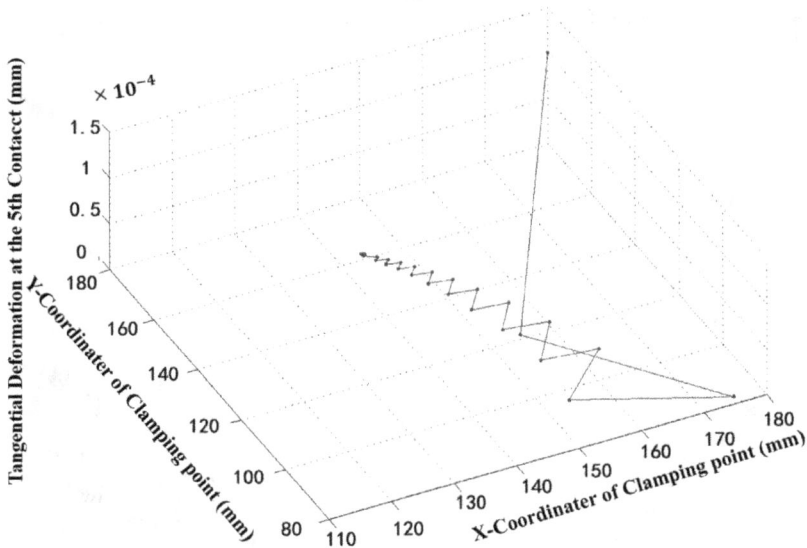

(5b) Tangential deformation change with clamping positions at 5th contact.

Fig. 11.4. (*Continued*)

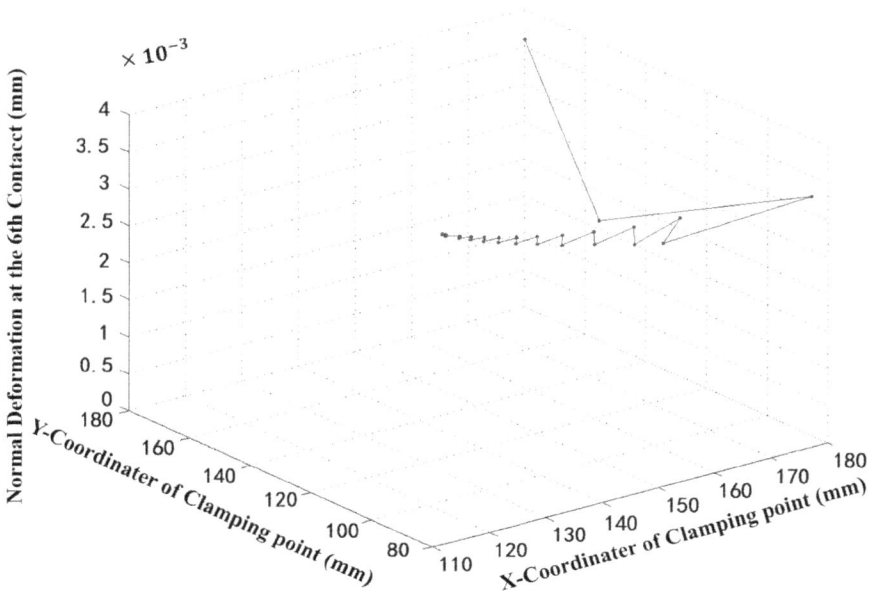

(6a) Normal deformation change with clamping positions at 6th contact.

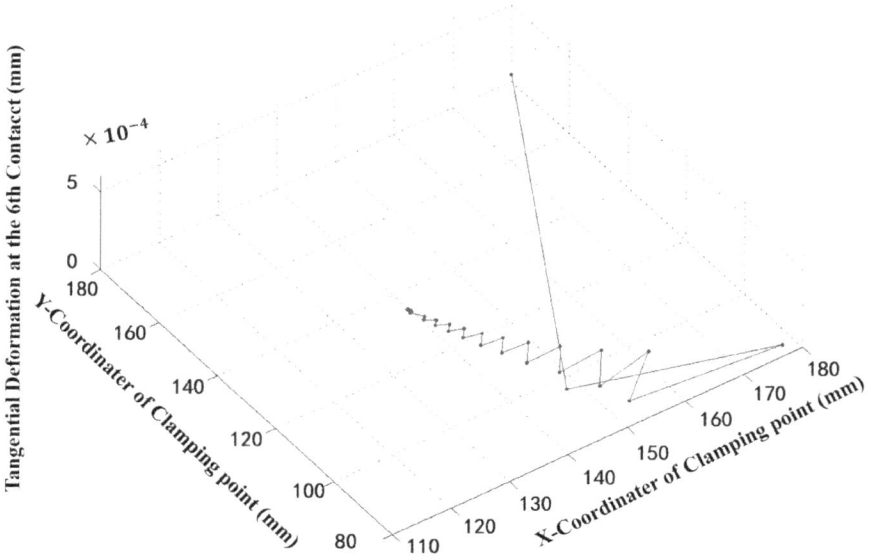

(6b) Tangential deformation change with clamping positions at 6th contact.

Fig. 11.4. (*Continued*)

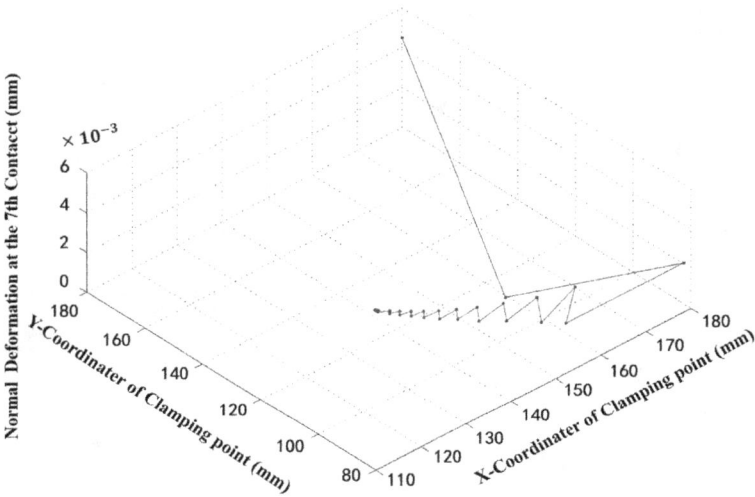

(7a) Normal deformation change with clamping positions at 7th contact.

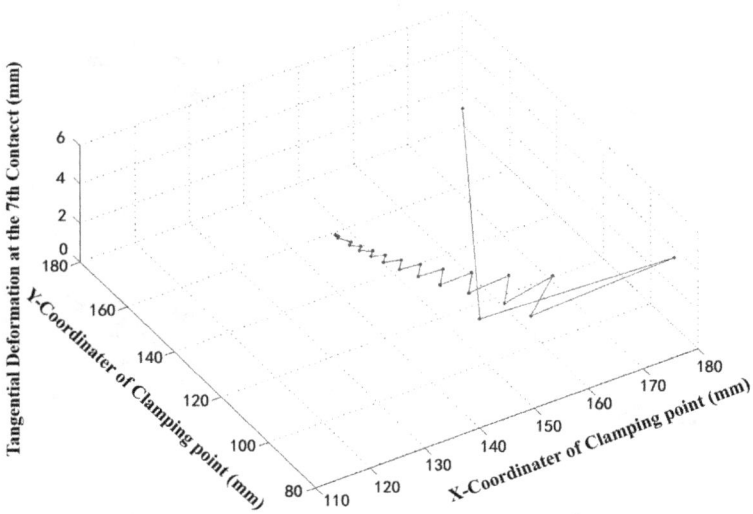

(7b) Tangential deformation change with clamping positions at 7th contact.

Fig. 11.4.   (*Continued*)

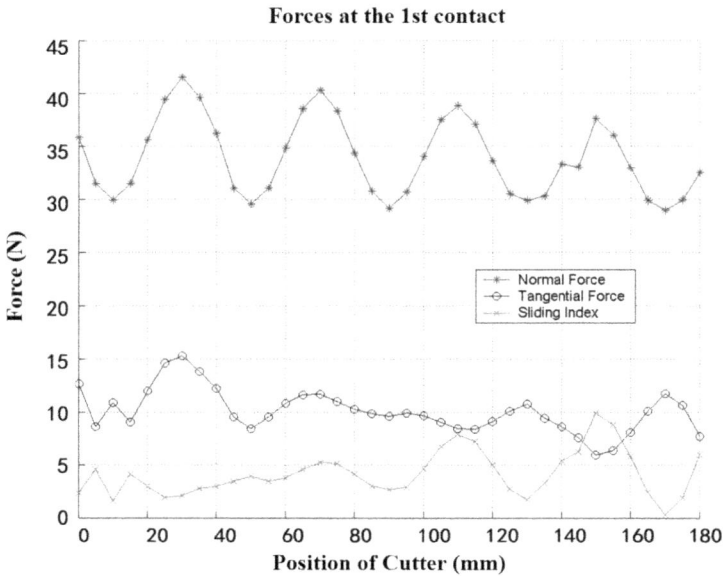

Fig. 11.5. Contact force and sliding index at the 1st contact.

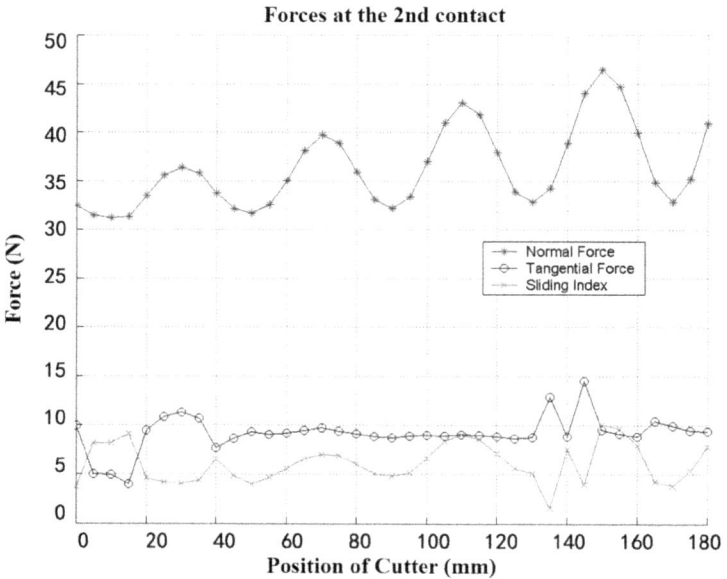

Fig. 11.6. Contact force and sliding index at the 2nd contact.

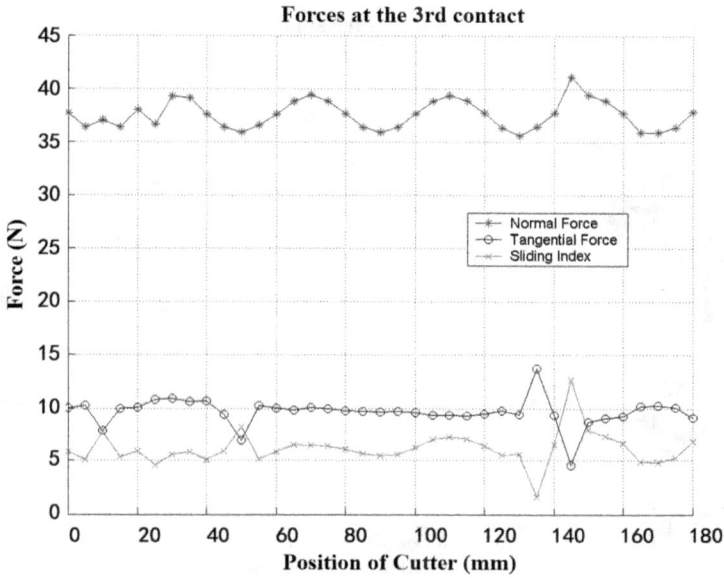

Fig. 11.7. Contact force and sliding index at the 3rd contact.

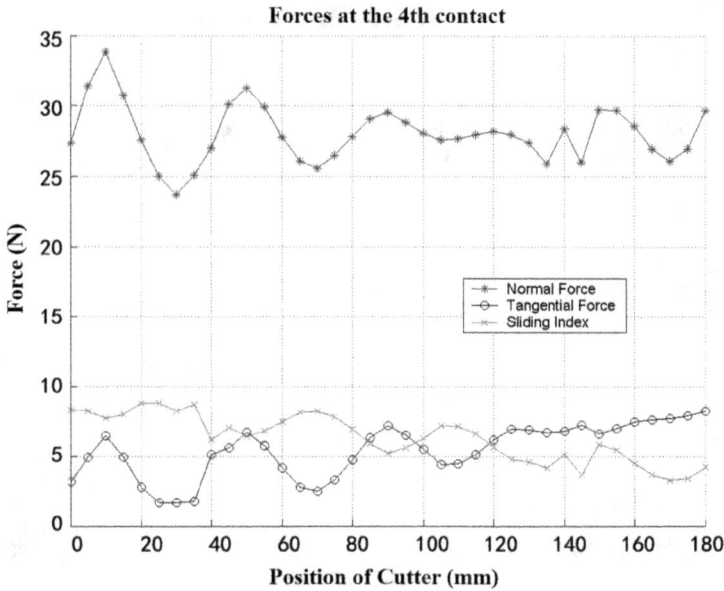

Fig. 11.8. Contact force and sliding index at the 4th contact.

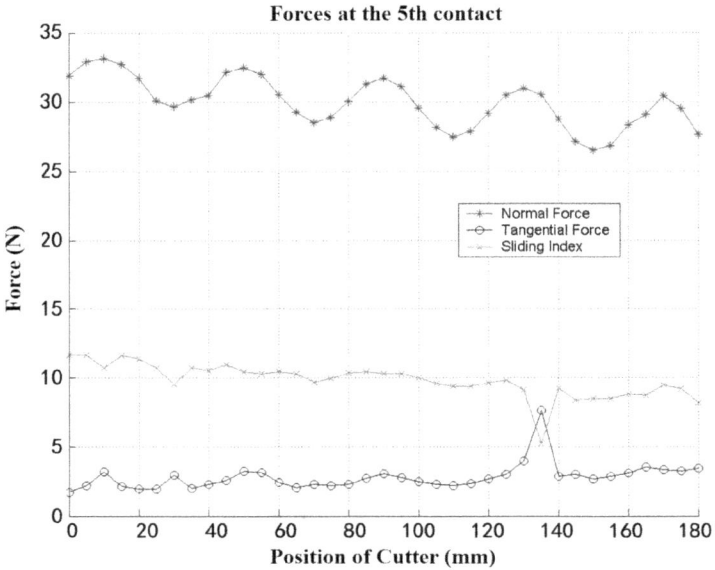

Fig. 11.9. Contact force and sliding index at the 5th contact.

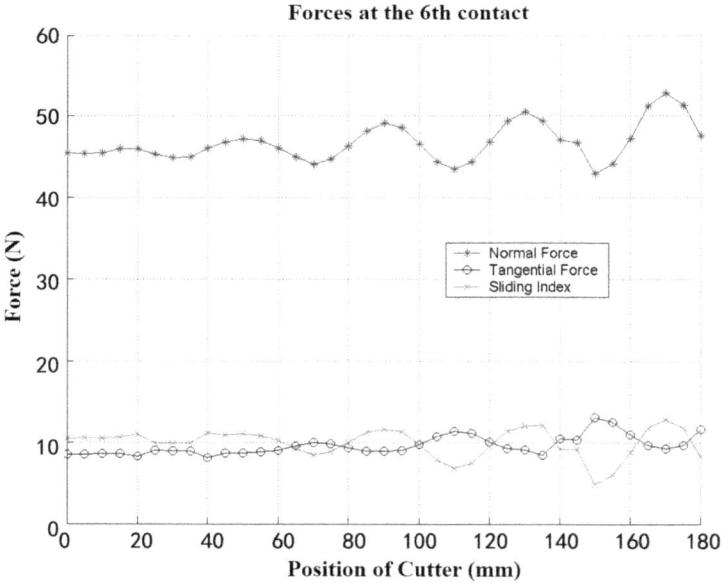

Fig. 11.10. Contact force and sliding index at the 6th contact.

Forces at the 7th contact

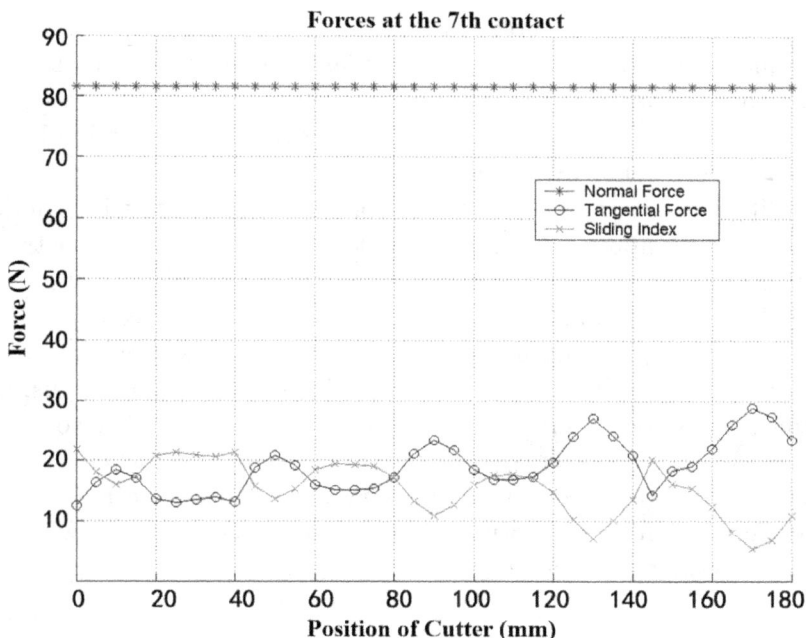

Fig. 11.11.   Contact force and sliding index at the 7th contact.

11.5   Summary

The clamping planning problem is one of the fundamental issues in fixture design. This chapter focuses on the determination of an optimal clamping scheme, including the magnitudes and the positions of the clamping forces in a workpiece-fixture system. We model the contacts between the workpiece and the fixture locators/clamps as locally elastic contacts with friction, whereas the workpiece and fixture elements are otherwise treated as rigid bodies. The optimization goal is to minimize the norm of the elastic deformations at these contacts. The problem is formulated as a constrained nonlinear programming with force/torque equilibrium, elastic deformation compatibility, and dry frictional constraints. Using the exterior penalty function method, we transform the constrained nonlinear optimization into an unconstrained nonlinear programming of nonlinear least squares. Consequently, the optimal magnitudes and positions of clamping forces are obtained by using the Levenberg–Marquardt method. The proposed planning method is illustrated with an example.

The modeling of the locator/clamp-workpiece (fingertip-object) contact is crucial to robotic fixturing/grasping, contact analysis, and stability evaluation. The model of contact compliance with friction derived in Chapter 7 establishes a tractable relationship between the small locator/clamp/workpiece (fingertip/object) displacements and changes in contact forces arising from these displacements. The modeling and planning methods presented in this chapter apply to other systems with the characteristic of passive contacts. Thus, the framework proposed in this chapter could be readily applied to power grasps or enveloping grasps in robotic manipulation, where passive contacts are combined with active contacts to achieve a broad class of manipulation tasks. Optimal planning of the active forces will play an important role in determining the success of these tasks.

References

[1]   Asada H., By A. B., Kinematic analysis of workpart fixturing for flexible assembly with automatically reconfigurable fixtures. *IEEE Journal of Robotics and Automation*, 1(2), pp. 86–93, 1985.

[2]   Bicchi A. On the closure properties of robotic grasping. *International Journal of Robotics Research*, 14(4), pp. 319–334, 1995.

[3]   Cai W. Hu S. J. and Yuan J. X. Deformable sheet metal fixturing: Principles, algorithms, and simulation. *ASME Transactions-Journal of Manufacturing Science and Engineering*, 118, pp. 318–324, 1996.

[4]   Fang B., DeVor R. E. and Kapoor S. G. An elastodynamic model of frictional contact and its influence on the dynamics of a workpiece-fixture system. *ASME Transactions-Journal of Manufacturing Science and Engineering*, 123, pp. 481–489, 2001.

[5]   Fox R. L. *Optimization Methods for Engineering Design*, Addison-Wesley Publishing Company, 1971.

[6]   Goyal S., Ruina A., and Papadopoulos J. Planar Sliding with Dry Friction, Part 1, Limit surface and moment function. *Wear*, 143, pp. 307–330, 1991.

[7]   Zheng Y., Rong Y., Hou Z. A finite element analysis for stiffness of fixture units. *ASME Transactions-Journal of Manufacturing Science and Engineering*, 127, pp. 429–432, 2005.

[8]   Johnson K. L. *Contact Mechanics*, Cambridge University Press, New York, 1985.

[9]   Kao I. and Cutkosky M. R. Dextrous manipulation with compliance and sliding. *International Journal of Robotics Research*, 12(1), pp. 20–40, 1992.

[10] Kulankara K., Satyanarayana S., and Melkote S. N. Iterative fixture layout and clamping force optimization using the genetic algorithm. *ASME Transactions-Journal of Manufacturing Science and Engineering*, 124, pp. 119–125, 2002.

[11] Vallapuzha S., De Meter E. C., Choudhuri S., and Khetan P. R. An investigation into the use of spatial coordinates for the genetic algorithm based solution of the fixture layout optimization problem. *International Journal of Machine Tools and Manufacture*, 42(2), pp. 265–275, 2002.

[12] Wu N. H., and Chan K. C. Genetic algorithm based approach to optimal fixture configuration. *Computers and Industrial Engineering*, 31(3-4), pp. 919–924, 1996.

[13] Lee S. H. and Cutkosky M. R. Fixture planning with friction. *ASME Transactions-Journal of Engineering for Industry*, 113, pp. 320–327, 1991.

[14] Li B. and Melkote S. N. An elastic contact model for the prediction of workpiece-fixture contact forces in clamping. *ASME Transactions-Journal of Manufacturing Science and Engineering*, 121, pp. 485–493, 1999.

[15] Li B. and Melkote S. N. Fixture clamping force optimisation and its impact on workpiece location accuracy. *International Journal of Advanced Manufacturing Technology*, 17, pp. 104–113, 2001.

[16] Lin Q., Burdick J., and Rimon E. Constructing minimum deflection fixture arrangements using frame invariant norms. *IEEE Transactions on Automation Science and Engineering*, 3(3), pp. 272–286, 2006.

[17] Marin R. A. and Ferreira P. M. Optimal placement of fixture clamps: Maintaining form closure and independent regions of form closure. *ASME Transactions-Journal of Manufacturing Science and Engineering*, 124, pp. 676–685, 2002.

[18] Marin R. A. and Ferreira P. M. Optimal placement of fixture clamps: Minimizing the maximum clamping forces. *ASME Transactions-Journal of Manufacturing Science and Engineering*, 124, pp. 686–694, 2002.

[19] Mason M. and Salisbury J. K. *Robot Hands and the Mechanics of Manipulation*, MIT Press, Cambridge, MA, 1985.

[20] Murray R. M., Li Z. and Sastry S. S. *A Mathematical Introduction to Robotic Manipulation*, CRC Press, Boca Raton, FL, 1994.

[21] Scales L. E. *Introduction to Non-Linear Optimization*, Macmillan Publishers Ltd, London, 1985.

[22] Tao Z. J., Kumar A. S. and Nee A. Y. C. Automatic generation of dynamic clamping forces for machining fixtures. *International Journal of Production Research*, 37(12), pp. 2755–2776, 1999.

[23] Pelinescu D. M. and Wang M. Y. Multi-objective optimal fixture layout design. *Robotics and Computer-Integrated Manufacturing*, 18(5–6), pp. 365–372, 2002.

[24] Wang M. Y. and Pelinescu D. M. Optimizing fixture layout in a point-set domain. *IEEE Transactions on Robotics and Automation*, 17(3), pp. 312–323, 2001.

[25] Wang Y. T., Kumar V. Simulation of mechanical systems with multiple frictional contacts. *ASME Transactions-Journal of Mechanical Design*, 116, pp. 571–580, 1994.

[26] Xiong C. H., Li Y. F., Ding H. and Xiong Y. L. On the dynamic stability of grasping. *International Journal of Robotics Research*, 18(9), pp. 951–958, 1999.

[27] Xiong C. H., Li Y. F., Rong Y. and Xiong Y. L. Qualitative analysis and quantitative evaluation of fixturing. *Robotics and Computer Integrated Manufacturing*, 18(5–6), pp. 335–342, 2002.

[28] Xiong C. H., Wang M. Y., Tang Y. and Xiong Y. L. On prediction of passive contact forces of workpiece-fixture systems. *Proceedings IMechE Part B: Journal of Engineering Manufacture*, 219(B3), pp. 309–324, 2005.

[29] Xiong C., Rong Y., Koganti R., Zaluzec M. J. and Wang N. Geometric variation prediction in automotive assembling. *Assembly Automation*, 22(3), 260–269, 2002.

[30] Xiong Y. L., Ding H., and Wang M. Y. Quantitative analysis of inner force distribution and load capacity of grasps and fixtures. *ASME Transactions-Journal of Manufacturing Science and Engineering*, 124, pp. 444–455, 2002.

[31] Xydas N. and Kao I. Modeling of contact mechanics and friction limit surface for soft fingers with experimental results. *International Journal of Robotics Research*, 18(9), pp. 941–950, 1999.

[32] Kang Y., Rong Y., and Yang J. C. Computer-aided fixture design verification. Part 3. Stability Analysis. *International Journal of Advanced Manufacturing Technology*, 21(10–11), pp. 842–849, 2003.

[33] Liu Y. H., Lam M. L., Ding D. A complete and efficient algorithm for searching 3-D form-closure grasps in the discrete domain. *IEEE Transactions on Robotics*, 20(5), pp. 805–816, 2004.

[34]   Gopalakrishnam K., Goldberg K., Bone G. M., Zaluzec M. J., Koganti R., Pearson R., Deneszczuk P. A. Unilateral fixtures for sheet-metal parts with holes. *IEEE Transactions on Automation Science and Engineering*, 1(2), pp. 110–120, 2004.

[35]   Tan E. Y. T., Kumar A. S., Fuh J. Y. H., Nee A. Y. C. Modeling, analysis, and verification of optimal fixturing design. *IEEE Transactions on Automation Science and Engineering*, 1(2), 121–133, 2004.

[36]   Zhong W., Hu S. J. Modeling machining geometric variation in a N-2-1 fixturing scheme. *ASME Transactions-Journal of Manufacturing Science and Engineering*, 128, pp. 213–219, 2006.

Chapter 12

Design and Implementation of Anthropomorphic Hand for Replicating Human Grasping Functions

Designing an anthropomorphic hand with a limited number of actuators to replicate the grasping functions of the human hand presents a challenging problem. This chapter explores a comprehensive theory for the design of such an anthropomorphic hand, aiming to endow it with natural grasping abilities. To analyze the grasping mechanism of the human hand in daily living, a grasping experimental paradigm is established. The study investigates the movement relationships among joints within a digit, among digits in the human hand, and the postural synergic characteristics of the fingers during grasping. A design principle for the anthropomorphic mechanical digit is developed to reproduce the digit grasping movement of the human hand. Additionally, a design theory for the kinematic transmission mechanism, which can be embedded into the palm of the anthropomorphic hand to replicate the postural synergic characteristics of the fingers using a limited number of actuators, is proposed. Formulating the design method for the anthropomorphic hand to replicate human grasping functions is a key focus. The effectiveness of the proposed design method for the anthropomorphic hand is validated through grasping experiments.

12.1   Introduction

Over the past few decades, numerous researchers have dedicated their efforts to understanding the fundamental aspects of multi-fingered robotic hands [1–6]. Impressive developments have been achieved, resulting in multifingered robotic hands like the Stanford/ JPL hand [2] and the Utah/MIT hand [7], where each joint is actuated independently. More recent designs of fully actuated hands can be found in Refs. [8–10]. However, these sophisticated designs are not suitable for use as prosthetic hands, as the abundance of actuators makes them challenging to control through an intuitive human-machine interface. In response to the complexity of modern mechanical hands, special attention has been given to reducing the number of actuators.

Coupling multiple joints in a mechanical hand is a method to reduce the number of actuators [11]. By employing joint coupling design, the hand can adapt passively to the physical properties of objects, thus potentially decreasing the requirement for complex sensing and control. This leads to an easier and more reliable operation of the hand [12,13]. However, the underlying principles to determine the most suitable joint couplings remain largely unexplored.

Underactuation, that is, the number of actuators is less than the degrees of freedom of the hand, is another way to decrease the number of actuators in a mechanical hand [14,15]. A tendon-driven robotic finger is developed based on an anatomical model of a human finger [16]. The optimization of tendon routing in robotic fingers is described in Ref. [17], and the optimum tendon routing can produce fingers with force-production capabilities that can exceed that of human hands [17]. A design method is proposed for tendon-driven mechanisms with active and passive tendons based on kinematic analyses [18], and connected motion of the DIP and PIP joints is achieved [19]. A prototype of soft gripper with tendon-pulley differential transmission is presented by Hirose and Umetani [20]. Due to its self-adaptability, the differential mechanisms are widely used in robotic hands [21] and prosthesis [22]. Force capabilities of underactuated fingers are analyzed, and an ejection phenomenon leading to unstable grasps due to uncontrollable force distribution may be found in Ref. [23]. The implementation of stable precision grasps by underactuated grippers is investigated by means of simple design modifications of the digital phalanges of the fingers [24]. A method for first-order form-closure analysis of underactuated grasps and a necessary and sufficient condition on the minimum number of unilateral constraints required for the first-order form-closure are proposed in Ref. [25]. Up to now, almost all fingers in underactuated prosthetic hands flex actively around one direction by means of a differential mechanism actuated by a DC motor. When the motor releases the driving mechanism, torsion springs in the joints extend the finger. This transmission reduces the number of actuators to only one in each finger, making it possible to use only six actuators in a hand, including the one for thumb's opposing motion [26, 27]. Such a scheme is adopted by a soft hand [28] and many commercial prosthetic hands, such as i-Limb hand, Bebionic hand, and Vincent hand. It means that the prosthetic hands have self-adaptive grasping ability but cannot grasp an object in a natural-looking manner like its human counterpart. Meanwhile, six actuators are still too many to be controlled via human–machine interface [29]. Coupling multiple fingers with one actuator, such as in Refs. [30, 31], leads to significant reduction in grasping dexterity. How to replicate human hand-grasping functions with a few actuators is still an open question.

This chapter is to introduce a general theory for designing an anthropomorphic hand with a few actuators and endowing the hand with natural grasping functions. The design principle is based on the analysis result of the human hand motions, i.e., movement relationship among joints in a digit, movement relationship among digits in a hand, and postural synergic characteristic of the fingers. A new mechanical design method is formulated for anthropomorphic hands with a few actuators. As a result, an anthropomorphic hand, called X-hand, has been developed by utilizing such a design method. Meanwhile, abundant experiments have been carried out to demonstrate the effectiveness of the developed robotic hand under various grasping types, grasping forces, and holding capabilities.

12.2   Grasping Mechanism and Postural Synergic Characteristic of Human Hand

In order to design an anthropomorphic hand to replicate the grasping functions of the human hand, it is necessary to measure the joint angles of hand fingers by the data capture device, such as the CyberGlove II. The measured joints for thumb are called the interphalangeal (IP), the metacarpophalangeal (MCP), and the carpometacarpal (CMC) joint. The joints for other fingers are called distal interphalangeal (DIP), proximal interphalangeal (PIP), and metacarpophalangeal (MCP) joints, as shown in Fig. 12.1. The obtained movement data of the CMC, MCP, and IP joints of the thumb, and the MCP, PIP, and DIP joints of the fingers can be described in Fig. 12.3. The 33 different grasp types, as shown in

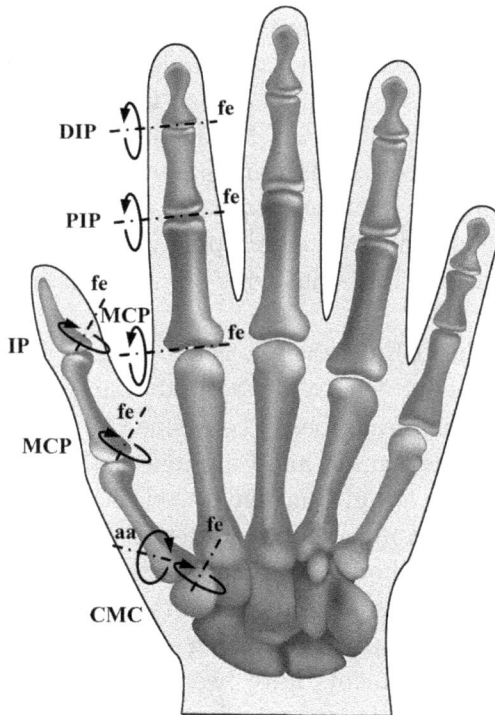

Fig. 12.1.   Joints and kinematic model of human hand.

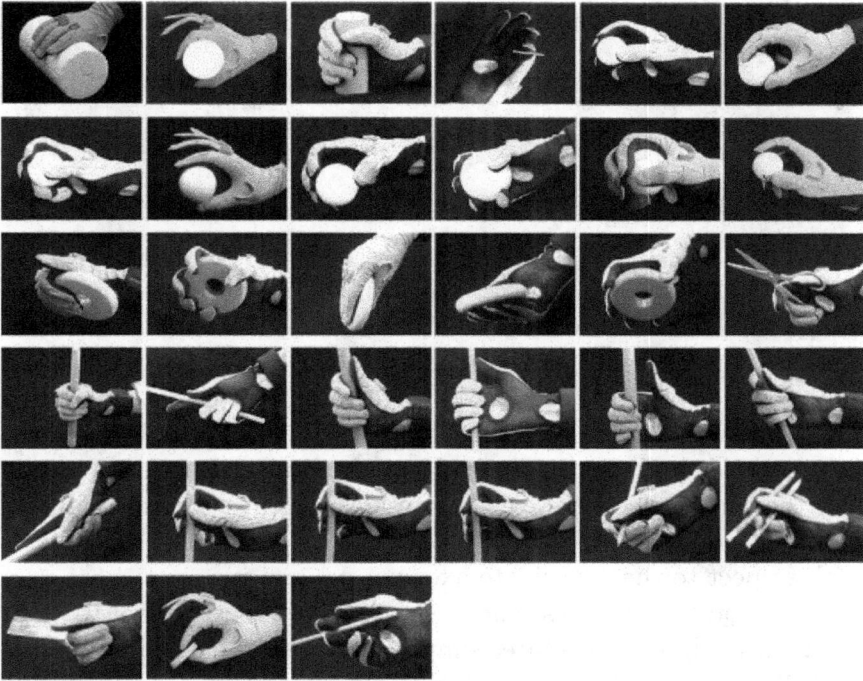

Fig. 12.2. Grasp types and final grasping configuration of human hand.

Fig. 12.2, are chosen in the grasping experimental paradigm to measure joints angles.

The joint movement data can be expressed in a matrix $\mathbf{Q}_d \in \Re^{n \times m}$ where n denotes the number of joints in a digit and m represents the number of discrete joint angle during the grasping from the initial configuration to the final configuration for the 33 grasp types. The ith row, jth column element q_{ij} of the matrix $\mathbf{Q}_d \in \Re^{n \times m}$ is the ith joint angle in a digit at the jth discrete point.

12.3 Movement Relationship among Joints in a Digit

By using the principal component analysis (PCA) on the matrix $\mathbf{Q}_d$, the complicated motion of every digit can be represented as follows:

$$\mathbf{q} = \bar{\mathbf{q}} + \beta_1 \mathbf{p}_1 + \beta_2 \mathbf{p}_2 + \beta_3 \mathbf{p}_3 \tag{12.1}$$

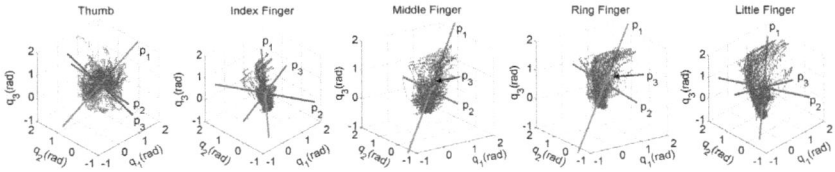

Fig. 12.3.   Joint movements of digits.

where $\mathbf{q} = \begin{pmatrix} q_1 & q_2 & q_3 \end{pmatrix}^T \in \Re^{3 \times 1}$ represents the joint motion of a digit. q_1, q_2 and q_3 denote the MCP, PIP, and DIP joint angles (CMC joint angle around the axis-fe, MCP, and IP joint angles for the thumb). The second subscript j of the joint angle q_{ij} is omitted for simplifying the expression. $\overline{\mathbf{q}} \in \Re^{3 \times 1}$ is the average of $\mathbf{q}$. $\mathbf{p}_1$, $\mathbf{p}_2$, and $\mathbf{p}_3$ are the first, second, and third principal component vectors which are in fact the three orthogonal eigen directions (eigenvectors) of the joint motion of a digit (see Fig. 12.3). β_1, β_2, and β_3 are the coordinates with respect to the coordinate axes $\mathbf{p}_1$, $\mathbf{p}_2$, and $\mathbf{p}_3$.

Equation (12.1) shows that the complicated motion of every digit can be decomposed into three simple movements along three orthogonal directions, which means that the motion of digits is composed of three eigen movements along the eigen directions although the motion of the human hand during grasping is very complex. The eigenvalues σ_i ($i = 1, 2, 3$) of the covariance matrix $\mathbf{Q}_d \mathbf{Q}_d^T$ corresponding to the eigenvectors $\mathbf{p}_i$ ($i = 1, 2, 3$) have the relationship: $\sigma_1 \geq \sigma_2 \geq \sigma_3$. The larger the eigenvalue, the greater the variation of the movement data in the corresponding eigen direction, which means that the range of the digital motion along the eigen direction with greater variation of the movement data is larger. Using the eigenvalues σ_i ($i = 1, 2, 3$), an index t_w to measure the cumulative percentage of total variation of the joint movement is defined as

$$t_w = \sum_{i=1}^{w} \sigma_i \Big/ \sum_{i=1}^{3} \sigma_i \qquad (12.2)$$

The joint movement of a digit in the space spanned by the eigen directions $\mathbf{p}_1, \ldots, \mathbf{p}_w$ is called the *primary motion* when $t_w \geq 80\%$, and the joint movement in the space spanned by other eigen directions except the former w eigen directions is referred to as the *secondary motion*. Calculating the index t_w on the basis of the PCA on the matrix $\mathbf{Q}_d$, we find that $t_1 > 80\%$ for all of the digits during grasping activities of daily living, which means that the eigen direction $\mathbf{p}_1$ can be referred to as the *primary motion* direction of the digit (in fact the movement forming the grasping pattern is generated by the primary movement of the digit), the movement in the plane spanned by the eigen directions $\mathbf{p}_2$ and $\mathbf{p}_3$ can be called the *secondary motion* of the digit which endows the digit with the adaptive ability to the shape of the grasped object. Thus, the *primary motion* $\mathbf{q}_{pm}$ and *secondary motion* $\mathbf{q}_{sm}$ of the digit can be represented as

$$\mathbf{q}_{pm} = \bar{\mathbf{q}} + \beta_1 \mathbf{p}_1 \tag{12.3}$$

$$\mathbf{q}_{sm} = \beta_2 \mathbf{p}_2 + \beta_3 \mathbf{p}_3 \tag{12.4}$$

Equation (12.3) also shows that there exist postural synergies among joints in a digit, which means that there exists an inherent coordinated movement relationship among joints in a digit determined by the proportional relationship of elements in the eigenvector $\mathbf{p}_1$. In other words, if the movement of the MCP joint in a digit is given, then the movement of the other joints such as the PIP and DIP joints can be determined. The motion process (namely, configuration) of the digit is described by the coordinate β_1 along the *primary motion* direction $\mathbf{p}_1$.

The characteristic of the joint movement data of digits can be interpreted by the constant density ellipsoid defined with the covariance matrix $\mathbf{Q}_d \mathbf{Q}_d^T$, the ellipsoid is represented as

$$(\mathbf{q} - \bar{\mathbf{q}})^T \left(\mathbf{Q}_d \mathbf{Q}_d^T \right)^{-1} (\mathbf{q} - \bar{\mathbf{q}}) = c^2 \tag{12.5}$$

where c is a constant.

The length $\sqrt{\sigma_i}$ $(i = 1, 2, 3)$ of the ith principal axis $\mathbf{p}_i$ of the ellipsoid reflects the joint motion capability of the digit along the eigen direction $\mathbf{p}_i$, which means that the motion capability of the digital joint along different directions can be described by the orientation and shape of the ellipsoid. Without loss of generality, Fig. 12.4 shows the constant density ellipsoid corresponding to the index finger.

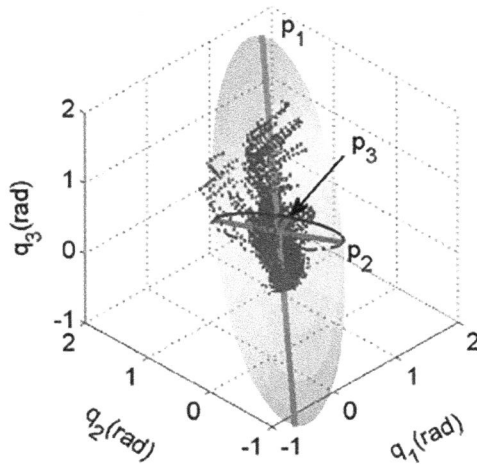

Fig. 12.4. Constant density ellipsoid corresponding to the index finger.

12.4   Movement Relationship among Digits in the Hand

In order to explore the grasping mechanism and movement characteristic of the human hand to design an anthropomorphic hand, it is necessary to investigate the movement relationship among digits in the human hand, based on the grasping data of six variables, that is, the MCP joints of four fingers (Index: $\mathbf{I}$, Middle: $\mathbf{M}$, Ring: $\mathbf{R}$, and Little: $\mathbf{L}$) and the thumb motions around the axis-fe ($\mathbf{T}_\mathrm{f}$) and the axis-aa ($\mathbf{T}_a$) in the CMC described in Fig. 12.1.

Further, we calculate Pearson correlation coefficients between different variables to measure the degree of movement dependencies among these joints. The bigger the correlation coefficient between two joints, the higher the degree of movement dependency between the two joints. Figure 12.5 gives the movement dependencies among the CMC joint of the thumb and the MCP joints of the fingers. Using the correlation coefficients as shown in Fig. 12.5 and the similarity measure, we can obtain the clustering dendrogram of the movement dependencies among digits in the human hand, as shown in Fig. 12.6.

From Fig. 12.6, it can be found that the motion generated by $\mathbf{T}_\mathrm{f}$ and $\mathbf{T}_\mathrm{a}$ belongs to one cluster, and the motion generated by the

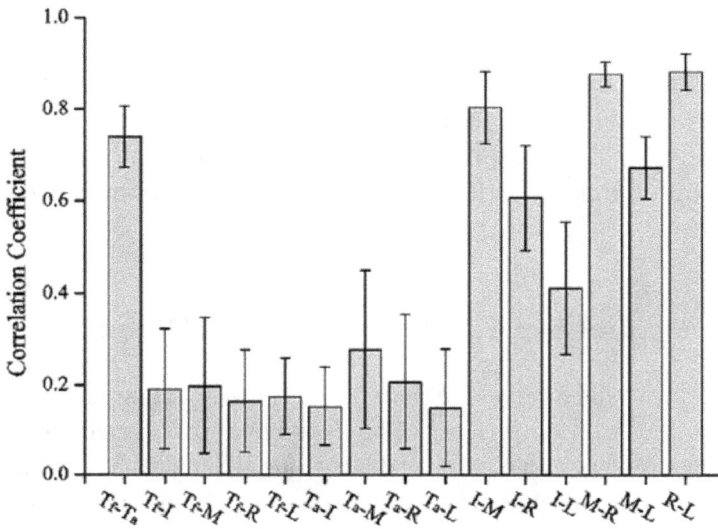

Fig. 12.5. Movement dependencies among the CMC joint of the thumb and the MCP joints of the fingers.

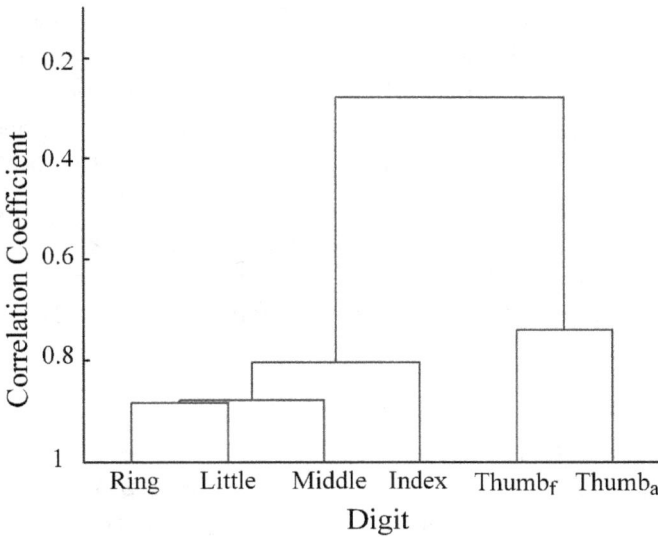

Fig. 12.6. Clustering dendrogram of the movement dependencies among digits.

MCP joints of the index, middle, ring, and little fingers belongs to another cluster, which connotes that the thumb moves independently of other four fingers in the grasping. The movements between $\mathbf{T}_f$ and $\mathbf{T}_a$ in the thumb are correlated, and the movements between index, middle, ring, and little fingers are also correlated. The degree of correlation among digits depends on the correlation coefficients in Fig. 12.6.

12.5   Postural Synergic Characteristic of Fingers

The results of the analysis of the movement relationship among digits indicate that the four fingers are in motion simultaneously, but not all of the four fingers move independently of one another in the grasping. The movement dependencies among the four fingers imply that there exist postural synergies in the four fingers. The postural synergies are called simply synergies or eigengrasps which is a small number of dominant postures to describe the entire act of grasp.

There exists inherent coordinated movement relationship (12.3) among joints in each finger. Thus, as long as the coordinated movement relationship of the MCP joints in the four fingers is found out, then postural synergies in the four fingers can be determined.

Using the singular value decomposition (SVD) on the motion data matrix $\mathbf{Q}_h = \begin{pmatrix} \mathbf{q}_1^1 & \mathbf{q}_1^2 & \mathbf{q}_1^3 & \mathbf{q}_1^4 \end{pmatrix}^T \in \Re^{4 \times m}$ of the MCP joints of the four fingers where $\mathbf{q}_1^1 \in \Re^{1 \times m}$, $\mathbf{q}_1^2 \in \Re^{1 \times m}$, $\mathbf{q}_1^3 \in \Re^{1 \times m}$, and $\mathbf{q}_1^4 \in \Re^{1 \times m}$ are the MCP joint angle series of the index, middle, ring, and little fingers, respectively (obtained in Section 12.2), that is, q_1 in the previous sections, we can obtain

$$\mathbf{Q}_h = \mathbf{U}\mathbf{L}\mathbf{S}^T \tag{12.6}$$

where $\mathbf{U} = \begin{pmatrix} \mathbf{u}_1 & \mathbf{u}_2 & \mathbf{u}_3 & \mathbf{u}_4 \end{pmatrix} \in \Re^{4 \times 4}$ with $\mathbf{u}_i \in \Re^{4 \times 1}$ and $\mathbf{S} = \begin{pmatrix} \mathbf{s}_1 & \mathbf{s}_2 & \mathbf{s}_3 & \mathbf{s}_4 \end{pmatrix} \in \Re^{m \times 4}$ with $\mathbf{s}_i \in \Re^{m \times 1}$, $i = 1, 2, 3, 4$; each of the matrices $\mathbf{U}$ and $\mathbf{S}$ has orthonormal columns so that $\mathbf{U}^T \mathbf{U} = \mathbf{I}_4$, $\mathbf{S}^T \mathbf{S} = \mathbf{I}_4$ ($\mathbf{I}_4 \in \Re^{4 \times 4}$ is an identity matrix). $\mathbf{L}$ is a (4×4) diagonal matrix with ith diagonal element $l_i^{1/2}$ ($i = 1, 2, 3, 4$) which is the singular value of the matrix $\mathbf{Q}_h$. The related matrices can be found in Table 12.1.

Table 12.1.

Variables	$\mathbf{U} = \begin{bmatrix} \mathbf{u}_1 & \mathbf{u}_2 & \mathbf{u}_3 & \mathbf{u}_4 \end{bmatrix}$	$\mathbf{L} = \mathrm{diag}\left\{ l_1^{\frac{1}{2}}, l_2^{\frac{1}{2}}, l_3^{\frac{1}{2}}, l_4^{\frac{1}{2}} \right\}$
Results	$\begin{bmatrix} -0.35 & 0.76 & -0.31 & -0.45 \\ -0.44 & 0.38 & 0.34 & 0.74 \\ -0.52 & -0.25 & 0.66 & -0.48 \\ -0.64 & -0.46 & -0.60 & 0.13 \end{bmatrix}$	$\begin{bmatrix} 42.35 & 0 & 0 & 0 \\ 0 & 11.77 & 0 & 0 \\ 0 & 0 & 6.83 & 0 \\ 0 & 0 & 0 & 3.69 \end{bmatrix}$

Equation (12.6) can be rewritten as

$$\Delta \mathbf{q}_h = \xi_1 \mathbf{u}_1 + \xi_2 \mathbf{u}_2 + \xi_3 \mathbf{u}_3 + \xi_4 \mathbf{u}_4 \tag{12.7}$$

where $\Delta \mathbf{q}_h \in \Re^{4 \times 1}$ is a column of the matrix $\mathbf{Q}_h$ which represents a posture departure from the initial configuration where all fingers are in the natural extension state in the human hand configuration space. $\mathbf{u}_1$, $\mathbf{u}_2$, $\mathbf{u}_3$, and $\mathbf{u}_4$ are the four orthonormal eigen directions which are called the first, second, third, and fourth postural synergies (also called eigengrasps or eigenvectors) of the four fingers (see Fig. 12.7). The proportional relationship of four elements in the eigenvector $\mathbf{u}_i$ ($i = 1, 2, 3, 4$) determines the inherent coordinated movement characteristic (namely, postural synergic characteristic) of four fingers. ξ_i is the coordinate along the coordinate axe $\mathbf{u}_i$; its value represents the scalar weight of the eigengrasp $\mathbf{u}_i$ contributed to the corresponding natural grasping posture in the configuration space of the human hand ($i = 1, 2, 3, 4$). The range of the coordinate ξ_i is $[\xi_i^{\min}, \xi_i^{\max}]$, $i = 1, 2, 3, 4$. Along each eigen direction, three postures, denoting the minimum, the average, and the maximum configuration, are given in Fig. 12.7. The corresponding coordinates are $\xi_i^{\min}$, ξ_i^{ave}, and $\xi_i^{\max}$, respectively. ξ_i^{ave} is the average of $\xi_i^{\min}$ and $\xi_i^{\max}$.

Equation (12.7) shows that the complicated grasp posture in the activities of daily living can be decomposed into four eigengrasps. In other words, the variety of grasp postures of the human hand is formulated by the linear combination of the four eigengrasps. In fact, using the singular value $l_i^{1/2}$ ($i = 1, 2, 3, 4$) of the matrix $\mathbf{Q}_h$, we can define the contribution rate of the different eigengrasps for the

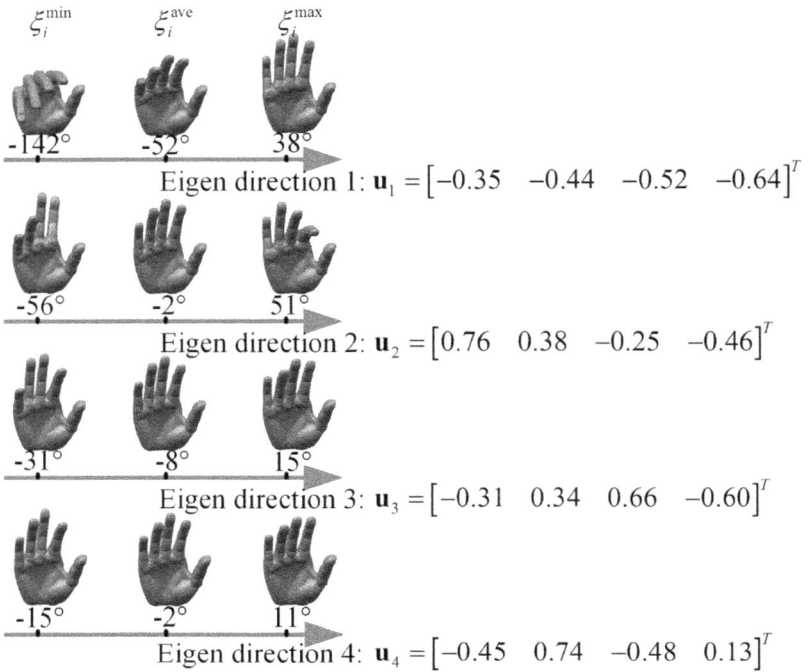

Fig. 12.7. Posture synergic characteristic of the four fingers where the thumb is fixed at its initial configuration. Only the movement of the MCP joints corresponding to the four fingers is considered.

motion reconstruction as

$$C_r^i = l_i \bigg/ \sum_{j=1}^{4} l_j, \quad (i = 1, \ldots, 4) \tag{12.8}$$

Using Eq. (12.8), we can find that the contribution rates of the four eigengrasps to form the actual motion of the human hand are 90%, 7%, 2%, and 1%, respectively, which means that the first and second eigengrasps play a dominant role in the motion reconstruction of the human hand. Therefore, the actual motion of the human hand can be reconstructed approximately by the linear combination of the first and second eigengrasps. The grasp motion spectra of the human hand can be described in the orthogonal 2-D coordinate frame composed of the eigen direction 1 and the eigen direction 2, as shown

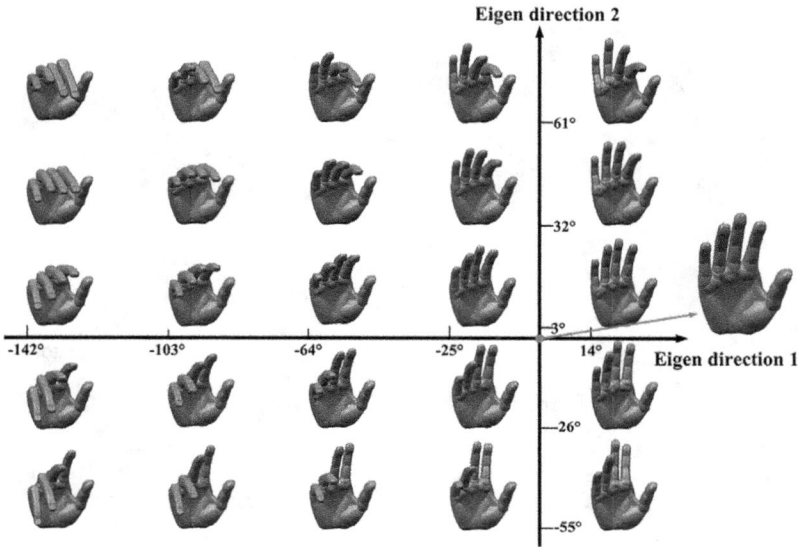

Fig. 12.8. Grasp motion spectra of the human hand. Note that, similar to Fig. 12.7, only four MCP joints are considered here.

in Fig. 12.8. The origin of the coordinate frame corresponds to the initial grasp posture where all fingers are in the extension state.

12.6   Design Principle of Desired Mechanical Digit

Based on the decomposition of human digit movements, a design principle of the desired mechanical digit can be described by the following equivalent conversion from human digits to mechanical digits, as shown in Fig. 12.9. In order to implement the related movement among joints in a human digit by a compact mechanical digit, the tendon-pulley transmission mechanism, as an independent module, is used in the designed mechanical digit. The primary motion is achieved via an actuator, and the secondary motion is implemented with mechanical compliance matching statistic parameters of human motion data.

On consideration of the elements having same sign in the 1st eigen motion $\mathbf{p}_1$ of the index finger, the tendon driving the joints

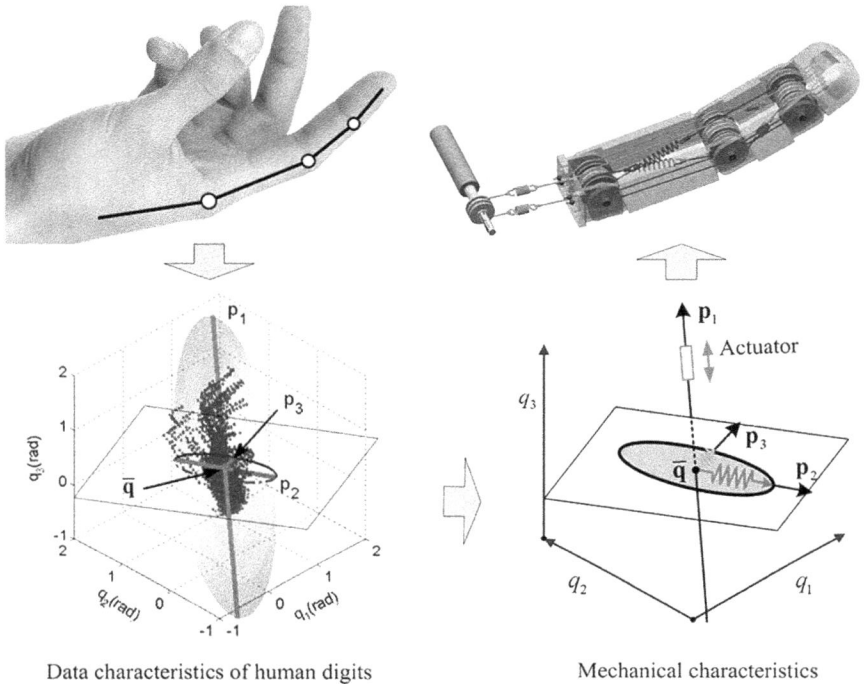

Fig. 12.9.   Equivalent conversion from human digits to mechanical digits.

of the index finger to generate the primary motion should be wrapped on the set of pulleys around the same direction as shown in Fig. 12.10(a).

In order to constrain the motion of the index finger along the eigen motion directions $\mathbf{p}_2$ and $\mathbf{p}_3$, we may embed mechanisms as shown in Figs. 12.10(b) and 12.10(c) in the finger mechanism in Fig. 12.10(a). Note that the wrapping directions of the tendons on the set of pulleys in Figs. 12.10(b) and 12.10(c) depend on the sign of the elements in the 2nd and 3rd eigen motions $\mathbf{p}_2$ and $\mathbf{p}_3$. Here assume that the mechanisms in Figs. 12.10(b) and 12.10(c) can constrain the motion along the eigen directions $\mathbf{p}_2$ and $\mathbf{p}_3$ respectively. How to determine the mechanism parameters in Figs. 12.10(b) and 12.10(c) will be given later. Thus, the mechanism to reproduce the primary motion $\mathbf{p}_1$ of the index finger under the effect of the actuator a can be designed

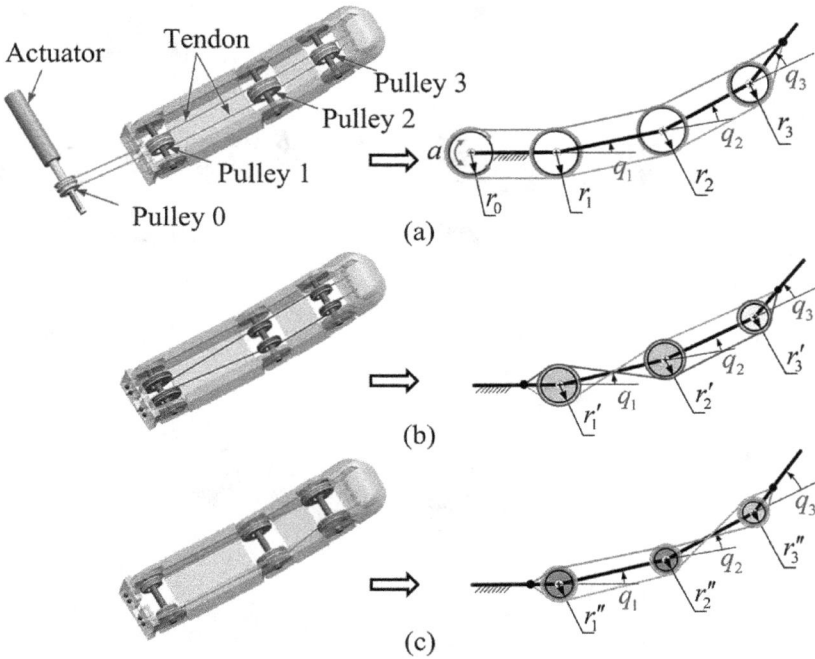

Fig. 12.10. Structure of mechanical digit.

as shown in Fig. 12.11 after embedding mechanisms as shown in Figs. 12.10(b) and 12.10(c) in the mechanism in Fig. 12.10(a). If we implant the springs in the mechanisms shown in Figs. 12.10(b) and 12.10(c), then the resultant finger mechanism can be designed as shown in Fig. 12.12.

Considering the deformation of the springs implanted in the mechanical digit in Fig. 12.12 and neglecting the effect of the actuator a, we can find the following movement relationship:

$$\boldsymbol{\Delta}_e = \mathbf{R}_m \cdot \boldsymbol{\Delta q} \tag{12.9}$$

where $\boldsymbol{\Delta}_e = \begin{bmatrix} \Delta_1 & \Delta_2 & \Delta_3 \end{bmatrix}^T$ represents the deformation vector of the springs, Δ_i ($i = 1, 2, 3$) denotes the deformation of a couple of the springs implanted in the tendons, $\boldsymbol{\Delta q} = \begin{bmatrix} \Delta q_1 & \Delta q_2 & \Delta q_3 \end{bmatrix}^T$ denotes the displacement of the joint angles of the mechanical digit with $\Delta q_1 = q_1 - q_{10}$, $\Delta q_2 = q_2 - q_{20}$, and $\Delta q_3 = q_3 - q_{30}$, q_{i0} ($i = 1, 2, 3$)

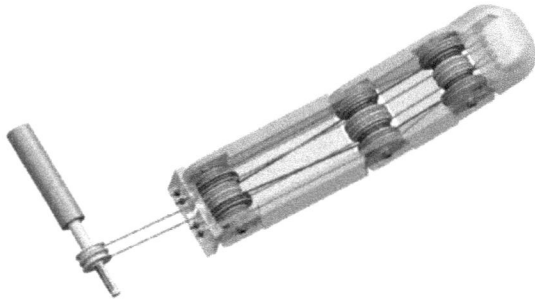

Fig. 12.11. Mechanism to reproduce primary motion of index finger.

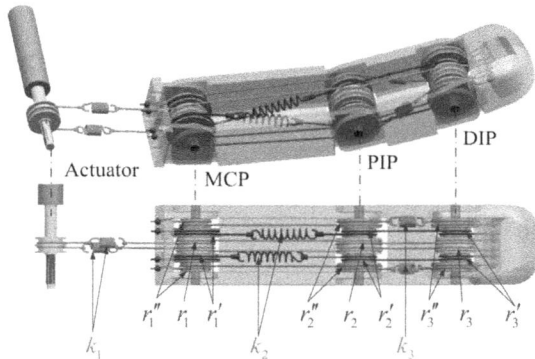

Fig. 12.12. Mechanism to reproduce primary and secondary motion of index finger.

represents the ith initial joint angle of a digit. The geometrical transformation matrix $\mathbf{R}_m$ is as follows:

$$\mathbf{R}_m = \begin{bmatrix} r_1 & r_2 & r_3 \\ r_1' & -r_2' & -r_3' \\ r_1'' & r_2'' & -r_3'' \end{bmatrix} \quad (12.10)$$

where r_i $(i = 1, 2, 3)$ is the radius of the ith pair of pulleys in the tendon-pulley transmission mechanism in Fig. 12.10(a), $r_i'(i = 1, 2, 3)$ is the radius of the ith pair of pulleys in the mechanism in Fig. 12.10(b), and r_i'' $(i = 1, 2, 3)$ is the radius of the ith pair of pulleys in the mechanism in Fig. 12.10(c).

The elastic potential stored in the mechanical digit is represented as

$$E_p = (1/2)\boldsymbol{\Delta}_e^T \mathbf{K}_\Delta \boldsymbol{\Delta}_e \qquad (12.11)$$

where

$$\mathbf{K}_\Delta = \mathrm{diag}\begin{bmatrix} k_1 & k_2 & k_3 \end{bmatrix} \qquad (12.12)$$

is the stiffness matrix of the springs implanted in the mechanical digit. As shown in Fig. 12.12, k_1 is the stiffness of a pair of springs implanted in the actuator tendons, k_2 is the stiffness of a pair of springs implanted in the mechanism in Fig. 12.10(b), and k_3 is the stiffness of a pair of springs implanted in the mechanism in Fig. 12.10(c).

From the joint eigen motion of a human digit (12.1), we have

$$\boldsymbol{\Delta}\mathbf{q} = \mathbf{P}\boldsymbol{\kappa} \qquad (12.13)$$

where $\mathbf{P} = \begin{bmatrix} \mathbf{p}_1 & \mathbf{p}_2 & \mathbf{p}_3 \end{bmatrix} \in \Re^{3\times 3}$ and $\boldsymbol{\kappa} = \begin{bmatrix} \kappa_1 & \kappa_2 & \kappa_3 \end{bmatrix}^T \in \Re^{3\times 1}$. The mechanical digit aims to reproduce the motion of the human digit in the eigen motion space spanned by the eigenvectors $\mathbf{p}_i$, $(i = 1, 2, 3)$.

Substituting Eqs. (12.9) and (12.13) into (12.11) yields

$$E_p = (1/2)\boldsymbol{\kappa}^T \left(\mathbf{P}^T \mathbf{R}_m^T \mathbf{K}_\Delta \mathbf{R}_m \mathbf{P} \right) \boldsymbol{\kappa} \qquad (12.14)$$

From Eq. (12.14), it can be found that the stiffness matrix of the mechanical digit can be described in the eigen motion space as follows:

$$\mathbf{K}_p = (\mathbf{R}_m \mathbf{P})^T \mathbf{K}_\Delta (\mathbf{R}_m \mathbf{P}) = \mathbf{P}^T \mathbf{K}_q \mathbf{P} \qquad (12.15)$$

Then, we can describe the compliance ellipsoid in the eigen motion space as follows:

$$\boldsymbol{\kappa}^T \left(\mathbf{K}_p^T \mathbf{K}_p \right) \boldsymbol{\kappa} = 1 \qquad (12.16)$$

In order to match the mechanical digit adaptive ability to the shape of the object with the adaptability of the human digit, the mechanical digit is to match its compliance ellipsoid with the constant density ellipsoid of human digit (as shown in Fig. 12.4).

Let $\Delta_1 = c_1\kappa_1$, $\Delta_2 = c_2\kappa_2$, and $\Delta_3 = c_3\kappa_3$, that is,

$$\Delta_e = \mathbf{c}\kappa \tag{12.17}$$

where $\mathbf{c} = \text{diag}\begin{bmatrix} c_1 & c_2 & c_3 \end{bmatrix} \in \Re^{3\times3}$ with the constants c_1, c_2, and c_3. Equation (12.17) means that the three tendon-pulley transmission mechanisms in Fig. 12.12 reproduce the motions of human digit along the three eigen direction $\mathbf{p}_1$, $\mathbf{p}_2$ and $\mathbf{p}_3$, respectively.

Comparing Eqs. (12.9) and (12.17), it can be found that

$$\mathbf{R}_m \cdot \Delta\mathbf{q} = \mathbf{c}\kappa \tag{12.18}$$

Substituting Eqs. (12.13) into (12.18), and considering $\mathbf{P}^{-1} = \mathbf{P}^T$, yields

$$\mathbf{R}_m\mathbf{P} = \mathbf{c}, \mathbf{R}_m = \mathbf{c}\mathbf{P}^T \tag{12.19}$$

In Eq. (12.19), we have the proportional relation of the pulley radii

$$r_1 : r_2 : r_3 = \left\|p_1^1\right\| : \left\|p_1^2\right\| : \left\|p_1^3\right\| \tag{12.20}$$

$$r_1' : r_2' : r_3' = \left\|p_2^1\right\| : \left\|p_2^2\right\| : \left\|p_2^3\right\| \tag{12.21}$$

$$r_1'' : r_2'' : r_3'' = \left\|p_3^1\right\| : \left\|p_3^2\right\| : \left\|p_3^3\right\| \tag{12.22}$$

where the symbol ":" indicates proportionality, and

$$c_1 = \sqrt{r_1^2 + r_2^2 + r_3^2} \tag{12.23}$$

$$c_2 = \sqrt{(r_1')^2 + (r_2')^2 + (r_3')^2} \tag{12.24}$$

$$c_3 = \sqrt{(r_1'')^2 + (r_2'')^2 + (r_3'')^2} \tag{12.25}$$

Substituting Eqs. (12.19) into (12.15) yields

$$\mathbf{K}_p = \mathbf{c}^T\mathbf{K}_\Delta\mathbf{c} = \text{diag}\begin{bmatrix} c_1^2 k_1 & c_2^2 k_2 & c_3^2 k_3 \end{bmatrix} \tag{12.26}$$

Thus, the compliance ellipsoid in the eigen motion space could be written as

$$\kappa^T \text{diag}\begin{bmatrix} c_1^4 k_1^2 & c_2^4 k_2^2 & c_3^4 k_3^2 \end{bmatrix} \kappa = 1 \tag{12.27}$$

Equation (12.27) shows that the orientation of the compliance ellipsoid of the desired mechanical digit matches the orientation of the constant density ellipsoid of the human digit. To match their shapes, the length proportional relation among the principal axes of the two ellipsoids must remain unchanged:

$$(1/c_1^2 k_1) : (1/c_2^2 k_2) : (1/c_3^2 k_3) = \sqrt{\sigma_1} : \sqrt{\sigma_2} : \sqrt{\sigma_3} \qquad (12.28)$$

Since the primary motion of the mechanical digit along the first eigen direction $\mathbf{p}_1$ is provided by the active actuator, we don't need to care about the length match between two ellipsoids in the first eigen motion direction $\mathbf{p}_1$, that is, the match condition (12.28) can be simplified as follows:

$$\frac{k_2}{k_3} = \frac{\sqrt{\sigma_3}\left((r_1'')^2 + (r_2'')^2 + (r_3'')^2\right)}{\sqrt{\sigma_2}\left((r_1')^2 + (r_2')^2 + (r_3')^2\right)} \qquad (12.29)$$

In summary, to design an anthropomorphic mechanical digit so that it can reproduce the digit grasping movement of the human hand, its geometrical and physical parameters must satisfy the conditions (12.20)–(12.22) and (12.29) which constitute the design principle of all desired anthropomorphic mechanical digits.

12.7   Simplification of Desired Mechanical Digit Design

The original mechanical digit shown in Fig. 12.12 can reproduce the human grasping movement because it satisfies the conditions (12.20)–(12.22) and (12.29). However, from Fig. 12.12, it can be found that there are six pulleys arranged orderly on each joint axis. It is well known that it is difficult to embed all the pulleys in a compact mechanical digit whose size must be similar to a human digit. We have to simplify the mechanical digit structure so that all of the tendon-pulley mechanisms may be accommodated in the anthropomorphic artificial digit. Particularly, we shorten the compliance coupling chains between joints and decrease the number of pulleys on the joint axis, that is, the compliance coupling chains from joint 2 to joint 3 shown in Fig. 12.10(b) and from joint 1 to joint 2 shown in Fig. 12.10(c) are canceled, and the number of pulleys on each

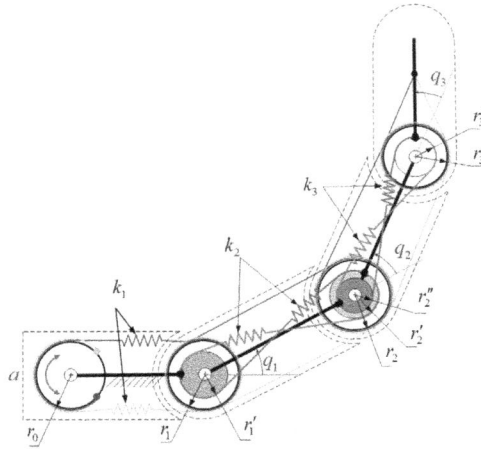

Fig. 12.13.   Schematic diagram of the simplified mechanical digit.

joint axis among joint 1, joint 2, and joint 3 shown in Fig. 12.12 is decreased to 4 from 6. The schematic diagram of the simplified mechanical digit is shown in Fig. 12.13. We can obtain the joint movement $\Delta\mathbf{q}$ generated by the actuator a from Fig. 12.13 as follows:

$$\Delta\mathbf{q} = \begin{bmatrix} \Delta q_1 \\ \Delta q_2 \\ \Delta q_3 \end{bmatrix} = \frac{ar_0}{r_1 r'_2 r'_3 + r'_1 r_2 r'_3 + r'_1 r''_2 r_3} \begin{bmatrix} r'_2 r'_3 \\ r'_1 r'_3 \\ r'_1 r''_2 \end{bmatrix} \qquad (12.30)$$

where r_0 is the radius of the actuator pulley in Fig. 12.10(a). Thus, to reproduce the primary motion of the human digit, the geometrical parameters of the simplified mechanical digit must satisfy the proportionality $\Delta q_1 : \Delta q_2 : \Delta q_3 = p_1^1 : p_1^2 : p_1^3$, that is,

$$r'_2/r'_1 = p_1^1/p_1^2, r'_3/r''_2 = p_1^2/p_1^3 \mid \qquad (12.31)$$

Similarly, the relationship between the deformation of the springs implanted in the simplified mechanical digit shown in Fig. 12.13 and the corresponding displacement of the joint angles of the mechanical digit can be represented as

$$\Delta_e = \mathbf{R}' \cdot \Delta\mathbf{q} \qquad (12.32)$$

where the geometrical transformation matrix $\mathbf{R}'$ is as follows:

$$\mathbf{R}' = \begin{bmatrix} r_1 & r_2 & r_3 \\ r_1' & -r_2' & 0 \\ 0 & r_2'' & -r_3' \end{bmatrix} \tag{12.33}$$

For the simplified mechanical digit shown in Fig. 12.13, the stiffness matrix $\mathbf{K}_q$ in the joint space is changed to as

$$\mathbf{K}_q = \begin{bmatrix} r_1^2 k_1 + (r_1')^2 k_2 & r_1 r_2 k_1 - r_1' r_2' k_2 & r_1 r_3 k_1 \\ r_1 r_2 k_1 - r_1' r_2' k_2 & r_2^2 k_1 + (r_2')^2 k_2 + (r_2'')^2 k_3 & r_2 r_3 k_1 - r_2'' r_3' k_3 \\ r_1 r_3 k_1 & r_2 r_3 k_1 - r_2'' r_3' k_3 & r_3^2 k_1 + (r_3')^2 k_3 \end{bmatrix} \tag{12.34}$$

Let $\mathbf{M}_\eta = \mathbf{K}_p^T \mathbf{K}_p$ with $\mathbf{K}_p = \mathbf{P}^T \mathbf{K}_q \mathbf{P}$, we have

$$\mathbf{M}_\eta = \mathbf{P}^T \mathbf{K}_q^T \mathbf{K}_q \mathbf{P} \tag{12.35}$$

The compliance ellipsoid described in the eigen motion space can be represented as

$$\boldsymbol{\kappa}^T \mathbf{M}_\eta \boldsymbol{\kappa} = 1 \tag{12.36}$$

In fact, the eigen motion space $\mathbf{P} \in \Re^{3\times3}$ can be broken into two subspaces $\mathbf{P}_p \in \Re^{3\times1}$ and $\mathbf{P}_s \in \Re^{3\times2}$, that is, $\mathbf{P} = \begin{bmatrix} \mathbf{P}_p & \mathbf{P}_s \end{bmatrix}$, and $\mathbf{P}_p = \mathbf{p}_1$, $\mathbf{P}_s = \begin{bmatrix} \mathbf{p}_2 & \mathbf{p}_3 \end{bmatrix}$. Thus, the symmetrical matrix $\mathbf{M}_\eta$ can be rewritten as

$$\mathbf{M}_\eta = \begin{bmatrix} \mathbf{M}_p & \mathbf{M}_{ps} \\ \mathbf{M}_{sp} & \mathbf{M}_s \end{bmatrix} \tag{12.37}$$

where $\mathbf{M}_p = \mathbf{P}_p^T \mathbf{K}_q^T \mathbf{K}_q \mathbf{P}_p$, $\mathbf{M}_{ps} = \mathbf{P}_p^T \mathbf{K}_q^T \mathbf{K}_q \mathbf{P}_s$, $\mathbf{M}_{sp} = \mathbf{P}_s^T \mathbf{K}_q^T \mathbf{K}_q \mathbf{P}_p$, and $\mathbf{M}_s = \mathbf{P}_s^T \mathbf{K}_q^T \mathbf{K}_q \mathbf{P}_s$.

However, generally speaking, the symmetrical matrix $\mathbf{M}_\eta$ is not diagonal, which means that the compliance ellipsoid defined in Eq. (12.36) is not consistent with the constant density ellipsoid defined in Eq. (12.5), as shown in Fig. 12.14, thus the grasping movement of the simplified mechanical digit doesn't match with the human digit.

To guarantee that the simplified mechanical digit can reproduce the digit grasping movement of the human hand, the primary motion

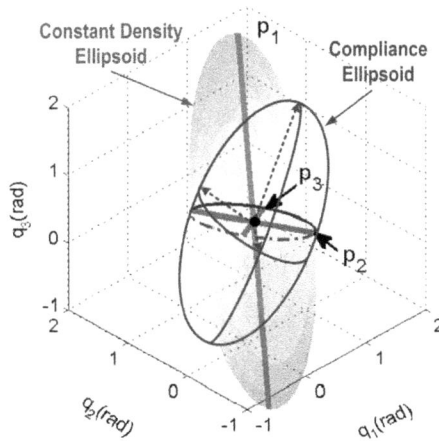

Fig. 12.14.   Compliance ellipsoid of the simplified mechanical digit.

and the secondary motion of the simplified mechanical digit must match with the human digit. First, the geometrical parameters of the simplified mechanical digit must satisfy the condition (12.31) so that the primary motion matching can be achieved. Second, two ellipses, namely (a) the ellipse formed by the compliance ellipsoid intersecting with the plane spanned by the 2nd and 3rd eigen motions $\mathbf{p}_2$ and $\mathbf{p}_3$ and (b) the ellipse formed by the constant density ellipsoid intersecting with the same plane, must match each other's shape, as shown in Fig. 12.14, so that the simplified mechanical digit can reproduce the secondary motion of the human digit, which means that the matrix $\mathbf{M}_s$ must be diagonal, and its diagonal element is inversely proportional to the square of the length of the principal axes of the intersecting ellipse,

$$\begin{cases} \mathbf{M}_s^{ij} = 0, \text{ for } i \neq j, \ i = 1, \ 2; \ j = 1, \ 2 \\ \frac{1}{\mathbf{M}_s^{11}} : \frac{1}{\mathbf{M}_s^{22}} = \sigma_2 : \sigma_3 \end{cases} \tag{12.38}$$

Especially, when $\mathbf{M}_{ps}$ is null matrix besides satisfying the conditions (12.31) and (12.38), that is, $\mathbf{M}_{ps} = \mathbf{0}$, the compliance of the mechanical digit is decoupled, and the mechanical digit becomes the desired one described in Section 12.6. Thus, conditions (12.31) and (12.38) constitute the design principle of simplified anthropomorphic mechanical digit as shown in Fig. 12.13, and guarantee that the simplified digit can be embedded in anthropomorphic hands and does not lose the information of human digit movement.

12.8 Kinematic Transmission Mechanism Design

The results shown above indicate that the actual grasping motion $\mathbf{Q}_h$ of the human hand can be reconstructed approximately by the linear combination of the first eigen grasp $\mathbf{u}_1$ and the second eigen grasp $\mathbf{u}_2$, which implies that Eq. (12.6) can be represented approximately as follows:

$$\hat{\mathbf{Q}}_h = \mathbf{W}\boldsymbol{\sigma}, \quad \mathbf{W} = \begin{bmatrix} l_1^{\frac{1}{2}} & \mathbf{u}_1 & l_2^{\frac{1}{2}} & \mathbf{u}_2 \end{bmatrix}, \quad \boldsymbol{\sigma} = \begin{bmatrix} \mathbf{v}_1 & \mathbf{v}_2 \end{bmatrix}^T \quad (12.39)$$

where $\hat{\mathbf{Q}}_h = \begin{bmatrix} \hat{\mathbf{q}}_1^1 \cdots \hat{\mathbf{q}}_1^4 \end{bmatrix}^T$ is considered as the motion output series of the MCP joints of the four fingers of the hand to be designed, $\boldsymbol{\sigma} \in \Re^{2 \times m}$ can be regarded as the motion input series, and the matrix $\mathbf{W} \in \Re^{4 \times 2}$ is referred to as the kinematic transmission matrix which is determined for the grasping activities of daily living.

Equation (12.39) shows that the fingers of the mechanical hand can reproduce the grasping movement of the human fingers by using the 2-D actuator $\boldsymbol{\sigma}$ exerting on a kinematic transmission mechanism (namely, matrix $\mathbf{W}$).

The matrix $\mathbf{W}$ can be decomposed into two transmission matrices $\mathbf{D} \in \Re^{4 \times 4}$ and $\mathbf{S}' \in \Re^{4 \times 2}$ as follows:

$$\mathbf{W} = \mathbf{D}\mathbf{S}' \quad (12.40)$$

where

$$\mathbf{D} = \begin{bmatrix} s_{11} + s_{12} & & & \\ & s_{21} + s_{22} & & \\ & & s_{31} + s_{32} & \\ & & & s_{41} + s_{42} \end{bmatrix}$$

$$\mathbf{S}' = \begin{bmatrix} \frac{s_{11}}{s_{11}+s_{12}} & \frac{s_{12}}{s_{11}+s_{12}} \\ \frac{s_{21}}{s_{21}+s_{22}} & \frac{s_{22}}{s_{21}+s_{22}} \\ \frac{s_{31}}{s_{31}+s_{32}} & \frac{s_{32}}{s_{31}+s_{32}} \\ \frac{s_{41}}{s_{41}+s_{42}} & \frac{s_{42}}{s_{41}+s_{42}} \end{bmatrix}$$

and $s_{ij}(i = 1, 2, 3, 4; j = 1, 2)$ are the elements of the matrix $\mathbf{W}$.

Substituting Eqs. (12.40) into (12.39) yields

$$\hat{\mathbf{Q}}_h = \mathbf{DI}, \quad \mathbf{I} = \mathbf{S}'\sigma \tag{12.41}$$

where $\mathbf{I} = \begin{bmatrix} \mathbf{I}_1^T & \cdots & \mathbf{I}_4^T \end{bmatrix}^T \in \Re^{4 \times m}$ is a connection matrix with $\mathbf{I}_i \in \Re^{1 \times m}$.

12.9   Design of motion distribution mechanism

It should be noted that the matrix $\mathbf{S}'$ has some characteristics, for example, the sum of all elements in each row is one although the value of every element perhaps is different. The kinematic relation transmitted by the matrix $\mathbf{S}'$ can be explained as follows: When two motion inputs are exerted along two parallel lines, the motion output position determined by the transmission matrix $\mathbf{S}'$ must be in the connecting line of the two corresponding motion input positions, and the motion output position is also determined with respect to the two motion input positions. Thus, when we design a sliding and guide bar mechanism as shown in Fig. 12.15, its parameters must satisfy the condition

$$d_i^1/d_0 = s_{i1}/(s_{i1} + s_{i2}) \quad \text{and} \quad d_i^2/d_0 = s_{i2}/(s_{i1} + s_{i2}) \tag{12.42}$$

so that the mechanism can implement the motion distribution relationship

$$\mathbf{I}_i = (s_{i1}/(s_{i1} + s_{i2}))\,\boldsymbol{\sigma}_1 + (s_{i2}/(s_{i1} + s_{i2}))\,\boldsymbol{\sigma}_2 \tag{12.43}$$

where d_0 is the horizontal distance between two input hinges, d_i^1 denotes the horizontal distance between the first input hinge and the ith output hinge, and $d_i^2 = d_0 - d_i^1$ denotes the horizontal distance between the second input hinge and the ith output hinge. d_i^1 is set as a signed value to ensure position uniqueness ('+' is set when the second output hinge and input hinge are on the same side of the first output hinge; '−' is set when the second output hinge and input hinge are on the different side of the first output hinge).

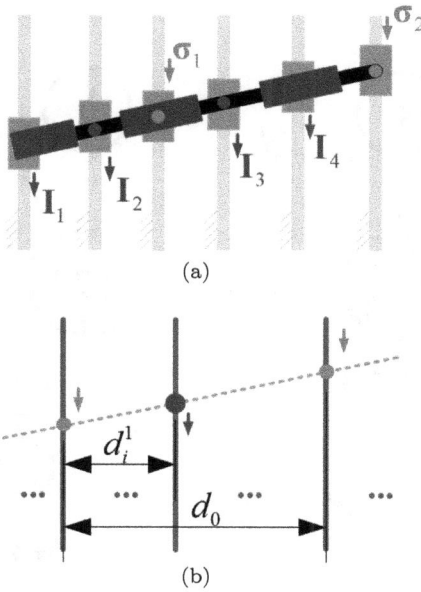

Fig. 12.15.   Motion distribution mechanism.

The designed mechanism (see Fig. 12.15) is referred to as the motion distribution mechanism, and its distribution coefficients are $s_{i1}/(s_{i1} + s_{i2})$ and $s_{i2}/(s_{i1} + s_{i2})$ $(i = 1, \ldots, 4)$, respectively, which must be implemented by the mechanism parameters d_0, d_i^1, and d_i^2 (see Eq. (12.42)).

12.10   Design of motion scaling mechanism

Moreover, the matrix $\mathbf{D}$ in Eq. (12.41) is a diagonal matrix; it makes inputs and outputs form one-to-one match. The ith element $s_{i1} + s_{i2}$ of the matrix $\mathbf{D}$ builds up a mapping relationship from the ith motion input $\mathbf{I}_i$ to the ith motion output $\hat{\mathbf{q}}_1^i$ $(i = 1, \ldots, 4)$. When we design a tendon-pulley mechanism as shown in Fig. 12.16, it must ensure the relationship between the inputs and outputs as follows:

$$\hat{\mathbf{q}}_1^i = (s_{i1} + s_{i2})\, \mathbf{I}_i, \quad (i = 1, \ldots, 4) \tag{12.44}$$

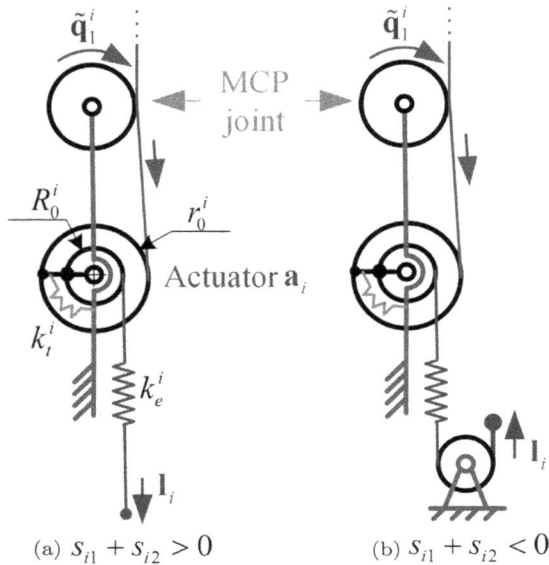

Fig. 12.16.   Motion scaling mechanism.

It should be noted that each top pulley in Fig. 12.16 must be fixed with the corresponding MCP joint so that the finger can obtain the desired motion $\hat{\mathbf{q}}_1^i$. The bottom two pulleys in Fig. 12.16 should be installed on the axis of the actuator. Because the four fingers have different motion ability in the grasping activities of daily living, using such two pulleys will facilitate the different motion implementation of fingers. In order to match the grasping compliance between the anthropomorphic hand and the human hand, we use the extension and torsion springs in the tendon-pulley mechanism as shown in Fig. 12.16.

Considering the deformations of springs and the torque balance condition on the bottom pulleys, we can obtain the following relationship from Fig. 12.16:

$$\mathbf{I}_i = R_0^i \left(1 + k_t^i / \left(\left(R_0^i \right)^2 k_e^i \right) \right) \mathbf{a}_i \qquad (12.45)$$

where k_e^i and k_t^i are the stiffness of the extension and torsion springs of the ith actuator respectively. R_0^i is the radius of the interior pulley of the ith actuator. $\mathbf{a}_i \in \Re^{1 \times m}$ is the driving series of the ith mechanical digit actuator.

Considering the simplified mechanical digit shown in Fig. 12.13, and using Eq. (12.30), we obtain

$$\hat{\mathbf{q}}_1^i = (r_0^i/b_i)\mathbf{a}_i \tag{12.46}$$

where r_0^i is the radius of the exterior pulley of the ith actuator. b_i is related to the ith finger parameters (namely, radii of pulleys) which is represented as

$$b_i = \left(r_1 r_2' r_3' + r_1' r_2 r_3' + r_1' r_2'' r_3\right) / \left(r_2' r_3'\right) \tag{12.47}$$

Note that the radii of pulleys in each finger have different values. Here we omit the finger label i on the radii of pulleys without causing confusion.

Combining Eqs. (12.45) and (12.46) yields

$$\hat{\mathbf{q}}_1^i = (r_0^i/b_i) \cdot \left(R_0^i k_e^i / \left(k_t^i + \left(R_0^i\right)^2 k_e^i\right)\right) \mathbf{I}_i \tag{12.48}$$

Comparing Eqs. (12.44) and (12.48), we obtain

$$(r_0^i/b_i) \cdot (R_0^i k_e^i/(k_t^i + (R_0^i)^2 k_e^i)) = s_{i1} + s_{i2} \tag{12.49}$$

Equation (12.49) shows that, to design the desired tendon-pulley mechanism, we need to synthetically consider its parameters and the mechanical finger's parameters and ensure all of them satisfy condition (12.49) so that the designed mechanism can implement the motion scaling relationship (12.44).

Such a mechanism is referred to as the motion scaling mechanism; its scaling coefficient is $s_{i1} + s_{i2}$ $(i = 1, \ldots, 4)$.

Furthermore, if $(R_0^i)^2 k_e^i \gg k_t^i$, then Eq. (12.49) can be simplified further as

$$r_0^i/R_0^i = b_i(s_{i1} + s_{i2}) \tag{12.50}$$

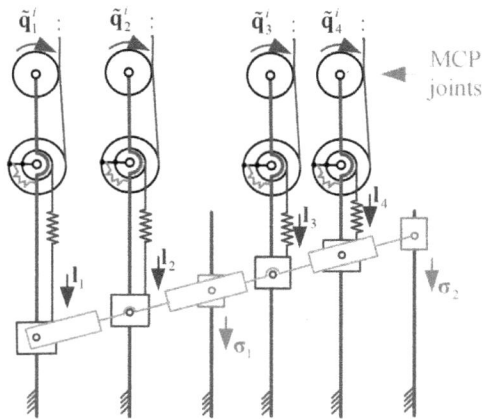

Fig. 12.17. Final kinematic transmission mechanism.

12.11 Design of motion combining mechanism

From the perspective of mathematical form, the transmission matrix $\mathbf{S}$ is the product of matrices $\mathbf{D}$ and $\mathbf{S}'$. Thus, the mechanical implementation of the matrix $\mathbf{S}$ means that the input of the motion scaling mechanism corresponding to each finger must be connected serially to the related output branch of the motion distribution mechanism.

The final mechanism that implements mechanically the transmission matrix $\mathbf{S}$ is shown in Fig. 12.17. Thus, we can use lower dimensional motion (namely, the two actuators' motion $\boldsymbol{\sigma}_1$ and $\boldsymbol{\sigma}_2$) to reconstruct synergically the higher dimensional motion (namely, the grasp motion $\hat{\mathbf{q}}_1^1$, $\hat{\mathbf{q}}_1^2$, $\hat{\mathbf{q}}_1^3$, and $\hat{\mathbf{q}}_1^4$) via the mechanism, as shown in Fig. 12.17.

12.12 Optimization of Transmission Mechanism

As mentioned above, the designed kinematic transmission mechanism can coordinate the four fingers to reproduce synergistically the anthropomorphic grasping movement. However, the size of the mechanism perhaps is either too large to be embedded into the palm, or some size such as the distance between two hinges may be too small to install the mechanical elements in the narrow space because the mechanical elements cannot be made too small. Thus, we have to optimize the kinematic transmission mechanism so that it can be embedded into the palm on the premise of ensuring that the anthropomorphic grasping motion $\hat{\mathbf{Q}}_h$ remains unchanged.

Giving a nonsingular transformation matrix $\mathbf{T}$, Eq. (12.39) can be rewritten as

$$\hat{\mathbf{Q}}_h = \mathbf{W}^*\boldsymbol{\sigma}^* = \mathbf{W}\boldsymbol{\sigma} \quad \text{with} \quad \mathbf{W}^* = \mathbf{WT}, \quad \boldsymbol{\sigma}^* = \mathbf{T}^{-1}\boldsymbol{\sigma}$$

$$(12.51)$$

In order to simplify the calculation, the matrix $\mathbf{T}$ is defined as

$$\mathbf{T} = \begin{bmatrix} f_1 \cos \varphi_1 & f_2 \cos \varphi_2 \\ -f_1 \sin \varphi_1 & -f_2 \sin \varphi_2 \end{bmatrix} \tag{12.52}$$

with four unknown parameters f_1, f_2, φ_1, and φ_2 which are related to the parameters of the new kinematic transmission mechanism.

It can be found from Eq. (12.51) that the anthropomorphic grasping motion $\hat{\mathbf{Q}}_h$ can be generated by a new kinematic transmission mechanism (namely, the new kinematic transmission matrix $\mathbf{W}^*$) via the two actuators' input $\boldsymbol{\sigma}^*$.

The matrix $\mathbf{W}^*$ needs to be implemented by the optimized new kinematic transmission mechanism. From the formula above, it can be found that the motion $\hat{\mathbf{Q}}_h$ remains unchanged although the transmission matrix and the actuators' input series are adjusted, which means that it is feasible to optimize the mechanism parameters so that the mechanism can be embedded into the palm on the premise of ensuring that the anthropomorphic grasping motion $\hat{\mathbf{Q}}_h$ remains unchanged.

The matrix $\mathbf{T}$ is naturally formed when the parameters of the mechanism are adjusted. How to reasonably determine the parameters of the matrix $\mathbf{T}$ is the key to optimize the new kinematic transmission mechanism.

From Eqs. (12.42) and (12.43), we can find that, if $s_{i2} = 0$ or $s_{i2}/(s_{i1}+s_{i2}) = 1$, then the motion input path will overlap the motion output path in the motion distribution mechanism, which means that the number of the fixed leaders will decrease to 4 from 6 (2 input and 4 output leading lines). This leads to a compact transmission mechanism design. Note that any two paths of the four motion outputs overlapping with the paths of two motion inputs cannot influence the locations of the four sliders and the ability to ensure the new transmission mechanism reproduces the anthropomorphic grasping motion $\hat{\mathbf{Q}}_h$. Here the sliders connecting the middle and little fingers are set as input sliders (see Fig. 12.20); we have

$$s_{42}^* = 0 \quad \text{and} \quad s_{22}^*/\left(s_{21}^* + s_{22}^*\right) = 1 \tag{12.53}$$

where s_{42}^*, s_{21}^*, and s_{22}^* are the elements of the matrix $\mathbf{W}^*$ which are related to the parameters f_1, f_2, φ_1, and φ_2 of the matrix $\mathbf{T}$.

According to Eqs. (12.51), (12.52), and (12.53), we obtain

$$\begin{cases} s_{i1}^* = (s_{i1}\cos\varphi_1 - s_{i2}\sin\varphi_1)\,f_1 = x_{i1}f_1 \\ s_{i2}^* = (s_{i1}\cos\varphi_2 - s_{i2}\sin\varphi_2)\,f_2 = x_{i2}f_2 \end{cases} \tag{12.54}$$

where $\varphi_1 = \arctan\left(s_{21}/s_{22}\right) = -76.7°$, $\varphi_2 = \arctan\left(s_{41}/s_{42}\right) = 78.6°$. Since φ_1 and φ_2 are constants, $x_{iu} = s_{i1}\cos\varphi_u - s_{i2}\sin\varphi_u$ $(i = 1,\ldots,4,\ u = 1,2)$ are also constants. Now the number of parameters required to be optimized changes to 2 from 4, that is, f_1 and f_2.

12.12.1   Objective Function

In order to make sure that the kinematic transmission mechanism can be embedded into the palm, it is important to keep the size of the mechanism as small as possible. Here we define that the optimization objective is to minimize the maximal horizontal distance between any two hinges in the motion distribution mechanism, that is,

$$\min\left\{\max_{i,j}\ d_{ij}\right\} \tag{12.55}$$

where d_{ij} is the horizontal distance between the ith hinge and jth hinge, $i,j = 1,\ldots,4$, and $i \neq j$.

Equation (12.42) shows that the elements of the transmission matrix of the motion distribution mechanism determine the proportional relationship of distances between any two hinges. For the new transmission matrix $\mathbf{W}^*$ and mechanism, we have

$$d_{ij} = d_0 \left| s'^*_{i2} - s'^*_{j2} \right|, \quad \text{and} \quad d^{min}_{allow} = d_0 \left(\min_{\substack{i,j \\ i \neq j}} \left\{ \left| s'^*_{i2} - s'^*_{j2} \right| \right\} \right)$$

(12.56)

then d_{ij} can be further represented as

$$d_{ij} = \frac{\left| s'^*_{i2} - s'^*_{j2} \right|}{\min\limits_{\substack{i,j \\ i \neq j}} \left\{ \left| s'^*_{i2} - s'^*_{j2} \right| \right\}} d^{min}_{allow}$$

(12.57)

where $s'^*_{i2} = s^*_{i2}/(s^*_{i1} + s^*_{i2})$ and $s'^*_{j2} = s^*_{j2}/(s^*_{j1} + s^*_{j2})$ are the elements of the matrix $\mathbf{S}'^*$ decomposed from the matrix $\mathbf{W}^*$ with the elements s^*_{ij} $(i = 1, \ldots, 4; \; j = 1, 2)$. d^{min}_{allow} is the allowable minimal distance between any two adjacent hinges. Considering the manufacturability and mechanical assembly ability, here d^{min}_{allow} is set as 14 mm (including the length of the linear bearing 10 mm, the hinge diameter 3 mm, and the interval between the linear bearing and hinges 1 mm), as shown in Fig. 12.20(c).

Thus, substituting Eq. (12.54) into Eq. (12.57), the objective function (12.55) can be rewritten as

$$\min \frac{\max\limits_{\substack{i,j \\ i \neq j}} \left\{ \left| \frac{x_{i2}(f_2/f_1)}{x_{i1} + x_{i2}(f_2/f_1)} - \frac{x_{j2}(f_2/f_1)}{x_{j1} + x_{j2}(f_2/f_1)} \right| \right\}}{\min\limits_{\substack{i,j \\ i \neq j}} \left\{ \left| \frac{x_{i2}(f_2/f_1)}{x_{i1} + x_{i2}(f_2/f_1)} - \frac{x_{j2}(f_2/f_1)}{x_{j1} + x_{j2}(f_2/f_1)} \right| \right\}} d^{min}_{allow}$$

(12.58)

12.12.2 Constraints

Due to the palm space limitation, whether the slider stroke P_i in the motion distribution mechanism or the scaling coefficient in the motion scaling mechanism must be restricted.

For example, if the slider stroke P_i is too large, then the palm size of the mechanical hand will be much larger than the size of the normal adult palm. Conversely, if the slider stroke P_i is too small, then the force transmission effect of the mechanical hand will be affected which will result in a small MCP joint torque. Thus, the slider stroke P_i must be in a reasonable range, namely

$$P_{allow}^{min} \leq P_i \leq P_{allow}^{max}, \quad i = 1, \ldots, 4 \tag{12.59}$$

where P_{allow}^{min} and P_{allow}^{max} are the allowable minimal and maximal values of the slider stroke which are set as 14 mm and 28 mm, respectively.

Since $P_i = \hat{q}_1^{range}/(s_{i1}^* + s_{i2}^*)$, Eq. (12.59) can be rewritten as

$$P_{allow}^{min} \leq \hat{q}_1^{range}/(s_{i1}^* + s_{i2}^*) \leq P_{allow}^{max} \tag{12.60}$$

where $\hat{q}_1^{range}$ is the motion range of the MCP joint of the mechanical hand; normally, we set $\hat{q}_1^{range} = \pi/2$.

Substituting Eq. (12.54) into Eq. (12.60), we can obtain

$$\frac{P_{allow}^{min}}{P_{allow}^{max}} \leq \frac{x_{i1} + x_{i2}\,(f_2/f_1)}{x_{j1} + x_{j2}\,(f_2/f_1)} \leq \frac{P_{allow}^{max}}{P_{allow}^{min}} \tag{12.61}$$

For the motion scaling mechanism, the radiuses of the scaling pulleys cannot be too small to be manufactured. Certainly, the size of the scaling pulleys cannot be too large to be embedded into the mechanical palm. Thus, the radius ratio constraints of the scaling pulleys can be described as

$$\gamma^0 \leq r_0^i/R_0^i \leq 1/\gamma^0 \tag{12.62}$$

where γ^0 is the allowable limit value of the radius ratio of the scaling pulleys. Considering the limit of the pulleys' radii in 4 mm $\sim$ 8 mm, γ^0 is set as 0.5.

According to Eq. (12.50), we can rewrite Eq. (12.62) on the new kinematic transmission mechanism as follows:

$$\gamma^0 \leq b_i\,(s_{i1}^* + s_{i2}^*) \leq 1/\gamma^0 \tag{12.63}$$

Substituting Eq. (12.54) into Eq. (12.63), we can obtain

$$\frac{b_j}{b_i} \times (\gamma^0)^2 \leq \frac{x_{i1} + x_{i2}\,(f_2/f_1)}{x_{j1} + x_{j2}\,(f_2/f_1)} \leq \frac{b_j}{b_i} \times \frac{1}{(\gamma^0)^2} \tag{12.64}$$

12.12.3 Optimization Problem Formulation

$$xx^* = \underset{xx \in D}{\arg\min} \left\{ \frac{\underset{\substack{i,j \\ i \neq j}}{\max} \left| \frac{x_{i2}xx}{x_{i1}+x_{i2}xx} - \frac{x_{j2}xx}{x_{j1}+x_{j2}xx} \right|}{\underset{\substack{i,j \\ i \neq j}}{\min} \left| \frac{x_{i2}xx}{x_{i1}+x_{i2}xx} - \frac{x_{j2}xx}{x_{j1}+x_{j2}xx} \right|} d_{allow}^{min} \right\} \quad (12.65)$$

subject to
$$\left\{ \begin{array}{l} \frac{x_{i1}+x_{i2}xx}{x_{j1}+x_{j2}xx} - \frac{P_{allow}^{max}}{P_{allow}^{min}} \leq 0 \\[2mm] \frac{P_{allow}^{min}}{P_{allow}^{max}} - \frac{x_{i1}+x_{i2}xx}{x_{j1}+x_{j2}xx} \leq 0 \\[2mm] \frac{x_{i1}+x_{i2}xx}{x_{j1}+x_{j2}xx} - \frac{b_j}{b_i}\frac{1}{(\gamma^0)^2} \leq 0 \\[2mm] \frac{b_j}{b_i}\left(\gamma^0\right)^2 - \frac{x_{i1}+x_{i2}xx}{x_{j1}+x_{j2}xx} \leq 0 \end{array} \right. \left| \begin{array}{l} i,j = 1, \ldots, 4, \\ \text{and } i \neq j. \end{array} \right.$$

The problem of reasonably determining the matrix **T** can be transformed as a constrained optimization problem. Note that there exists the variable f_2/f_1 either in the objective function (12.58) or in the constraints (12.61) and (12.64). Let $xx = f_2/f_1$, then we can transform further the problem into a constrained optimization problem (12.65) with only one design variable xx.

Although the new optimization problem is still a constrained nonlinear one, we can use the enumeration method to solve the problem, that is, we can search the optimal solution with different step lengths in the feasible domain D of the design variable xx. By solving the problem (12.65), the width of the motion distribution mechanism is minimized, and the new kinematic transmission mechanism can be embedded into the anthropomorphic palm.

12.13 Grasping Experiments of Anthropomorphic Hand

The anthropomorphic mechanical digit as shown in Fig. 12.13 has 10 parameters which are seven geometrical parameters, r_1, r_2, r_3, r_1', r_2', r_2'', r_3', and three physical parameters, k_1, k_2, k_3. The conditions (12.31) and (12.38), which guarantee that the simplified mechanical digit can reproduce the digital grasping movement of the human hand, constitute four constraint equations. Considering the manufacturability of the designed elements, we can choose a group of feasible parameters satisfying the constraint equations of the conditions (12.31) and (12.38) in the design process. In fact, we may also obtain the optimal parameters by optimizing some performance such as loading ability, stability, etc., which is an issue to be further studied in the future.

Now we determine the parameters of the mechanical index finger whose motion characteristic is shown in Table 12.2. Given $r_2' = 6\,\text{mm}$ and $r_2'' = 4\,\text{mm}$, we can obtain $r_1' = 6.8\,\text{mm}$ and $r_3' = 7.6\,\text{mm}$ via Eq. (12.31). Then, given $k_1 = 40\,\text{N/mm}$ and $r_1 = 4.0\,\text{mm}$, the relationship among r_2, r_3, k_2, and k_3 is determined by the condition (12.38), as shown in Fig. 12.18. A group of feasible parameter values is chosen in the constraint (12.38) (see Fig. 12.18) as $r_2 = 5.8\,\text{mm}$, $r_3 = 4.0\,\text{mm}$, $k_2 = 6.5\,\text{N/mm}$, $k_3 = 24.2\,\text{N/mm}$.

Without loss of generality, the parameters of the other digits can also be determined in a similar way. Based on the radii of pulleys in fingers, we can calculate the results of b_i in (12.47) which are shown in Table 12.3. According to Eqs. (12.61) and (12.64), the constraint of xx (i.e., f_2/f_1) could be obtained as $0.96 \leq xx \leq 2.42$. As stated above, enumeration method is used to solve the constrained nonlinear problem (12.65). The step length is finally set as 0.01. It can be found that the optimal solution $xx^* = 2.12$. The objective function could be described as a curve with respect to the variable xx, as shown in Fig. 12.19. Actually, giving the optimal solution xx^*, we can search a pair of feasible values f_1 and f_2 according to the dimensional

Table 12.2.   Eigen motion of index finger.

Eigen motion direction	$\mathbf{P}_1 = \begin{pmatrix} p_1^1 \\ p_1^2 \\ p_1^3 \end{pmatrix}$	$\mathbf{P}_2 = \begin{pmatrix} p_2^1 \\ p_2^2 \\ p_2^3 \end{pmatrix}$	$\mathbf{P}_3 = \begin{pmatrix} p_3^1 \\ p_3^2 \\ p_3^3 \end{pmatrix}$
	$= \begin{pmatrix} 0.61 \\ 0.70 \\ 0.37 \end{pmatrix}$	$= \begin{pmatrix} 0.79 \\ -0.50 \\ -0.35 \end{pmatrix}$	$= \begin{pmatrix} 0.06 \\ 0.50 \\ -0.86 \end{pmatrix}$
Contribution rate	91%	8%	1%

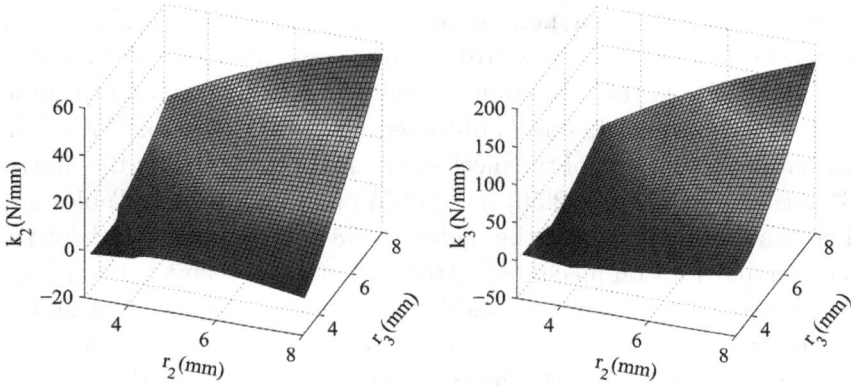

Fig. 12.18. Relationship among r_2, r_3, k_2, and k_3.

Table 12.3. Key numerical results.

b_i	$\mathbf{S} = [l_1^{\frac{1}{2}}\mathbf{u}_1 \quad l_2^{\frac{1}{2}}\mathbf{u}_2]$	$\mathbf{T}$	$\mathbf{s}^*$
$b_1 = 0.0482$	$\begin{bmatrix} -14.7219 & 8.9509 \\ -18.7162 & 4.4245 \\ -22.1298 & -2.9976 \\ -27.1493 & -5.4605 \end{bmatrix}$	$\begin{bmatrix} -0.0011 & -0.0020 \\ -0.0047 & 0.0101 \end{bmatrix}$	$\begin{bmatrix} -0.0258 & 0.1201 \\ 0 & 0.0826 \\ 0.0388 & 0.0147 \\ 0.0561 & 0 \end{bmatrix}$
$b_2 = 0.0482$			
$b_3 = 0.0500$			
$b_4 = 0.0640$			

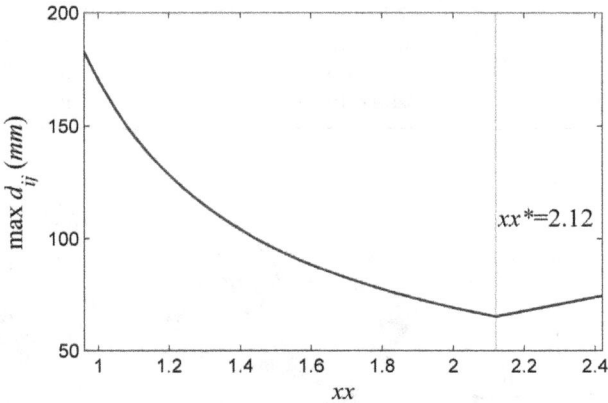

Fig. 12.19. The curve of the objective function with respect to the variable xx.

constraints of the palm, i.e., the constrains (12.61) and (12.64). When f_1 reaches -0.0049, f_2 is -0.0104. Thus, based on Eq. (12.52), matrix $\mathbf{T}$ is finalized. At the same time, the matrix $\mathbf{S}^*$ which is used to determine the final parameters is obtained, as shown in Table 12.3. The structure parameters of the mechanism are determined by the matrix $\mathbf{S}^*$ according to Eqs. (12.50) and (12.57) and are listed in Table 12.4. The maximum stroke of the sliders is 28 mm and the width of the motion distribution mechanism (the distance between the two hinges which are at the ends) reaches 60.7 mm (when the modified matrix $\mathbf{S}'^*$ reserves one decimal fraction). The final structure is shown in Fig. 12.20. Therefore, the mechanism could be embedded into an anthropomorphic palm leaving space for the actuating module of the thumb.

Based on the mechanical parameters calculated above, an anthropomorphic hand has been designed with a size similar to a human

Table 12.4. Final mechanical structure parameters.

Parameters	Values	Parameters	Values
r_0^1	7.8 mm	R_0^1	4.0 mm
r_0^2	7.5 mm	R_0^2	4.4 mm
r_0^3	7.5 mm	R_0^3	7.0 mm
r_0^4	6.6 mm	R_0^4	7.5 mm
d_{12}	14.0 mm	d_{23}	32.7 mm
d_{34}	14.0 mm	d_{14}	60.7 mm

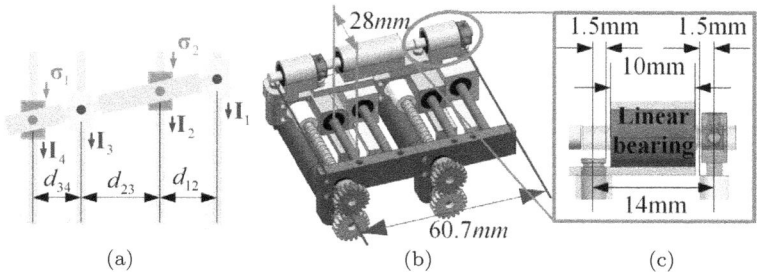

Fig. 12.20. Motion distribution mechanism structure.

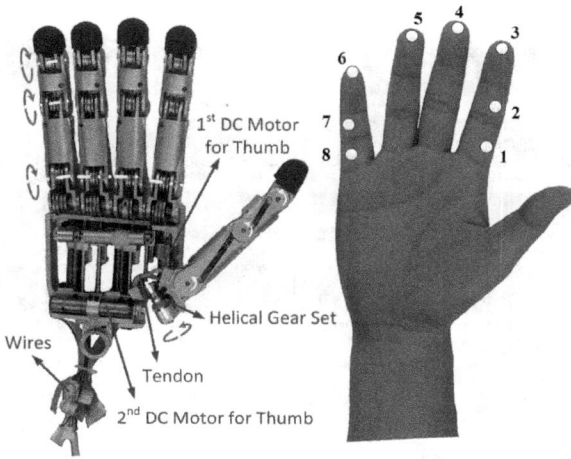

Fig. 12.21. The designed X-hand and human hand in the same scale.

hand. The hand is named as X-hand, as shown in Fig. 12.21. A fine cooperation of the thumb and the other four fingers plays a crucial role in grasping. This cooperation can be examined through the Kapandji test. In the Kapandji test, a human subject usually has to touch eight easily identifiable locations on his/her fingers with the tip of his/her thumb. The eight locations are shown in Fig. 12.21. Based on the Kapandji test, feasible placements of the thumb and the fingers are adopted in our robotic hand, as shown in Fig. 12.21. Two DC motors (Maxon DCX10L, gear GPX10 16:1, and encoder ENX10) are used to drive the kinematic transmission mechanism so that the four fingers generate anthropomorphic grasping movement. As mentioned in Section 12.4, because the thumb moves independently of the four fingers during the grasping of a human hand, the thumb mechanism of the X-hand is designed independently of the kinematic transmission mechanism. There are two DC motors in the thumb mechanism of the X-hand. The abduction movement and adduction movement of the thumb are realized by the 1st DC motor, and the thumb flexion movement and extension movement by the 2nd DC motor, as shown in Fig. 12.21. Thus, there are a total of four DC motors in the X-hand. A few drivers in the X-hand provide the potential feasibility

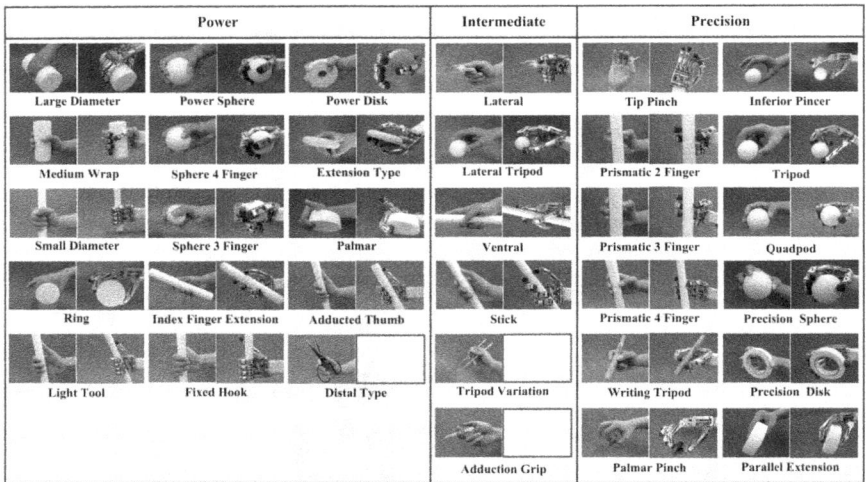

Power			Intermediate	Precision	
Large Diameter	Power Sphere	Power Disk	Lateral	Tip Pinch	Inferior Pincer
Medium Wrap	Sphere 4 Finger	Extension Type	Lateral Tripod	Prismatic 2 Finger	Tripod
Small Diameter	Sphere 3 Finger	Palmar	Ventral	Prismatic 3 Finger	Quadpod
Ring	Index Finger Extension	Adducted Thumb	Stick	Prismatic 4 Finger	Precision Sphere
Light Tool	Fixed Hook	Distal Type	Tripod Variation	Writing Tripod	Precision Disk
			Adduction Grip	Palmar Pinch	Parallel Extension

Fig. 12.22. Anthropomorphic grasping tasks carried out by X-hand. The process to form these grasps is shown in the supplementary movies.

of the biosignal (such as myoelectrical signal) based control for the X-hand.

In order to evaluate the anthropomorphic grasping ability of the X-hand, here we define the anthropomorphic grasping ability as the capacity to replicate the natural movement in the grasping activities of daily living. Since Feix taxonomy as shown in Fig. 12.2 reflects the grasping activities, we choose it as the grasping test tasks of the X-hand. Figure 12.22 demonstrates the ability of the X-hand to achieve these grasps.

12.14　Summary

This chapter focuses on the development of a general design theory for anthropomorphic hands, aiming to achieve the ambitious goal of replicating the natural grasping functions of the human hand using a specially designed hand with only a few actuators controlled by bio-signals. The key concept in designing an anthropomorphic hand is to reduce the complexity of the higher-dimensional grasp motion space of the human hand in daily activities through feature extraction. Subsequently, the grasp motion in the higher-dimensional space is reconstructed using the motion of the few actuators in the lower-dimensional space (dimension reduction space) facilitated by the anthropomorphic mechanical hand. The design approach presented in this chapter can also be extended to other robotic systems intended to imitate human or animal natural behaviors using a limited number of actuators controlled by bio-signals.

References

[1] Cutkosky M. R. *Robotic Grasping and Fine Manipulation.* Kluwer Academic Publishers, Boston, MA, USA, 1985.

[2] Mason M. T. and Salisbury J. K. *Robot Hands and the Mechanics of Manipulation.* MIT Press, Cambridge, MA, USA, 1985.

[3] Murray R. M., Li Z. X., and Sastry S. S. *A Mathematical Introduction to Robotic Manipulation.* CRC Press, Inc., Boca Raton, FL, USA, 1994.

[4] Xiong C. H., Li Y. F., Ding H., and Xiong Y. L. On the dynamic stability of grasping. *International Journal of Robotics Research*, 18(9), pp. 951–958, 1999.

[5] Xiong C. H., Ding H., and Xiong Y. L. *Fundamentals of Robotic Grasping and Fixturing.* World Scientific Publishing Co Pte Ltd, Singapore, 2007.

[6] Xiong C. H., Wang M. Y., and Xiong Y. L. On clamping planning in workpiece-fixture systems. *IEEE Transactions on Automation Science and Engineering*, 5(3), pp. 407–419, 2008.

[7] Jacobsen S. C., Wood J. E., Knutti D. F., and Biggers K. B. The UTAH/M.I.T. dextrous hand: Work in progress. *International Journal of Robotics Research*, 3(4), pp. 21–50, 1984.

[8] Martin J. and Grossard M. Design of a fully modular and backdrivable dexterous hand. *International Journal of Robotics Research*, 33(5), pp. 783–798, 2014.

[9] Palli G., Melchiorri C., Vassura G., Scarcia U., Moriello L., Berselli G., and et al. The DEXMART hand: Mechatronic design and experimental evaluation of synergy-based control for human-like grasping. *International Journal of Robotics Research*, 33(5), pp. 799–824, 2014.

[10] Grebenstein M., Chalon M., Friedl W., Haddadin S., Wimbock T., Hirzinger G., and et al. The hand of the DLR Hand Arm System: Designed for interaction. *International Journal of Robotics Research*, 31(13), pp. 1531–1555, 2012.

[11] Pons J. L., Rocon E., Ceres R., Reynaerts D., Saro B., Levin S., and *et al.* The MANUS-HAND dextrous robotics upper limb prosthesis: Mechanical and manipulation aspects. *Autonomous Robots*, 16(2), pp. 143–163, 2004.

[12] Dollar M. and Howe R. D. Joint coupling design of underactuated hands for unstructured environments. *International Journal of Robotics Research*, 30(9), pp. 1157–1169, 2011.

[13] Dollar M. and Howe R. D. The highly adaptive SDM hand: Design and performance evaluation. *International Journal of Robotics Research*, 29(5), pp. 585–597, 2010.

[14] Birglen L., Gosselin C. M., and Laliberté T. *Underactuated Robotic Hands.* Springer, Berlin, Germany, 2008.

[15] Dechev N., Cleghorn W. L., and Naumann S. Multiple finger, passive adaptive grasp prosthetic hand. *Mechanism and Machine Theory*, 36(10), pp. 1157–1173, 2001.

[16] Shirafuji S., Ikemoto S., and Hosoda K. Development of a tendon-driven robotic finger for an anthropomorphic robotic hand. *International Journal of Robotics Research*, 33(5), pp. 677–693, 2014.

[17] Inouye J. M. and Valero-Cuevas F. J. Anthropomorphic tendon-driven robotic hands can exceed human grasping capabilities following optimization. *International Journal of Robotics Research*, 33(5), pp. 694–705, 2014.

[18] Ozawa R., Kobayashi H., and Hashirii K. Analysis, classification, and design of tendon-driven mechanisms. *IEEE Transactions on Robotics*, 30(2), pp. 396–410, 2014.

[19] Ozawa R., Hashirii K., and Kobayashi H. Design and control of underactuated tendon-driven mechanisms. In: *Proceedings of IEEE International Conference on Robotics and Automation (ICRA)*, pp. 1522–1527, 2009.

[20] Hirose S. and Umetani Y. The development of soft gripper for the versatile robot hand. *Mechanism and Machine Theory*, 13(3), pp. 351–359, 1978.

[21] Guo R., Nguyen V., Niu L., and Bridgwater L. Design and analysis of a tendon-driven, under-actuated robotic hand. In: *ASME International Design Engineering Technical Conferences and Computers and Information in Engineering Conference*, V05AT08A095, 2014.

[22] Massa B., Roccella S., Carrozza M. C., and Dario P. Design and development of an underactuated prosthetic hand. In: *Proceedings of IEEE International Conference on Robotics and Automation*, Washington, DC, USA, pp. 3374–3379, 2002.

[23] Birglen L. and Gosselin C. M. Kinetostatic analysis of underactuated fingers. *IEEE Transactions on Robotics and Automation*, 20(2), pp. 211–221, 2004.

[24] Kragten G. A., Baril M., Gosselin C., and Herder J. L. Stable precision grasps by underactuated grippers. *IEEE Transactions on Robotics*, 27(6), pp. 1056–1066, 2011.

[25] Krut S., Begoc V., Dombre E., and Pierrot F. Extension of the form-closure property to underactuated hands. *IEEE Transactions on Robotics*, 26(5), pp. 853–866, 2010.

[26] Carrozza M., Cappiello G., Micera S., Edin B., Beccai L., and Cipriani C. Design of a cybernetic hand for perception and action. *Biological Cybernetics*, 95(6), pp. 629–644, 2006.

[27] Dalley S. A., Wiste T. E., Withrow T. J., and Goldfarb M. Design of a multifunctional anthropomorphic prosthetic hand with extrinsic actuation. *IEEE/ASME Transactions on Mechatronics*, 14(6), pp. 699–706, 2009.

[28] Deimel R. and Brock O. A novel type of compliant and underactuated robotic hand for dexterous grasping. *International Journal of Robotics Research*, 35(1-3), pp. 161–185, 2016.

[29] Belter J. T., Segil J. L., Dollar A. M., and Weir R. F. Mechanical design and performance specifications of anthropomorphic prosthetic hands: A review. *Journal of Rehabilitation Research and Development*, 50(5), pp. 599–618, 2013.

[30] Cipriani C., Controzzi M., and Carrozza M. C. The SmartHand transradial prosthesis. *Journal of NeuroEngineering and Rehabilitation*, 8(29), pp. 1–13, 2011.

[31] Wiste T. E., Dalley S. A., Varol H. A., and Goldfarb M. Design of a multigrasp transradial prosthesis. *Journal of Medical Devices*, 5(3), p. 031009, 2011.

Chapter 13

Mechanical Implementation of Kinematic Synergy for Continual Grasping Generation

The postural synergy presented in Chapter 12 is extracted from quantities of joint trajectory. To reproduce different grasping tasks, it is necessary to determine the temporal weight sequences of each synergy, from the pre-grasp phase to the grasp phase. Additionally, a zero-offset posture needs to be pre-set before initiating any grasp. This chapter utilizes kinematic synergies extracted from angular velocity profiles to design the motion generation mechanism. These kinematic synergies, derived from quantities of grasp tasks, are implemented through the proposed eigen cam group in the tendon space. The process of fully extending the hand from the initial posture only requires an average rotation of the two eigen cam groups for one cycle, enabling a continuous grasp. Changing the grasp pattern is achieved by specifying new transmission ratio pairs for the two eigen cam groups. A hand prototype, illustrating the proposed design principle, has been developed, and grasping experiments were conducted to demonstrate the feasibility of this design method. The potential applications of this approach include prosthetic hands controlled by classified patterns extracted from bio-signals.

13.1   Introduction

When hand performing any complex task, all finger joints are closely coordinated over time. It is a particular challenge to control each joint of an anthropomorphic hand to precisely replicate the kinematic and dynamic characteristics of the time-varying posture of a human hand in a complex task [1, 2]. The most complex point is how to generate so many channels of control signals to coordinate the joints. In many cases, only very limited channels are available, such as the prosthetic hand controlled by bio-signal. Thus, under this situation the functional versatility and motion controllability seem to be incompatible for the anthropomorphic hand.

Since underactuated mechanism can reach the tradeoff between a versatile function and a few input channels, designing an anthropomorphic hand based on the underactuation principle attracts lots of attention and many types of underactuated mechanisms are proposed in the literature to address this issue. Some are based on linkages [3–5], some are based on tendon-actuated mechanisms [6–8], while others are based on flexible actuators [9–11]. Due to the self-adaptation to the object grasped, the underactuated hands have good compliance with limited inputs [12]. However, an underactuated hand cannot produce fruitful postures in free space because the finger joints are not independently controlled. Unlike the general underactuated mechanism, the dimensionality reduction-based techniques in the literature toward finding a lower-dimensional representation of the original grasp data, which also provides a potential way to address this problem. The dimensionality reduction methods generally consist of nonlinear and linear types. The most recent representatives of nonlinear approaches include the Gaussian process latent variable model [13] and unsupervised kernel regression [14]. The most used linear approach is Principal Component Analysis (PCA). The principal component or synergy [15–22] observed from the human hand-grasping activities couples all joints to form a specified static or time-varying posture. Any posture or continual time-varying posture can be approximated by a linear combination of a few significant synergies. Compared with the nonlinear approaches, the PCA method is more feasible to be implemented by the real-time algorithm [23–27] and drive mechanism [28].

In literature, much work has been done in applying the principle of synergy to deal with the motion generation of anthropomorphic hands via a few inputs. These inputs are the weight for combining several static posture synergies. For example, two [23] and three [24] postural synergies are used to control twenty-four actuators in the ACT hand. Other examples include the use of two synergies in the DLR II hand [25], four synergies in the SAH hand [26], two synergies in the UB Hand IV [27], etc. Matrone *et al.* used the X-Y coordinates of a mouse cursor on the screen [29] or two channels of sEMG from a human wrist [30] as two control inputs for the CyberHand. Differing from the software regulation in the above works, the postural synergy was implemented via a designed mechanism in tendon space [28].

The common characteristic of the motion generation of these anthropomorphic hands is that the used postural synergy is static and has no temporal characteristics. Thus, the implementation of a time-varying posture for a particular grasping movement is dependent on carefully specifying the time-varying weights of each synergy. The general procedure for determining the value of weights for a particular grasping movement is as follows: First, adjust the hand to the zero-offset posture in which the weights of each synergy are equal to zero. Second, open the hand to the fully extension posture. Third, close the hand to the suitable grasping posture. Finally, compute the corresponding weights at the two boundary postures. The intermediate temporal values of weight for this grasping movement are solved by linearly interpolating the weights corresponding to these three postures assuming an appropriate time interval [27]. The hand should be in the zero-offset posture before starting any grasping movement. The temporal weights from simple interpolation cannot ensure that it precisely duplicates the practical values, which affects the hand's performance. Moreover, the temporal sequence of weights depends on the prespecified grasp configuration for the considered object, which affects the hand's performance in an unstructured environment. In most cases, what can be provided before grasp is the grasp pattern but not the final grasp configuration. Thus, continual motion generation based on static postural synergy requires well-defined conditions that may limit its potential application.

To address these issues, this chapter investigates the mechanical implementation of the angular velocity synergy and proposes the general design principle of eigen cam group to implement such type of synergy. The mechanical implementation of synthesizing and regulating synergic velocity are also studied. Based on the design principle, this chapter proposes a prototype of an anthropomorphic hand whose continual grasp from the initial posture to the final grasp posture only requires averagely rotating the two input shafts one cycle under a particular transmission ratio.

13.2   Angular Velocity Synergy

Before giving the details of the design method, we begin with a brief overview of the mathematical form of angular velocity synergy. If n joints are considered and the period of grasp or manipulation is averagely sampled t_s times in one trial, the angular velocity evolvement of the hand joints at the duration can be represented as

$$\mathbf{v} = \begin{bmatrix} \omega_1(t_1) & \cdots & \omega_1(t_s) & \cdots & \omega_n(t_1) & \cdots & \omega_n(t_s) \end{bmatrix} \qquad (13.1)$$

where the element $\omega_i(t_j)$ is the ith joint velocity at the t_jth sample (t_1 represents the first sample at the initial time and t_s represents the final sample at the end time). Given N trials for grasping or manipulating different objects, a velocity matrix can be defined by

$$\mathbf{V} = \begin{bmatrix} \mathbf{v}_1 \\ \vdots \\ \mathbf{v}_N \end{bmatrix} \qquad (13.2)$$

where $\mathbf{v}_i (i = 1 \cdots N)$ is the ith trial of grasping or manipulation. This joint velocity matrix can be rewritten as the product of three smaller matrices by Singular Value Decomposition (SVD), which is illustrated by Eq. (13.3). Each row of the third matrix $\mathbf{S}$ is called a principal component (or called a synergy). The singular values are diagonally arranged from largest to smallest in the second matrix:

$$\mathbf{V} = \mathbf{U} \, \mathrm{diag}\{\lambda_1 \cdots \lambda_m\} \, \mathbf{S} \tag{13.3}$$

According to the importance of elements of the diagonal matrix, the matrix $\mathbf{V}$ can be approximated by $\hat{\mathbf{V}}$ if the first k principal components account for more than 80% of the total variance of the entire trial data:

$$\mathbf{V} \approx \hat{\mathbf{V}} = \mathbf{W}^k \mathbf{S}^k \tag{13.4}$$

where $\mathbf{W}^k = \mathbf{U}_k \mathrm{diag}\{\lambda_1 \cdots \lambda_k\}$, $\mathbf{U}_k$ is the first k columns of matrix $\mathbf{U}$, and $\mathbf{S}^k$ is the first k rows of matrix $\mathbf{S}$. In fact, according to our experimental data from human grasping movement, seeing the following subsection, the first two rows of the matrix $\mathbf{S}$ can account for 88% of the total variance. We think these two principal components are sufficient to account for the variance of the entire trial data. Thus, Eq. (13.4) can be written as

$$\mathbf{V} \approx \mathbf{W}^2 \mathbf{S}^2$$

$$= \begin{bmatrix} w_1^1 & w_1^2 \\ \vdots & \vdots \\ w_q^1 & w_q^2 \\ \vdots & \vdots \\ w_N^1 & w_N^2 \end{bmatrix} \begin{bmatrix} s_1^1(t_1) & \cdots & s_1^1(t_s) & \cdots & s_n^1(t_1) & \cdots & s_n^1(t_s) \\ s_1^2(t_1) & \cdots & s_1^2(t_s) & \cdots & s_n^2(t_1) & \cdots & s_n^2(t_s) \end{bmatrix}$$

$$\tag{13.5}$$

where the scalars of w_i^1 and w_i^2 corresponding to the ith row of matrix $\mathbf{V}$ are weights associated with the two principal components. Each principal component, i.e., the angular velocity synergy, is averagely sampled t_s times. To further explicit representation, each row of $\mathbf{V}$, i.e., $\mathbf{v}_i$, corresponding to one trial can be written as

$$\mathbf{v}_i \approx [\, w_i^1 \quad w_i^2\,]\mathbf{S}^2 \qquad (13.6)$$

or rewritten in the following form:

$$
\begin{bmatrix}
\omega_1(t_1) & \cdots & \omega_1(t_s) \\
& \vdots & \\
\omega_n(t_1) & \cdots & \omega_n(t_s)
\end{bmatrix}_i
$$

$$
\approx w_i^1
\begin{bmatrix}
s_1^1(t_1) & \cdots & s_1^1(t_s) \\
& \vdots & \\
s_n^1(t_1) & \cdots & s_n^1(t_s)
\end{bmatrix}
+ w_i^2
\begin{bmatrix}
s_1^2(t_1) & \cdots & s_1^2(t_s) \\
& \vdots & \\
s_n^2(t_1) & \cdots & s_n^2(t_s)
\end{bmatrix}
$$

$$(13.7)$$

where each row of the matrices at the two sides of Eq. (13.7) corresponds to one joint velocity evolvement at the duration of the hand movement. The reduced form of Eq. (13.7) for the ith joint is given by Eq. (13.8):

$$\omega_i\big|_{t_1}^{t_s} \approx w_i^1 \, s_i^1\big|_{t_1}^{t_s} + w_i^2 \, s_i^2\big|_{t_1}^{t_s} \qquad i = 1, \ldots, n \qquad (13.8)$$

Although the elements of each angular velocity synergy are discretely represented in Eq. (13.7) due to discrete sampling, it must be noted that each element is a continually varying function of time. This characteristic of velocity synergy is different from static postural synergy whose element is a constant scalar value. The static postural synergy just represents the principal motion direction in the joint space, and it cannot describe how to coordinate joints across time. The velocity synergy represents the eigen evolvement along the motion time. It just needs to specify one single weight for each velocity synergy to construct the continual time-varying posture for a complete manipulation. When initial hand posture is specified as the naturally full-extended fingers, the hand posture at each time during the motion period can be obtained by integrating angular velocity synergy.

In this chapter, we aim to develop the design method for reducing the control complexity and minimizing the quantity of input to implement the continual grasping movement. In the following sections, we

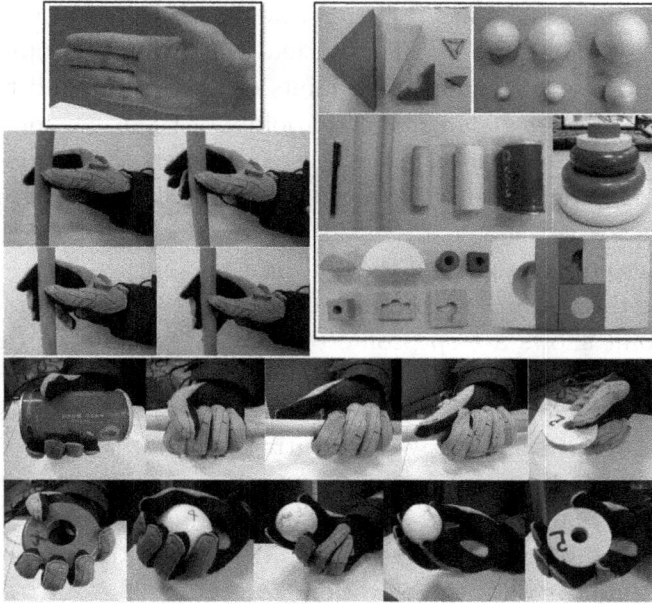

Fig. 13.1. The initial posture of each grasping movement (at left upper corner), the involved objects for grasping (at right corner), and the typical grasp patterns.

show how to achieve this goal by implementing the angular velocity synergy described by Eq. (13.7) in a mechanical manner.

13.3 Kinematic Synergy Extraction

Here we use the CyberGlove II to record the joint angle of the human hand while grasping objects with different shapes. Three healthy right-hand subjects wearing the CyberGlove are instructed to grasp each object shown in Fig. 13.1 for three times in appropriate grasp patterns. The initial posture of each grasp is with fully extended fingers and the abduction between thumb and index finger is in the comfortable status (see Fig. 13.1). Each grasping movement is averagely performed in three seconds including one second for the start and one second for the end of the grasp to ensure the zero velocity at the two ends. In the experiment, each object is grasped in different feasible types as many as possible. Finally, there are 322 grasping movements performed in total. Here, we only consider eleven sensors that correspond to the thumb rotation (TR), metacarpophalangeal (MCP) and interphalangeal (IP) joints of the thumb, and the MCP and proximal interphalangeal (PIP) joints of the other four fingers. These eleven joints can capture most characteristics of the human hand in grasping tasks. Note that the Distal Interphalangeal (DIP) joints are not considered because the flexion of the DIP joint of a finger is about two-thirds of that of the PIP joint.

Due to the resolution limit of the glove sensor, the raw trajectory of each joint shows a step profile that is not suitable for derivative operation. All the raw data from the glove sensors are first fitted by third-order spline with boundary constraints. Then, the resampled data are filtered using a low-pass Butterworth filter at a cut-off frequency of 2 Hz. Finally, the excess stationary region at the start and end stages is truncated and only the interval between onset time and mainly stopping time is reserved. Then the motion time is normalized to one second for velocity derivation.

The velocity synergies are extracted by applying the SVD procedure to the derived velocity data. Upon this result, the first and second synergies account for approximately 80% and 88% of the total variance, respectively. Figure 13.2 illustrates how sufficiency of synergy numbers account for the total variance of the entire training data. Error bars indicate standard deviation across subjects. Although the first three or more synergies can give more precision to account for the total variance, mechanically implementing more synergies will dramatically increase mechanism complexity compared with mechanical implementing two synergies. This is one of the points that we only employ the first two synergies to replicate the grasping movement. The waveforms of the first two synergies are shown in Fig. 13.3.

13.4   Mechanism Design

13.4.1   Cam Design Principle

The angular velocity profile of the element in one synergy is realized by a combined mechanism consisting of a cam, disc, and pulley. As shown in Fig. 13.4, the cam and disc are fixed together. The radius of

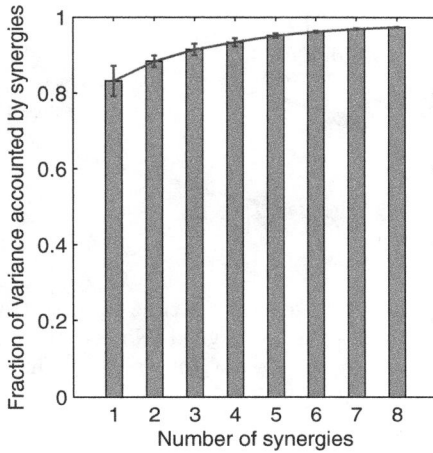

Fig. 13.2. Fraction of variance illustrating in bars is accounted by increasing number of synergies. Error bars indicate the standard and deviation across subjects.

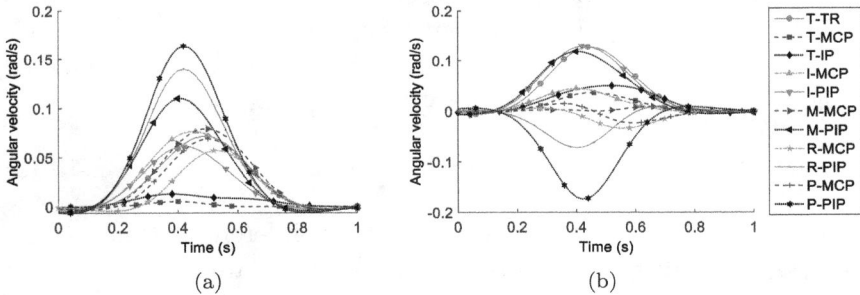

(a)	(b)

Fig. 13.3. Two most significant angular velocity synergies extracted from grasping trials. (a) The first significant angular velocity synergy, (b) the second significant angular velocity synergy. The abbreviated letters T, I, M, R, and P correspond to thumb, index finger, middle finger, ring finger, and pinky finger, respectively.

the cam is the function of the rotational angle. The radii of the cam and the disc at the initial rotational angle are equal. When the input shaft rotates at a constant angular velocity, the differential motion between the input angle $d\alpha$ and the output translation dy has the following relationship:

$$r_b(\alpha)d\alpha - r_a d\alpha = 2dy \tag{13.9}$$

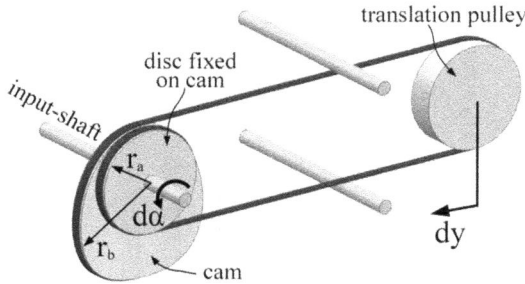

Fig. 13.4. Cam mechanism generating time-varying translation in tendon space. The axes of disc and cam are aligned. Note that there is only one cable in this mechanism with two end-tips of the cable are fixed on cam and disc, respectively.

Here, the dy is described in tendon space and it can map to joint space just by winding a cable around the axis of a revolutionary joint. In this subsection, we call the revolutionary joint *design-reference-finger-joint* whose radius and differential rotation are denoted by r_s and $d\alpha_s$, respectively. If the output translation dy in tendon space is applied to the *design-reference-finger-joint*, the dy can be represented as $dy = r_s d\alpha_s$. After taking it into Eq. (13.9), we get

$$r_b(\alpha)d\alpha - r_a d\alpha = 2r_s d\alpha_s \tag{13.10}$$

and dividing both sides of Eq. (13.10) by differential time dt, the expression about the radius of the cam is formulated as

$$r_b(\alpha) = 2r_s \frac{s(t)}{\bar{\omega}(t)_{0 \mapsto \alpha}} + r_a \tag{13.11}$$

where $s(t) = d\alpha_s/dt$ can be considered as the angular velocity of one joint in velocity synergy and $\bar{\omega}(t)_{0 \mapsto \alpha} = d\alpha/dt$ denotes the constant angular velocity of input axis in the range between 0 and α. In this subsection, we call $\bar{\omega}$ *input-reference-rotation*. According to Eq. (13.11), if the radii of the cam's disc and *design-reference-finger-joint*, i.e., r_a and r_s, are specified, respectively, the radius of the cam about the rotation angle α is particularly determined. This cam mechanism will ensure the velocity profile of *design-reference-finger-joint* follows one waveform of the velocity synergy when the input shaft rotates angle α from 0 radian at constant velocity $\bar{\omega}$ in one second. Since the velocity amplitudes in velocity synergy are generally small, the r_s is not the real radius of the finger joint but more like the scaling coefficient of the radius. In order to simplify the problem, the values of the r_a for all components of synergy are assumed to be the same. Similarly, the values of the r_s for all components of synergy are also assumed to be the same. Under these assumptions, the cam radius at angle α is dependent on the magnitude of the corresponding component of synergy at time t when the input axis rotates to angle α.

Considering that the contour of the cam should be continually smooth, we assume that the finger joint flexes from the initial angle to the target angle when the input shaft averagely rotates one cycle in one second. This assumption means that the maximum rotational angle of the input shaft is 2π radian and the *input-reference-rotation* $\bar{\omega}$ is $2\pi\,rad/s$. When the input shaft rotates back from 2π radian, the hand joints return to their initial position by the torsional springs installed at finger joint axes. For example, if the radii of the cam's disc r_a and *design-reference-finger-joint* r_s are set as $r_a = 10\,mm$ and $r_s = 150\,mm$, using the first angular velocity synergy shown in Fig. 13.3 as the values of s, the resulting radii of the cam and associated disc for the last ten joints are given in Fig. 13.5.

13.4.2 Cam Contour Correction

Note that some components of the second angular velocity synergy, such as R-MCP, R-PIP, P-MCP, and P-PIP, have obviously minus velocity profile which will result the corresponding cam with concave contour according to design formula Eq. (13.11) (see Fig. 13.6(a)). Since the concave contour brings non-continual contact between the cable and cam surface, the outputted velocity profiles will not follow profiles of the synergy. This problem can be solved by rotating the cam 180 degrees around the axis that is parallel to the direction of translation of the movable pulley and then changing the cable winding in a reverse manner (see the exampled cam 3 in Fig. 13.7).

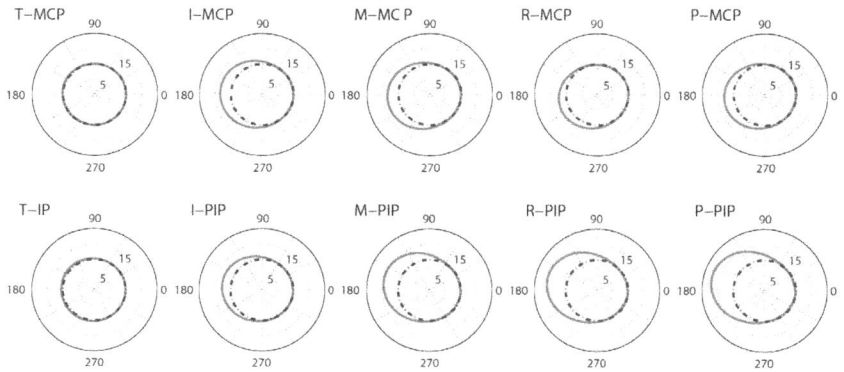

Fig. 13.5. The exampled cams and associated discs related to rotational angle in polar coordinate (radius-angle, unit: mm-degree) corresponding to the ten joints of the first kinematic synergy. The solid curves represent the cam contour and the dash-dot cycles represent the disc rigidly attached to cam.

The design formula for this type of cam is modified as follows:

$$r'_b(\alpha) = -2r_s \frac{s(t)}{\overline{\omega}(t)_{0 \mapsto \alpha}} + r_a \qquad (13.12)$$

where $s(t)$ corresponds to the special component which has obviously minus profile, such as R-MCP, R-PIP, P-MCP, and P-PIP in the second synergy. The corrected contour of the cams corresponding to these four components of the second synergy is shown in Fig. 13.6(b).

13.4.3   Time-Varying Posture Synthesis

In this subsection, we will use the cam mechanism proposed in the above section to synthesize the time-varying posture in the velocity profile represented by Eq. (13.7). The mechanism principle of angular velocity synthesis for one joint is shown in Fig. 13.8. This mechanism can be decomposed into two parts according to function: synergic velocity generator and ratio regulator. The synergic velocity generator is the above-proposed cam mechanism. The ratio regulator is similar to a cylindrical cone and has a continual varying diameter along the axis. The 3-D solid model of the ratio regulator is given in Fig. 13.9. The diameter ratio of the end section to the other section is

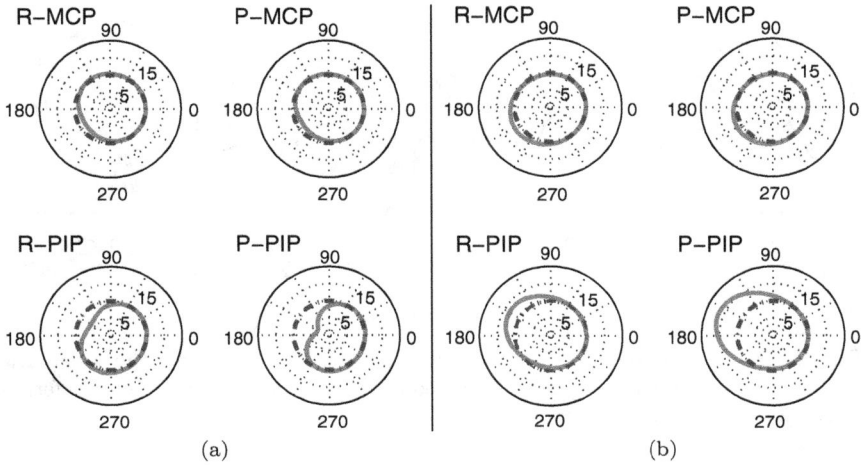

Fig. 13.6. Comparison of the cams contour in polar coordinate (radius-angle, unit: mm-degree) between (a) before correction and (b) after correction corresponding to the four components R-MCP, R-PIP, P-MCP, and P-PIP in the second synergy.

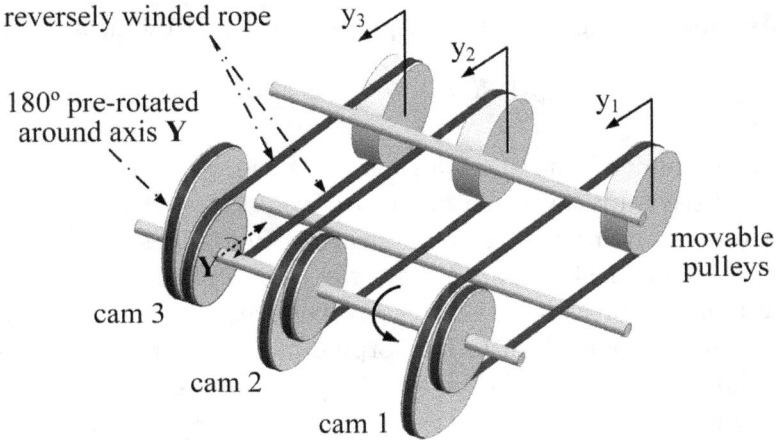

Fig. 13.7. Three exampled cams attach on the input shaft. Note that cam 3 is pre-rotated 180 degrees around axis Y paralleling to the translation direction of movable pulley, which is to generate obviously minus velocity profile of component in velocity synergy.

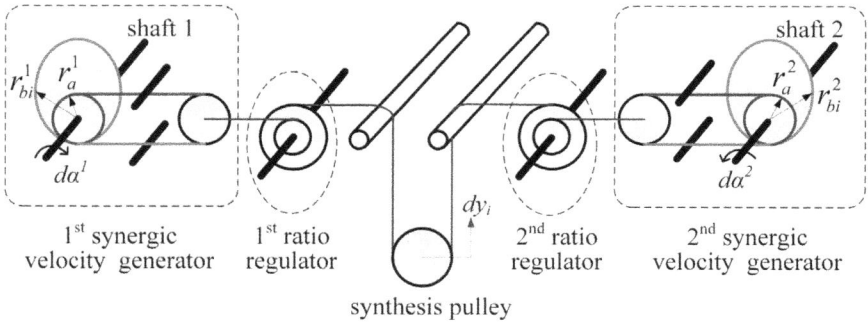

Fig. 13.8. Combining the two cam mechanisms to implement the kinematic synergy. The output velocity is synthesized in tendon space and can be transferred to the joint space through cable.

variable and can be viewed as an amplifier to scale the velocity outputted from the synergic velocity generator, thus, further to regulate the outputted velocity of synthesis pulley. In the following section, we call the diameter ratio of the end section to the other section transmission ratio.

As shown in Fig. 13.8, the output dy_i for the ith joint is given by

$$K^1(r_{bi}^1 da^1 - r_a^1 da^1) + K^2(r_{bi}^2 da^2 - r_a^2 da^2) = 4dy_i \qquad (13.13)$$

where K^1 and K^2 are the transmission ratio corresponding to the first and second ratio regulator, respectively. r_{bi}^1 and r_{bi}^2 are the cam radii corresponding to the components of first and second velocity synergy, respectively. r_a^1 and r_a^2 are the radii of cam associated disc corresponding to cam r_{bi}^1 and r_{bi}^2, respectively. If we apply dy_i to the ith finger joint of anthropomorphic hand, it can be reasonably formulated as $dy_i = r_i^o da_i^o$ in which r_i^o and da_i^o denote the radius and differential rotation of the ith finger joint, respectively. In order to simplify the velocity synthesis and focus on the motion analysis within the tendon space, we assume the radius of all the finger joints r_i^o are the same and equal to r^o. The Eq. (13.13) can be rewritten as the following form:

$$\frac{K^1 \omega^1}{4r^o}(r_{bi}^1 - r_a^1) + \frac{K^2 \omega^2}{4r^o}(r_{bi}^2 - r_a^2) = \omega_i^o \qquad (13.14)$$

where $\omega^1 = d\alpha^1/dt$, $\omega^2 = d\alpha^2/dt$ and $\omega_i^o = d\alpha_i^o/dt$. The ω^1 and ω^2 denote the rotational velocity of shafts 1 and 2, respectively. The ω_i^o denotes angular velocity of the ith hand joint. According to the cam design principle in above section, the ω^1 and ω^2 are both equal to the *input-reference-rotation* $\bar{\omega}$, i.e., 2π rad/s. From Eq. (13.11), we have

$$r_{bi}^1 - r_a^1 = 2r_s^1 \, s_i^1/\bar{\omega} \qquad (13.15)$$

$$r_{bi}^2 - r_a^2 = 2r_s^2 \, s_i^2/\bar{\omega} \qquad (13.16)$$

where s_i^1 and s_i^2 represent the ith component of the first and second velocity synergies in Eq. (13.7), respectively. Similarly, the r_s^1 and r_s^2 are the radius of *design-reference-finger-joint* corresponding to the first and second velocity synergy, respectively. After putting them into Eq. (13.14), the angular velocity of ith joint of the hand is given by

$$\omega_i^O = \frac{r_s^1}{2r^o}K^1 \cdot s_i^1 + \frac{r_s^2}{2r^o}K^2 \cdot s_i^2 \qquad i = 1, \cdots, 11 \qquad (13.17)$$

Comparing Eqs. (13.8) and (13.17), we basically implement the kinematic synergy synthesis via the proposed cam mechanism. To perform one particular grasping task from the initial fully extending posture, it only needs to specify the transmission ratio, i.e., K^1 and K^2, and then the two input shafts rotate one cycle at the speed of *input-reference-rotation* $\bar{\omega}$, i.e., 2π rad/s.

13.4.4 Design Parameters' Optimization

In the previous sections, we have made three assumptions about the design parameters to simplify the mechanism design: The radii r_s of *design-reference-finger-joint* corresponding to all components of one synergy are the same, the radii r_a of disc associated with cam corresponding to all components of one synergy are the same, and the joint radii r^o of all fingers are the same. Changing r_s and r_a will affect the size of cams. Proper r_s and r^o can limit the transmission ratio

motion-output cable

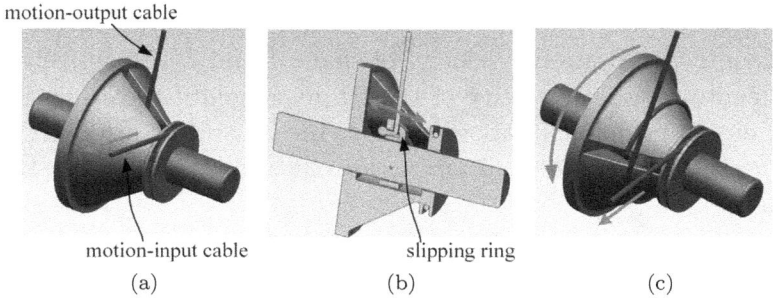

motion-input cable   slipping ring

(a)   (b)   (c)

Fig. 13.9. The 3-D model of the ratio regulator. (a) The initial configuration, (b) the section view, (c) the working configuration. The transmission ratio between the motion-input cable and motion-output cable can be freely adjusted by moving slipping ring on the axis when ratio regulator at initial configuration (corresponding to the initial hand posture).

in the lower range which brings the ratio regulator in compact size. In order to limit the size of the cams and ensure the ratio regulator with compact size, a simple optimization procedure corresponding to each velocity synergy can be formulated as

$$\min : f(r_s^k, r_a^k, r^o) = 2r_s^k \left| s_i^k \right|_{\max} / (2\pi) + r_a^k \quad k = 1,\, 2$$

subject to :

$$0 < K^k = \left| w^k \right|_{\max} / \frac{r_s^k}{2r^o} \leq \beta \qquad (13.18)$$

$$r^o{}_{\min} \leq r^o \leq r^o{}_{\max}$$

$$r_{a\,\min} \leq r_a^k \leq r_{a\,\max}$$

The meaning of the objective function is to limit the maximum radius of the cam corresponding to the joint with the most maximum velocity peak in kinematic synergy. The first constraint condition is to limit the transmission ratio in the proper range so that the size of the ratio regulator is compact.

Before optimization, several parameters should be specified first. The maximum absolute peak in two kinematic synergies, i.e., $\left| s_i^1 \right|_{\max}$ and $\left| s_i^2 \right|_{\max}$, can be determined from the velocity synergies shown in Fig. 13.3 and here their value are found to be 0.16 rad/s and 0.17 rad/s, respectively. The maximum absolute weight, i.e., $\left| w^1 \right|_{\max}$ and $\left| w^2 \right|_{\max}$, which can also be found from the SVD result of the motion capture data, are 56 and 18, respectively. Other boundary conditions

Table 13.1. Value of boundary condition for optimization.

Boundary parameters	Value
β	2
$r^{o}{}_{max}$, $r^{o}{}_{min}$	10 mm, 5 mm
$r_{a\,max}$, $r_{a\,min}$	15 mm, 10 mm

Table 13.2. The optimal result for the cam mechanism.

1st Kinematic synergy		2nd Kinematic synergy	
Parameters	Value	Parameters	Value
r_s^1	300 mm	r_s^2	102.5 mm
r_a^1	10.0 mm	r_a^2	10.0 mm
r^o	5.0 mm	r^o	5.0 mm

are given in Table 13.1. Using the boundary optimization procedure in Matlab, the optimal results of the design parameters under given constraints are obtained (see Table 13.2).

13.4.5   Prototype Design

According to the optimal result, the radii of the finger joints corresponding to the first and second synergy are the same value, which is consistent with the design assumption. The resultant cams can be assembled sequentially one by one on a common axis. These cams are fixed on the axis so that the grouped cams have the same angular velocity. We call the grouped cams eigen cam group. According to the optimized parameters, the prototypes of the two eigen cam groups are shown in Fig. 13.10. In order to change the transmission ratio conveniently, the conoid surface of the ratio regulator along the axis is separated into multi-grooves whose radius are increased one by one. The improved ratio regulator and its assembly are shown in Fig. 13.11. The number of grooves is not optimized here but just in the range that the selectable radius ratios can meet most of the grasp

Fig. 13.10.   Two eigen cam groups corresponding to the two angular velocity synergies in Fig. 13.3.

Fig. 13.11.   The 3-D solid model of the improved ratio regulator and local view of their assembly on a common axis.

task. According to Eq. (13.17), changing the transmission ratio will scale the magnitude of the synergetic velocity outputted from the cam mechanism, which will alter the maximum translation of cable winding around the synthesis pulley. It must be noted that this maximum translation cannot be changed just by varying the rotation speed of the input shaft because one cycle rotation of the cam fixes the output distance of the movable pulley in the cam mechanism no matter how much the rotation speed is.

The first version of the anthropomorphic hand prototype is developed, as shown in Fig. 13.12, for illustrating how to use the two eigen cam groups to generate the continual grasping movement. Two local views are also provided to illustrate the main components, i.e.,

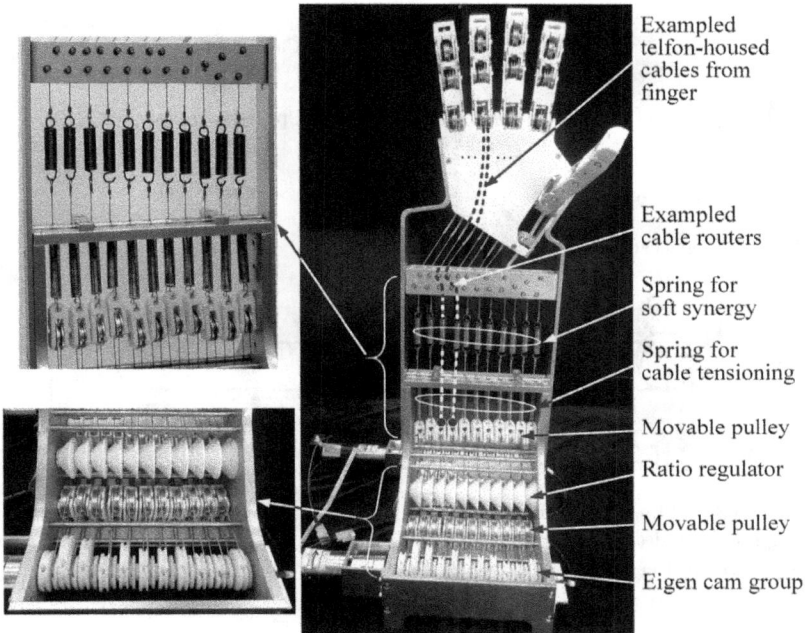

Fig. 13.12. The prototype of anthropomorphic hand based on the proposed principle of mechanical implementation of kinematic synergy. Two local views about the prototype are also given: The velocity synthesis mechanism and springs used to implement soft synergy and cables tensioning.

springs in the cable routers and velocity synthesis mechanism. The detailed cable router in the drive mechanism is shown in Fig. 13.13. There are 15 joints in the current hand prototype with five similar fingers and each finger has three flexion/extension joints. The typical finger mechanism with a cable router is shown in Fig. 13.14. The index, middle, ring, and little finger all have two active joints (proximal and medial joint) and one coupled joint (distal joint). As shown in Fig. 13.14, the tendon driving the medial joint passes through the axis of the proximal joint so that the coupled motion between the medial and proximal joint due to winded cable is suppressed. The thumb has three active joints in which the coupled joint shown in Fig. 13.14 is replaced by an active one. The torsion springs with stiffness 0.36 Nmm/deg and 0.65 Nmm/deg are mounted in the joint axis to guarantee the joint moving back when the cable

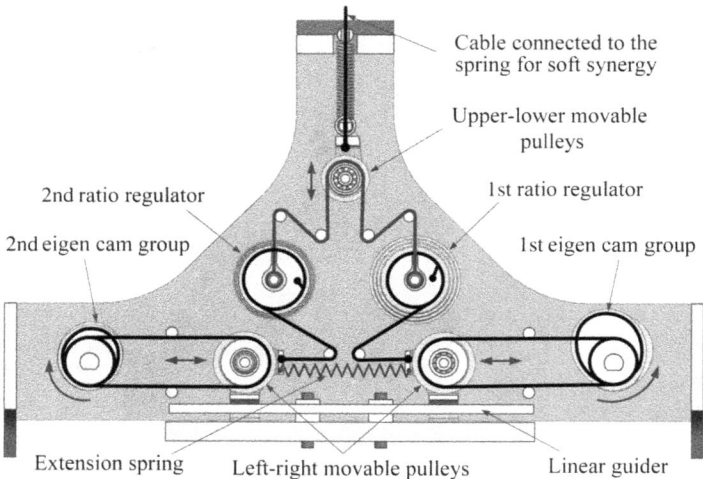

Fig. 13.13.   The cable router in the velocity synthesis mechanism. The extension spring connected the left and right moveable pulleys is used to tension the cables winding around cam mechanism.

is relaxed. The base of the thumb is fixed on the palm at an angle of 70 degrees to the palm from the side view and 30 degrees to the palm from the top view. We find this mounting manner is most appropriate for the thumb to perform precise grasping and power grasping in the current quantity of DoF. Under this framework, we map the rotation of the human thumb to the flexion/extension of the MCP joint of the robot thumb.

In order to assign compliant ability to the anthropomorphic hand in case the final envelope formed by the hand posture is not consistent with the object shape, extension springs are placed into the tendon route between the fingers and the drive mechanism (refer to the springs for soft synergy in Fig. 13.12). The function of actuation mechanism is more like the conception of *soft synergy*. In fact, these extension springs are crucial to the appropriate grasp because the drive mechanism is rigid and the stall of any finger joint during the input-shaft rotation will interrupt the motion of other joints and even deform the grasped object if no tendons are broken in this case.

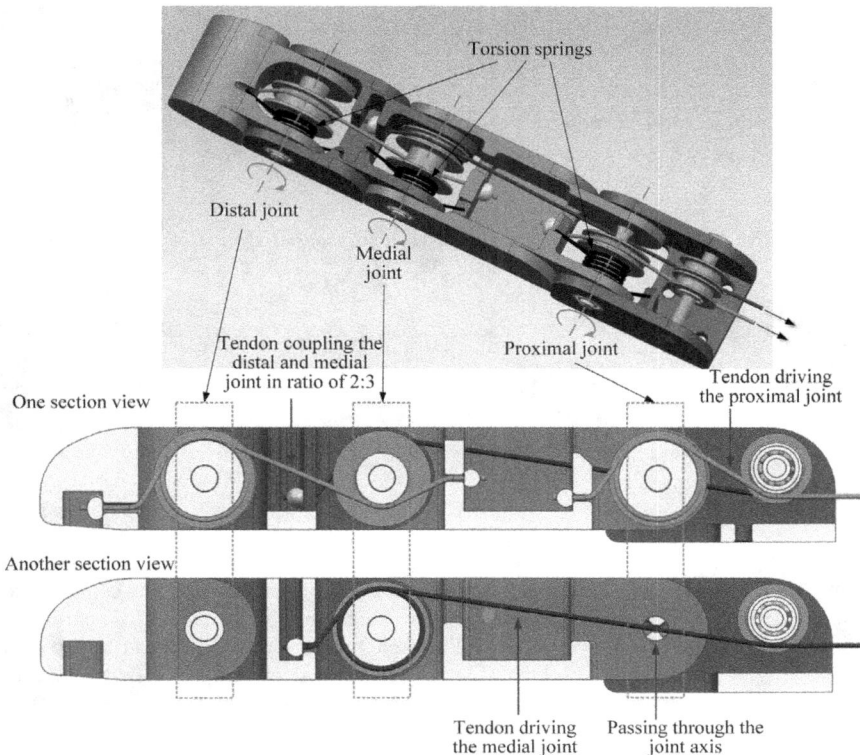

Fig. 13.14. The mechanism of one typical finger. Two section views are given to illustrate the involved cable routes.

For the two types of springs in the cable router of the prototype, as shown in Fig. 13.12, the stiffness of the springs implementing the function of soft synergy is not optimized and uniformly specified as 4 N/mm just for the appropriate grasping force. The lower ends of these springs are connected to the upper-lower movable pulleys. The other type of springs for tensioning the cables are suspended at the shelf and the other ends are also connected to the upper-lower movable pulleys. The stiffness of this type of spring is specified as 0.2 N/mm. Considering the input shafts have to provide enough torque to overcome the spring's extension force, the motor with a nominal output torque of 3.6 Nm is employed in the prototype.

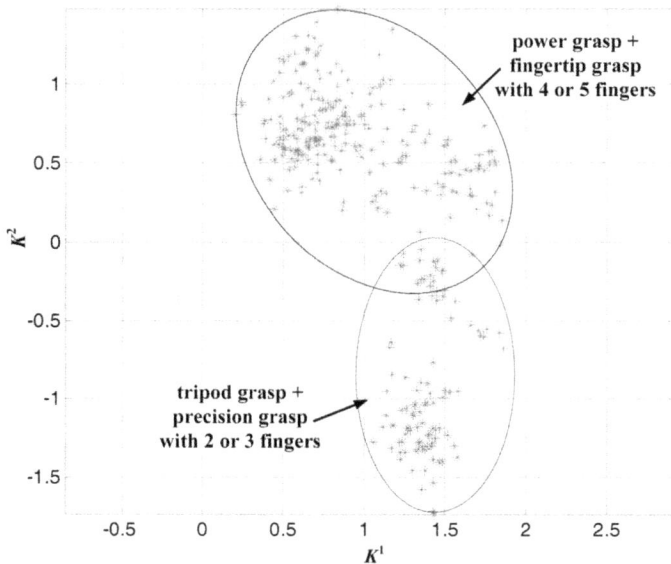

Fig. 13.15. The distribution of transmission ratio pair (K^1, K^2). According to the final hand shape and involved fingers' function, the transmission pairs are generally classified into two groups. The first one is called by power grasp and fingertip grasp with 4 or 5 fingers. These movements generally require almost all fingers roughly flexed as enveloped shape to hold object without precision. The second one is called by tripod grasp and precision grasping with 2 or 3 fingers. These movements generally require the thumb, index, and optional middle finger to grasp object by fingertips.

According to Eqs. (13.8) and (13.17), the transmission ratio is linearly proportional to the weight of the kinematic synergy. Based on weights that are computed from the SVD decomposition of experimental data, the distribution of the transmission ratio pair (K^1, K^2) is shown in Fig. 13.15. According to the trial data, the range of the ratio K^1 and K^2 is (0, 2) and (−1.5, 1.5), respectively. In the proposed hand prototype, the designed ranges for K^1 and K^2 are selected to be $(0.3, 2)$ and $(0.3, 1.5)$ considering the design feasibility of the ratio regulator. The minus value of K^2 in the range $(−1.5, −0.3)$ can be implemented by inversely rotating the second input shaft. Thus, except for a small region that cannot be reached, the designed ratio regulator can approximate most of the real ratios.

In our experiment, the grasp task and its related objects can be generally defined as a grasp pattern. All the grasp trials performed by the same grasp pattern can be grouped together. The corresponding transmission ratio pairs grouped on the plane can be represented by their average point. This average point, i.e., average transmission ratio pair, can be viewed as the typical representative of the grasp pattern. All typical transmission ratio pairs can be organized as grasp patterns and stored in the hand controller. In order to generate practical grasping movement, it is required only to specify the corresponding transmission ratio pair according to the grasp pattern and then averagely rotate the input shafts one cycle.

The prototype in the current version does not replicate all possible joints of the human hand, such as the thumb rotation and abduction between fingers. This prototype is just for illustration of the implementation possibility of the continual motion generation of the robotic hand whose joints are driven by the eigen cam groups. If more DoFs are added to the prototype, more cams will be involved and the robotic hand can realize more comprehensive grasping movement.

13.5 Simulation and Experiment

13.5.1 Simulation Based on Synergy Synthesis

In order to simply illustrate how the transmission ratio affects the temporal evolvement of hand posture, here we select four typical transmission ratios for each synergy. The corresponding time-varying hand postures for each transmission ratio are averagely sampled along the motion time by five snapshots (see Fig. 13.16).

The first kinematic synergy mainly generates the power grasping movement. A bigger transmission ratio means more flexion posture. The second kinematic synergy mainly generates the precision grasp.

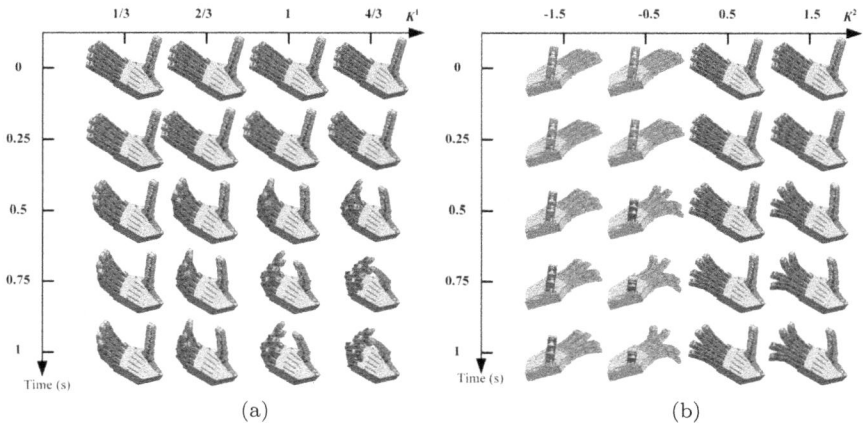

Fig. 13.16. The temporal sequence of hand postures at different transmission ratio corresponding to (a) the first synergy and (b) the second synergy, respectively.

Similarly, the flexion extent is in proportion to the value of the transmission ratio. Especially, the ring and little fingers inversely flex to the back of the palm when the transmission ratio is positive, while the thumb, index and middle finger flex inversely to the back of the palm when the transmission ratio is negative. This phenomenon indicates that the first kinematic synergy plays a significant role in grasp posture formation, especially the power grasping movement. While the second kinematic synergy just counteracts part of fingers flexion generated by the first kinematic synergy for producing complex movement, such as precision grasp. Due to mechanical constraints, no joint can inversely flex and thus there is no physical meaning to use the second velocity synergy alone. The rotation of the second eigen cam group should be always in company with the rotation of the first eigen cam group.

By combining the two kinematic synergies under different weights (transmission ratio), practical grasping movement can be approximately generated. Figure 13.17 gives four exampled resulting motions of the two kinematic synergies: grasping bottle with three fingers, power-grasping triangle, power-grasping cylinder, and precisely grasping sphere with three fingers. Due to the gap between the robot thumb and human thumb on DoF distribution, the thumb movement

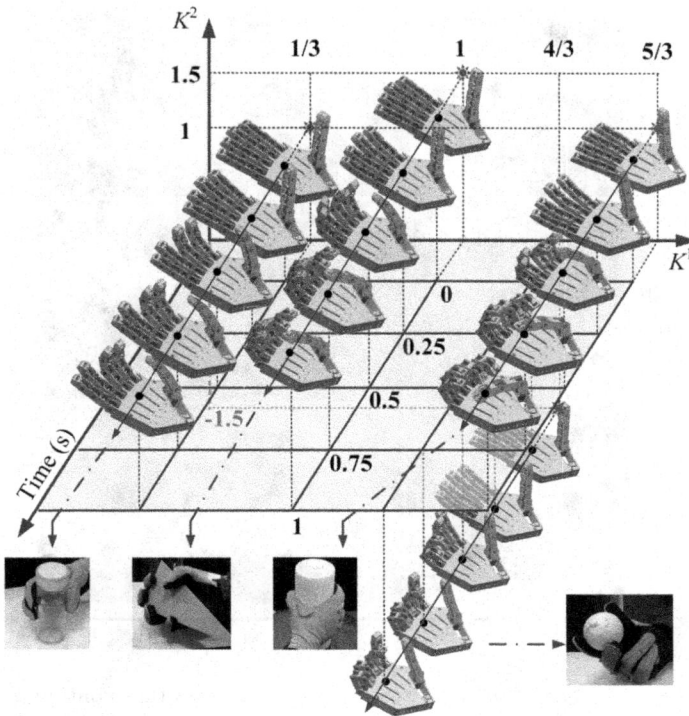

Fig. 13.17. The resultant temporal sequence of hand postures by combing the first and second synergy at different transmission ratios.

of the robotic hand is slightly different from that of the human thumb, especially in the movement of the power grasping cylinder. The joint trajectories for the movements under four transmission ratio pairs, i.e., $(4/3, -1.5), (1, 1.5), (5/3, 1)$, and $(1/3, 1)$, are given in Fig. 13.21. All joint ranges are mechanically limited in $[-8°, 115°]$, and the joint whose synthesis motion over this limitation will halt (see the P-PIP and M-PIP joints in Figs. 13.21(a) and 13.21(c), respectively).

As is shown in Fig. 13.21, different combinations of transmission ratios of the two kinematic synergies produce generally different hand grasp patterns. Although more complex grasping movements can be reproduced by using more kinematic synergies, it will dramatically increase the complexity of mechanism implementation.

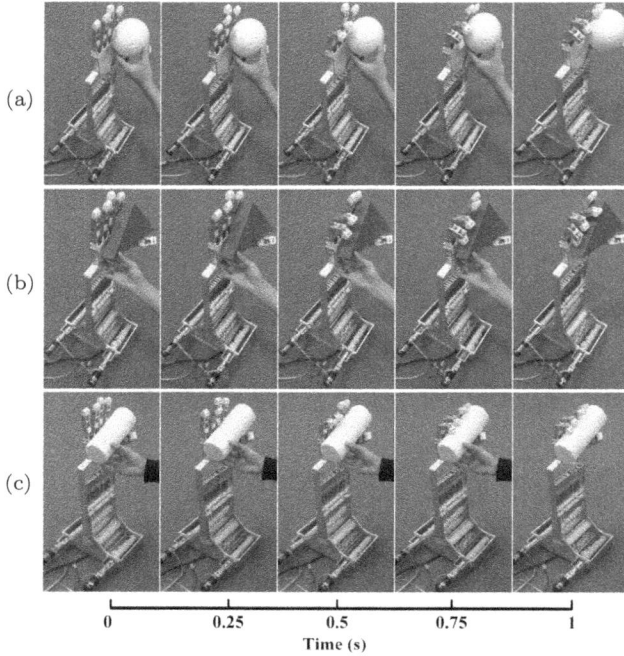

Fig. 13.18. The temporal posture sequence of the prototype continual grasping three objects in one second: (a) sphere, (b) triangle, and (c) cylinder. The corresponding transmission ratio pairs (K^1, K^2) are $(4/3, -1.5), (1, 1.5)$, and $(5/3, 1)$, respectively.

13.5.2 Grasping Experiments by the Prototype

Here we select three grasping tasks to validate the continual grasping ability: grasping a sphere, grasping a triangle, and grasping a cylinder. The initial posture for all tasks is the fingers in full extension. The transmission ratio pair of the three tasks is set to $(4/3, -1.5), (1, 1.5)$, and $(5/3, 1)$, respectively. The motion time is one second. When the two input motors averagely rotate one cycle, the temporal postures reach the final grasp configuration. Like the illustration in the above simulation, we also give five snapshots averagely sampled along the motion time for each task (see Fig. 13.18).

When the fingers freely move in space without contacting any objects, the hand posture variation and joint displacement at the end of cam rotation are only dependent on the specified transmission

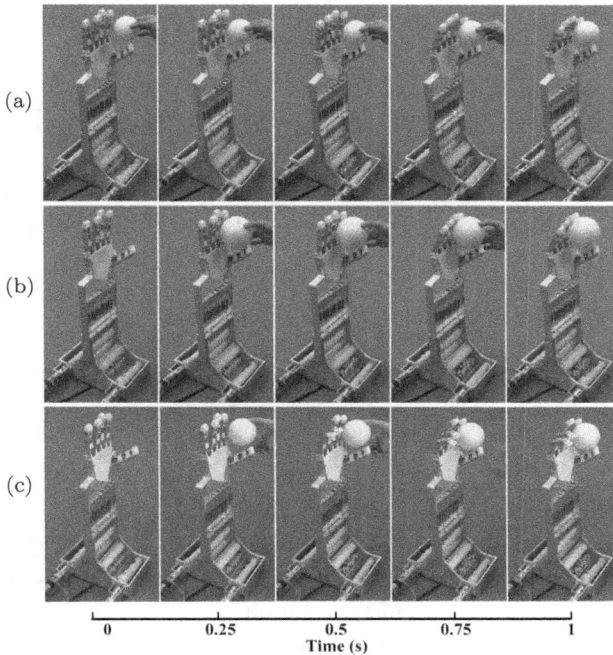

Fig. 13.19. The temporal sequence of the prototype grasping spheres with different diameter using precision grasp pattern. The transmission ratio pair for grasping sphere in (a) small diameter and (b) big diameter are both (1/3, 1). After minor changing the ratio pair to (2/3, 1), the grasp pattern therefore changes to (c).

ratio, which can be referred to in Fig. 13.17. Thus, the hand posture at the end of the cam rotation is impossible to actively change into another posture in free space for the given transmission ratio. However, if the hand moves in constrained space, i.e., grasping objects, the final hand posture will be not the posture formed in free space but dependent on the object shapes that are different in scale. This adaptability comes from the used springs implementing soft synergy (see Fig. 13.12). To illustrate this adaptability, the Fig. 13.19(a) and (b) give the experiment results in which the hand grasps two different spheres under the same transmission ratio pair (1/3, 1). Moreover, in order to clearly show the influence of transmission ratio variation on the grasp pattern, we make minor changes of the ratio pair to be

(2/3, 1), the resultant grasp pattern therefore gives some variation on ring and pinky fingers, which is shown in Fig. 13.19(c).

Additionally, the motion capability of the proposed hand can be generally evaluated by some metrics in the literature. However, for simple operation and quantitative evaluation consideration, the Modified Relative Mean Squared Error (mRMSE) with dimension-free is employed here to evaluate the posture difference between the robotic hand and human hand at each status along the motion time.

$$mRMSE = \sqrt{\frac{1}{n-1} \cdot \frac{\|\mathbf{P}_{rd} - \mathbf{P}_{hd}\|^2}{\|\mathbf{P}_{hd}\|^2}} \qquad (13.19)$$

where the $\mathbf{P}_{hd}$ and $\mathbf{P}_{rd}$ represent human and robotic hand posture respectively. The hand posture consisting of the considered joint angle is formulated as a vector. The joint angles of the robotic hand are measured by the VICON system (Oxford Metrics Ltd, UK). To use this evaluation index, the joint data of the robotic hand comes from the grasping movements in which no objects are put into the grasp region to ensure the fingers move freely to replicate the human hand posture. Moreover, in order to avoid the non-zero angle at initial status due to the difference in cable tensioning degree, we use the relative angle variation instead of the absolute angle for the evaluation. The four grasping movements shown in Fig. 13.17 are performed to evaluate the posture difference between the robotic and the human hand along motion time. The corresponding transmission ratio pairs for the robotic hand are specified according to the values shown in Fig. 13.17. Each grasp trial performed by the robotic hand is evaluated with four grasps performed by four subjects, respectively. Thus, the evaluation result of each grasp corresponds to four error curves. The evaluation results are shown in Fig. 13.20. As is shown, the posture difference between the robotic hand and human hand in the motion time is kept at a relatively low level (less than 20%). The most variation of the error is observed mainly in the first 0.6 second which covers the mainly flexing stage of the robotic hand fingers (refer to Fig. 13.21). Due to the angular velocity profile discrepancy between

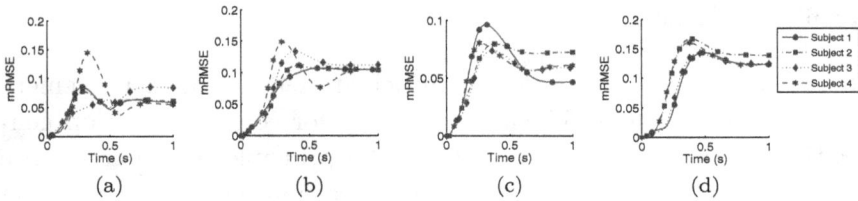

Fig. 13.20. The evaluation result of the time-varying posture difference between robotic hand moving in free space and human hand. For each grasp pattern from (a)–(d), the performance of the robotics hand is evaluated by mRMSE with the same grasps performed by four subjects. The correspondence between transmission ratio pair and grasp pattern is as follows: (a) $(4/3, -1.5)$ to tripod grasp sphere, (b) $(1, 1.5)$ to power grasp triangle, (c) $(5/3, 1)$ to power grasp cylinder, and (d) $(1/3, 1)$ to precision grasp bottle with two fingers.

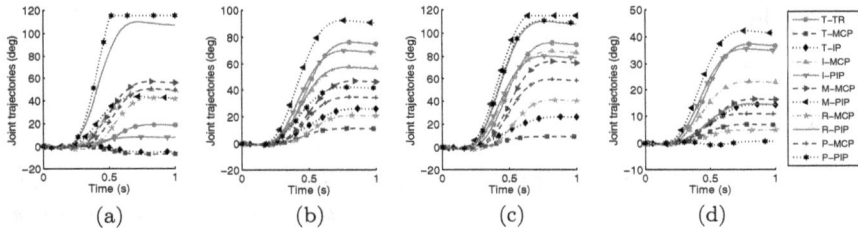

Fig. 13.21. The joint trajectories in free space under four transmission ratio pairs: (a) $(4/3, -1.5)$, (b) $(1, 1.5)$, (c) $(5/3, 1)$, and (d) $(1/3, 1)$, respectively.

the robotic and human hand, the accumulative error increases in the first half of the motion time. This motion reconstruction error mainly originates from the limited number of employed velocity synergies.

However, the simulation and hand-grasping experiments validate the proposed mechanical implementation principle of kinematic synergy on generating continual grasping movement. Although the motion period is considered as one second for one cycle in the design framework, the input shafts of the anthropomorphic hand prototype are allowed to rotate faster or slower than 2π rad/s. The motor rotating one cycle for a long time gives slow grasping, whereas the short time brings quick grasping.

13.6 Summary

This chapter presents a design principle of mechanical implementation of human hand kinematic synergy for continual motion generation of anthropomorphic hands. The principle does not require solving the temporal weight sequence of each synergy and presetting zero-configuration exited in current anthropomorphic hands based on the static postural synergy. The two most significant kinematic synergies observed in the angular velocity profile are extracted from quantities of human hand-grasping trials and implemented by the two eigen cam groups respectively. The continual grasping of the anthropomorphic hand is only dependent on the pre-specifying transmission ratio pair and the input motors just need to rotate averagely one cycle during the motion time.

References

[1] Ju Z. and Liu H. Human hand motion analysis with multisensory information. *IEEE/ASME Transactions on Mechatronics*, 19(2), pp. 456–466, 2014.

[2] Rombokas E., Malhotra M., Theodorou E. A., Todorov E., and Matsuoka Y. Reinforcement learning and synergistic control of the ACT hand. *IEEE/ASME Transactions on Mechatronics*, 18(2), pp. 569–577, 2013.

[3] Townsend W. The BarrettHand grasper — programmably flexible part handling and assembly. *Industrial Robot*, 27(3), pp. 181–188, 2000.

[4] Fukaya N., Asfour T., Dillmann R., and Toyama S. Development of a five-finger dexterous hand without feedback control: The TUAT/Karlsruhe humanoid hand. In: *Proceedings of IEEE/RSJ International Conference on Intelligent Robots and Systems*, Tokyo Big Sight, Japan, pp. 4533–4540, 2013.

[5] Dechev N., Cleghorn W. L., and Naumann S. Multiple finger, passive adaptive grasp prosthetic hand. *Mechanism and Machine Theory*, 36(10), pp. 1157–1173, 2001.

[6] Aukes D., Kim S., Garcia P., Edsinger A., and Cutkosky M. R. Selectively compliant underactuated hand for mobile manipulation. In: *Proceedings of IEEE International Conference on Robotics and Automation*, Minnesota, USA, pp. 2824–2829, 2012.

[7]   Dollar A. M. and Howe R. D. A robust compliant grasper via shape deposition manufacturing. *IEEE/ASME Transactions on Mechatronics*, 11(2), pp. 154–161, 2006.

[8]   Gosselin C., Pelletier F., and Laliberte T. An anthropomorphic underactuated robotic hand with 15 dofs and a single actuator. In: *Proceedings of IEEE International Conference on Robotics and Automation*, California, USA, pp. 749–754, 2008.

[9]   Bierbaum A., Schill J., Asfour T., and Dillmann R. Force position control for a pneumatic anthropomorphic hand. In: *Proceedings of IEEE/RAS International Conference on Humanoid Robots*, Paris, France, pp. 21–27, 2009.

[10] Gaiser I., Schulz S., Kargov A., Klosek H., Bierbaum A., Pylatiuk C., Oberle R., Werner T., Asfour T., Bretthauer G., and Dillmann R. A new anthropomorphic robotic hand. In: *Proceedings of IEEE/RAS International Conference on Humanoid Robots*, Daejeon, Korea, pp. 418–422, 2008.

[11] Schulz S., Pylatiuk C., and Bretthauer G. A new ultralight anthropomorphic hand. In: *Proceedings of IEEE International Conference on Robotics and Automation*, Seoul, Korea, pp. 2437–2441, 2001.

[12] Birglen L., Laliberté T., and Gosselin C. Grasping vs. manipulating. In: Siciliano B., Khatib O., and Groen F. (eds.), *Underactuated Robotic Hands*, Springer Tracts in Advanced Robotics, 40, Springer, Berlin, Germany, pp. 7–31, 2008.

[13] Romero J., Feix T., Ek C. H., Kjellstrom H., and Kragic D. Extracting postural synergies for robotic grasping. *IEEE Transactions on Robotics*, 29(6), pp. 1342–1352, 2013.

[14] Steffen J., Haschke R., and Ritter H. Towards dextrous manipulation using manipulation manifolds. In: *Proceedings of IEEE/RSJ International Conference on Robotics and Intelligent Systems*, Nice, France, pp. 2738–2743, 2008.

[15] Santello M., Flanders M., and Soechting J. F. Postural hand synergies for tool use. *Journal of Neuroscience*, 18(23), pp. 10105–10115, 1998.

[16] Mason C., Gomez J., and Ebner T. Hand synergies during reach-to-grasp. *Journal of Neurophysiology*, 86(6), pp. 2896–2910, 2001.

[17] Grinyagin I. V., Biryukova E. V., and Maier M. A. Kinematic and dynamic synergies of human precision-grip movements. *Journal of Neurophysiology*, 94(4), pp. 2284–2294, 2005.

[18] Vinjamuri R., Mingui S., Cheng-Chun C., Heung-No L., Sclabassi R. J., and Zhi-Hong M. Dimensionality reduction in control and coordination of the human hand. *IEEE Transactions on Biomedical Engineering*, 57(2), pp. 284–295, 2010.

[19]   Santello M., Flanders M., and Soechting J. F. Patterns of hand motion during grasping and the influence of sensory guidance. *Journal of Neuroscience*, 22(4), pp. 1426–1435, 2002.

[20]   Vinjamuri R., Sun M., Chang C.-C., Lee H.-N., Sclabassi R. J., and Mao Z.-H. Temporal postural synergies of the hand in rapid grasping tasks. *IEEE Transactions on Information Technology in Biomedicine*, 14(4), pp. 986–994, 2010.

[21]   Castellini C. and van der Smagt P. Evidence of muscle synergies during human grasping. *Biological Cybernetics*, 107(2), pp. 233–245, 2013.

[22]   Karnati N., Kent B. A., and Engeberg E. D. Bioinspired sinusoidal finger joint synergies for a dexterous robotic hand to screw and unscrew objects with different diameters. *IEEE/ASME Transactions on Mechatronics*, 18(2), pp. 612–623, 2013.

[23]   Rombokas E., Malhotra M., and Matsuoka Y. Task-specific demonstration and practiced synergies for writing with the ACT hand. In: *Proceedings of IEEE International Conference on Robotics and Automation*, Shanghai, China, pp. 5363–5368, 2011.

[24]   Zhang A., Malhotra M., and Matsuoka Y. Musical piano performance by the ACT Hand. In: *Proceedings of IEEE International Conference on Robotics and Automation*, Shanghai, China, pp. 3536–3541, 2011.

[25]   Wimbock T., Jahn B., and Hirzinger G. Synergy level impedance control for multifingered hands. In: *Proceedings of IEEE/RSJ International Conference on Intelligent Robots and Systems*, California, USA, pp. 973–979, 2011.

[26]   Rosell J., Suárez R., Rosales C., and Pérez A. Autonomous motion planning of a hand-arm robotic system based on captured human-like hand postures. *Autonomous Robots*, 31(1), pp. 87–102, 2011.

[27]   Ficuciello F., Palli G., Melchiorri C., and Siciliano B. Experimental evaluation of postural synergies during reach to grasp with the UB hand IV. In: *Proceedings of IEEE/RSJ International Conference on Intelligent Robots and Systems*, California, USA, pp. 1775–1780, 2011.

[28]   Brown C. Y. and Asada H. H. Inter-finger coordination and postural synergies in robot hands via mechanical implementation of principal components analysis. In: *Proceedings of IEEE/RSJ International Conference on Intelligent Robots and Systems*, San Diego, USA, pp. 2877–2882, 2007.

[29]   Matrone G., Cipriani C., Secco E., Magenes G., and Carrozza M. Principal components analysis based control of a multi-dof underactuated prosthetic hand. *Journal of NeuroEngineering and Rehabilitation*, 7(16), pp. 1–13, 2010.

[30]   Matrone G., Cipriani C., Carrozza M., and Magenes G. Real-time myoelectric control of a multi-fingered hand prosthesis using principal components analysis. *Journal of NeuroEngineering and Rehabilitation*, 9(40), pp. 1–13, 2012.

[31]   Cutkosky M. R. On grasp choice, grasp models, and the design of hands for manufacturing tasks. *IEEE Transactions on Robotics and Automation*, 5(3), pp. 269–279, 1989.

Index

www.ingramcontent.com/pod-product-compliance
Lightning Source LLC
Chambersburg PA
CBHW050634190326
41458CB00008B/2265